POPULATION ECOLOGY OF RAPTORS

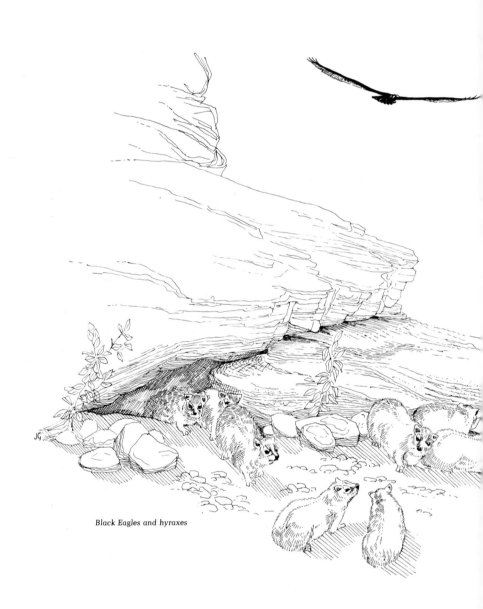

Black Eagles and hyraxes

Population Ecology of Raptors

by IAN NEWTON

ILLUSTRATIONS BY JIM GAMMIE

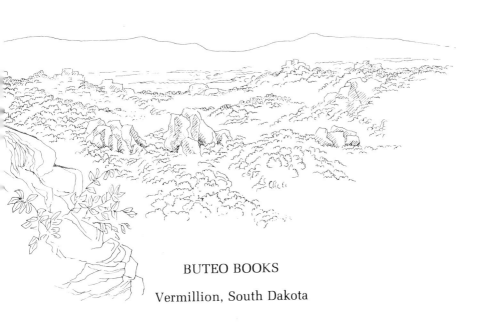

BUTEO BOOKS

Vermillion, South Dakota

ISBN 0–931130–03–4
Library of Congress Catalog Card Number 79–50279

Published in USA in 1979 by Buteo Books,
PO Box 481, Vermillion, South Dakota 57069 and
in Great Britain by T. & A. D. Poyser Ltd,
281 High Street, Berkhamsted, Hertfordshire, England

Text set in 10/11 pt VIP Melior, printed by photolithography,
and bound in Great Britain at The Pitman Press, Bath

Contents

List of Plates

List of Figures

Snake Eagle

List of Tables

11

Introduction

So many books have been written about birds-of-prey in recent years that to add another needs an especially good excuse. Mine is that I think this book offers something fresh and different that has not been attempted before. It is concerned with all aspects of population regulation in raptors, with their social behaviour, their dispersion, numbers, movements, breeding and mortality. These are fascinating and fast-developing fields, and my aim is to provide a synthesis of recently discovered facts from various parts of the world, which I hope will give a sound basis both for further research and for more effective conservation. I have tried to organise existing information into a coherent whole, interpret it critically, and at the same time pinpoint major gaps in knowledge. Throughout I have tried to write simply, in the hope that the book will be of value not only to the professional research worker, but also to anyone else whose enjoyment of these birds may be enriched by greater understanding.

The subjects of the book include the 287 species that comprise the order Falconiformes. This order contains five families of varying degrees of affinity, namely the Cathartidae (New World vultures and condors), the Pandionidae (Osprey), the Accipitridae (hawks, kites, buzzards, eagles and Old World vultures), the Sagittariidae (Secretary Bird) and the Falconidae (caracaras, falcons and falconets). All these

Illustration: Goshawk and hare

groups have the same specialisations for finding food, and for holding and tearing apart the bodies of other animals: acute vision, strong legs and feet equipped with sharp, curved claws, and a hooked beak. Yet they differ so much in the other details of their anatomy that they are almost certainly derived from more than one ancestral stock. They provide an example of convergent evolution – of unrelated animals growing to look like one another because they have the same way of life. In some respects, the convergence also applies to the owls, the nocturnal equivalents of the raptors which I shall not discuss. Throughout I have used the term 'raptors' only for members of the Falconiformes.

Our understanding of any basic ecological problem depends on the choice of an easy animal to study. No one who was interested solely in the general principles of population regulation would choose to work on diurnal birds-of-prey. Compared to most other birds, raptors usually nest at low density, often in inaccessible and remote places; and in many areas they are also liable to be shot or interfered with by our fellow men. All this helps to make for samples that to the statistician seem hopelessly small and hard to evaluate. Little wonder that, for many years, most of what we knew about raptor populations was contributed by a small number of devotees, often egg-collectors or falconers. In their quantitative aspects, most of these early scripts would not meet the scientific standards of today, but many are nonetheless full of wisdom and experience, and are well worth reading.

The last 10–15 years have seen an enormous expansion in research and conservation efforts devoted to raptors in many parts of the world. This has come mainly from a realisation of the drastic declines in the numbers of such birds since the last century, and of the poor and endangered status that many now have. For in the last hundred years, human activities have acted against the birds-of-prey as never before. Some of these actions were deliberate, and others were incidental to the growth in human population and material productivity. Four main factors have been involved. The first is the trend to more intensive land use which has increasingly destroyed habitats; in this category I include not only the conversion of natural and semi-natural areas into cultivated ones, but also the more intense use of existing farmland, which has often reduced prey-species to such a low level that the raptors which lived there 50 years ago can no longer do so. The second is deliberate persecution in the interests of game and stock rearing. This includes not only the huge culls from shooting and trapping by gamekeepers, but also those achieved unintentionally, in poisoning campaigns directed at other animals, such as wolves and foxes. The third, of which we are generally less aware in the western world, is the improved sanitation in many Asian and African cities, which has reduced the numbers of scavenging kites and vultures (though these were encouraged by the existence of cities in the first place). Fourth is the widespread use of DDT and other persistent organo-chlorine insecticides in agriculture, which by the nineteen-sixties had almost eliminated the populations of several species from large regions of Europe and North America. This was partly through causing direct mortality, and partly through causing the thinning and breakage of eggshells, which in turn lowered the breeding rate to a level at which it could no longer offset the natural mortality.

Such pesticides had their greatest impact on the bird-eating and fish-eating species. In the nineteen-sixties, the Peregrine *Falco peregrinus* disappeared completely as a breeder from the eastern United States, and by 1975 had declined by more than 90% in the western States and by about 50% in the forests and tundras of Alaska and northern Canada (Fyfe *et al* 1976). This last fact came as the greatest shock to many people: that Peregrines breeding in the last major wilderness in North America should be as contaminated as those elsewhere. The cause lay in Latin America, where these Peregrines and their arctic prey species wintered, and picked up the pesticides. It highlighted one of the most insidious effects of DDT and other organo-chlorines, that at sub-lethal level they could cause the steady decline of populations that bred in areas hundreds of kilometres distant from where these chemicals had been applied. In Europe, likewise, the Peregrine disappeared from large areas and declined in others; though by the mid nineteen-seventies, several local populations had recovered somewhat following restrictions in organo-chlorine use. Always of strong emotive appeal, the Peregrine has become for many people a symbol of concern for the environment. Through the well-being of its populations, it has assumed the role of an 'ecological barometer' of importance to everyone.

Thus research on the Peregrine and other raptors has done much to highlight the dangers of pollution, and to bring about restrictions in the use of organo-chlorine and mercurial pesticides. Against the powerful vested interest of the agro-chemical industry, the biologists who worked on these problems showed a degree of dedication, cooperation and integration of effort seldom met in ecological research. The case against DDT is probably as well substantiated now as it ever will be. It is based on a mass of detailed circumstantial field evidence from many species in several parts of the world, and on careful experimental work in laboratories. But there are still some who claim to be unconvinced, so strong is their motivation for wishing to believe otherwise.

In other respects, research on raptors has so far made little contribution to the mainstream of ecological thought. Few textbooks of ecology mention birds-of-prey, except incidentally as predators of other animals. After much hard fieldwork, raptor biologists have often found themselves merely confirming trends already established for easier and better studied birds. Yet raptors lend themselves especially well to studies of population regulation, and of the influence of food on breeding rates. Left alone, some of them show extreme stability of breeding population over long periods of years; and the role of spacing behaviour in limiting their densities in relation to food and other resources has already proved a fruitful field. Other raptor species fluctuate in numbers, in parallel with the numbers of their prey, and could provide more understanding of predator-prey relationships. The raptors also include some of the most extreme examples among birds of low reproductive rates, long deferred maturity and great longevity, and in these respects they parallel some large seabirds. As yet they are a very unevenly studied group. Some, such as the Peregrine, are now among the best-known birds in the world, but others, such as some tropical forest species, are among the least known, and even their nests and eggs have never been described.

The recent fate of birds like the Peregrine has also served to focus the attention of conservationists on raptors in a way that could not have been imagined only ten years ago. The idea of management for conservation is catching on in a big way, even in parts of Europe where, for several hundred years, the only management directed at raptors was aimed at total eradication. In North America, wilderness is still being developed, but for the first time with the long-term future of raptors in mind. The efforts of electric power companies to reduce eagle deaths from electrocution is another encouraging trend. A further break-through of recent years has been the successful breeding of Peregrines and other raptors in captivity, not just a few birds, but large numbers sufficient for release schemes. Thus it seems that the survival of endangered genotypes is ensured, at least in captivity, and that the production of enough birds for reintroduction to cleaned environments is a practical proposition. How successful the releases will be remains to be seen, but the omens are good.

The first two chapters in this book deal with the social organisation and spacing behaviour of raptors, and the next two with density regulation. There are then five chapters on various aspects of breeding, two on movements, one on mortality, and three on the effects on raptor populations of human persecution, organo-chlorine chemicals and other pollutants respectively. Lastly, there are two chapters dealing with conservation management and captive breeding. With this form of treatment it is inevitable that some topics arise in more than one chapter, but I have tried to reduce the repetition to a minimum and to cross-refer wherever possible. Nomenclature and taxonomy follow Brown & Amadon (1968). Scientific names are given at the first mention in the text, and there is a list of English common names and scientific names on page 322. References are cited in the customary way, but information given to me personally is acknowledged by a name only, with no date. To avoid breaking up the text too much, all tables are placed together at the end.

No one could work in this field without owing a lot to those who came before. As a boy, my imagination was fired by the writings of Seton Gordon on Golden Eagles and of J. H. Owen on Sparrowhawks, while later in life I was influenced by the writings of Derek Ratcliffe on Peregrines, and of the Craighead brothers on various raptors in the United States. In addition, I owe a lot to my friends, David Jenkins,

Derek Ratcliffe, Leslie Brown, Doug Weir and Nick Picozzi for much stimulating discussion, and especially to Mick Marquiss, who has also helpfully criticised the whole book in draft. I am grateful to the following for constructive comments on particular chapters: P. Frost (Chapter 11), Dr M. Fuller (Chapter 2), Dr & Mrs F. Hamerstrom and N. Picozzi (Chapter 1), Dr D. Jenkins (Chapters 1, 3, 9, 13 and 14), Dr R. Kenward (Chapters 11, 16 and 17), Dr D. Osborne (Chapter 15), Helen Snyder (Chapters 1 and 9), Dr I. Taylor (Chapter 6) and Dr P. Ward (Chapters 6, 7 and 11); to Sheila Adair and Liz Murray for drawing the diagrams and to Jim Gammie for his delightful illustrations of the birds themselves. I also thank my wife for encouragement, and for checking the whole book in draft. Finally, my general thinking on bird populations was influenced by six pleasant years at the Edward Grey Institute in Oxford, and by my teacher David Lack.

CHAPTER 1

Relationship between the sexes

In most kinds of birds, the male is bigger than the female, but in birds-of-prey, the female is usually the larger sex. In some raptor species the size difference is so slight as to be barely noticeable, while in others the female may weigh almost twice as much as her mate. This raises two questions: first why do raptors have this so-called 'reversed size dimorphism' (hereafter shortened to 'reversed dimorphism'), and second why is it more developed in some species than in others?

The fact that the female is larger than the male is evidently connected with the raptorial lifestyle, because it also occurs among owls and skuas, having arisen independently in all these groups by parallel evolution (Amadon 1959). If the raptors themselves are regarded as having arisen from more than one ancestral stock, then reversed dimorphism has evolved more than once in these birds too. Four out of the five families in the Order Falconiformes show it; and the exception contains only a single species (the Secretary Bird *Sagittarius serpentarius*), which hunts more like a stork than like a raptor. For reversed

Illustration: Hen Harriers

19

dimorphism to have evolved several times in unrelated birds of similar feeding habits strongly implies that it is an adaptive character, associated with this particular way of life.

The degree of size difference between the sexes is also linked with feeding habits. In the world as a whole, there are about 75 raptor species that each live almost exclusively on a single kind of prey – on fish, reptiles, birds, and so on, as the case may be. If these specialists are arranged according to diet, grouping all species of like food together, a remarkable relationship emerges (Figure 1). Dimorphism increases with the speed and agility of the prey. At one extreme, those vultures and condors that feed entirely from immobile carcasses show no consistent size difference between the sexes or, if they do, the male is slightly bigger than the female, as in non-raptorial birds. Next come the snail feeders, which show reversed dimorphism, but the female is only slightly larger than the male. The insect* feeders and the reptile

* Mainly sluggish or grounded forms, including grasshoppers, locusts, beetles and larvae of wasps and bees.

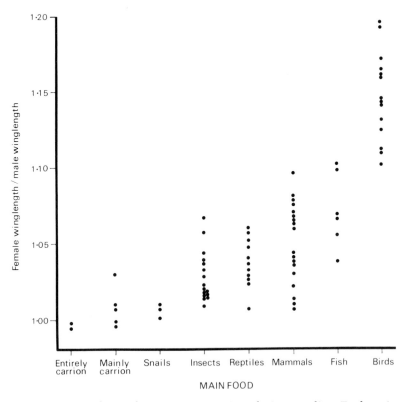

Figure 1. *Size-dimorphism in raptors in relation to diet. Each point represents one species, and only species which feed almost exclusively on one prey-type are included. The faster the prey, the greater the degree of dimorphism in the raptor.*

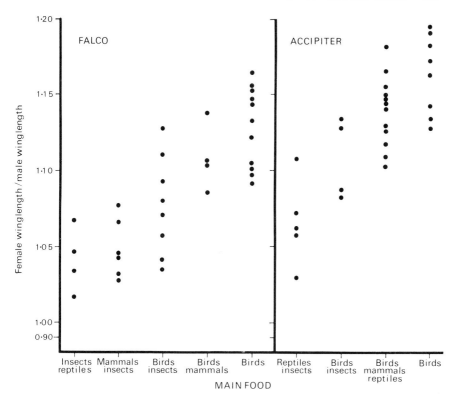

Figure 2. Dimorphism in relation to diet among species of the same genus. Each point represents one species. The same trends hold within genera, as among raptors as a whole.

feeders show somewhat greater dimorphism, followed by the mammal feeders and the fish feeders and then the bird feeders, which as a group are the most dimorphic of all. The extreme is shown by the small bird-eating accipiters in which the female is nearly twice as heavy as the male, namely the European Sparrowhawk *Accipiter nisus* and the North American Sharp-shinned Hawk *A. striatus*.

The link with feeding habits is apparent even among species in the same genus, if they take different foods (Figure 2). Among the falcons, for example, the insectivorous Lesser Kestrel *Falco naumanni* is less dimorphic than the rodent-eating Common Kestrel *F. tinnunculus*, and this species in turn is much less dimorphic than the bird-eating Merlin *F. columbarius* or Peregrine *F. peregrinus*. Likewise, among the accipiters, the frog-eating *A. soloensis* from eastern Asia and the reptile-eating *A. butleri* from the Nicobar Islands are both much less dimorphic than the bird-eaters mentioned above. Even in a single species, the degree of dimorphism may differ between races, according to diet; for example, the more insectivorous races of the American Kestrel *F. sparverius* are less dimorphic than the more rodent-eating ones (Storer 1966, Amadon 1975).

The relationship between diet and dimorphism is shown in Figure 1 only for the 25% of raptor species that have restricted diets, but it still holds for the majority of species that eat two or more kinds of animal. In such catholic species the amount of dimorphism is commensurate with the kinds of animals taken. Raptors that take a mixture of insects and birds, for example, are intermediate in dimorphism between the insect specialists and the bird specialists; while those that take a mixture of mammals and birds are intermediate between the mammal specialists and the bird specialists (Figure 3). Similarly among the vultures, those species that live on a mixture of carrion and living prey show more dimorphism than do those which live on carrion alone (Figure 1).

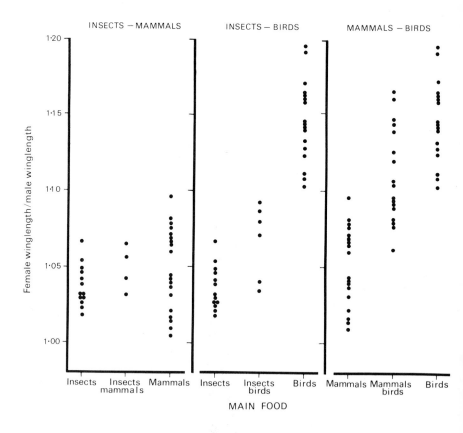

Figure 3. *Dimorphism in some general feeders compared with specialist feeders. Each point represents one species. In each of the three cases shown, the generalists are intermediate in dimorphism between two kinds of specialist.*

Two species of raptors eat a lot of bats, the New World Bat Falcon *Falco rufigularis* and the Old World Bat Hawk *Machaerhamphus alcinus*. The former takes many birds as well and is as dimorphic as some of the extreme bird specialists, while the latter takes fewer birds and is no more dimorphic than the average mammal feeder. I do not know why these species differ, since they both catch flying prey on the wing, but the falcon seems to hunt more by active chasing than the hawk does. The Old World Bat Hawk is the only obvious exception to the overall relationship in raptors between dimorphism and speed of prey.

Associated with the main trend is another; namely, for the more dimorphic species to take larger prey relative to their body size than the less dimorphic ones (excluding the carrion feeders). The snail-eating or insect-eating raptors capture prey weighing less than one-tenth of their own weight, but at the other extreme, some of the bird-eaters capture prey as heavy, or heavier, than themselves. Other species are intermediate. In other words the more dimorphic the raptor, the faster and the larger the prey it takes, relative to itself.

In Figures 1–3, I have depicted dimorphism by use of wing-lengths, taking the ratio of female wing-length over male wing-length as an index of the size difference between the sexes. It might have been better to use weights but, while wing-lengths of both sexes have been measured in almost all species, reliable weights were available for only a small proportion. Of course, weights show a greater divergence than wing-lengths – in the most highly dimorphic species, a 20% increase in wing-length from male to female is equivalent to almost a doubling in weight. But taking weights instead of wing-lengths would not have altered the overall relationship with diet.

So far as I am aware, this is the first time that anyone has drawn attention to the link between diet and dimorphism in this way, though several previous writers have discussed it in general terms, and particularly the extreme dimorphism associated with bird-feeding (Amadon 1959, 1975, Storer 1966). In a statistical study of the North American raptors, Snyder & Wiley (1976) found a significant correlation between the degree of dimorphism in different species and the percentage of birds in their diets (correlation coefficient (r) = 0·79). Their form of analysis was especially suited to raptors that have varied diets containing birds. But it could not be applied satisfactorily on a world scale, because many raptors outside North America have diets devoid of birds, yet still show considerable dimorphism if they eat fish or other fast-moving prey. Nonetheless, their findings on American raptors are consistent with the trends in world species shown in Figures 1–3. Likewise, among North American owls, Earhart & Johnson (1970) found that the extent of dimorphism was correlated with the proportion of vertebrates (as opposed to invertebrates) in the diet. Evidently the convergence between raptors and owls applies not only to the reversal of size dimorphism, but also to its extent.

In contrast, no consistent relationship exists between dimorphism and overall body-size, another possibility that has been suggested (Rensch 1950). In the falcons, it is the largest species that are most dimorphic (except for the Merlin), and in the accipiters the smallest; among raptors as a whole both large and small species include some slightly dimorphic kinds and some highly dimorphic ones. Nor does any consistent relationship exist between dimorphism in size and the degree of colour difference between the sexes; these two aspects are independent.

Explanations for the link between dimorphism and diet
Why is dimorphism so closely related to diet? In the gradation from hunting slow-moving to fast-moving prey, the numbers of competitors from other species declines. Carrion and insects, for example, are eaten by many different kinds of animals, from other insects to birds and mammals; so through competition, each such predator is likely to be restricted to a fairly narrow niche. But fast-moving prey, such as birds, are exploited by very few predators (mostly other birds), so that each such predator has a wider niche, and more scope for diet division within species. In other words, I suggest that it is the progressive reduction in competition from other animals, as one moves from exploiting slow-moving to fast-moving prey, that allows progressively greater food partitioning within species, the small sex taking the smaller items from the total prey spectrum and the large sex the larger items. So far as I know, this explanation has not previously been offered.

Other authors have proposed that the relationship between dimorphism and diet has to do with the difficulty of catching fast-moving, agile prey (Storer 1966, Earhart & Johnson 1970, Reynolds 1972). The idea is that, for successful predation, agility in the prey must be matched by similar agility in pursuer. One way that this could be achieved is for the raptor to be as close as possible in size to its prey, thus reducing the difference in manoeuvrability between them. This would account for the more dimorphic species taking relatively larger prey than the less dimorphic ones, and also for the increasing tendency, as prey speed increases, for the food niche to be partitioned, with the male taking small prey and the female large prey. These two explanations of the link between dimorphism and diet are not mutually exclusive.

A diet difference between the sexes has been noted in various species of accipiters, harriers and falcons, all of which are moderately or strongly dimorphic (Table 1, Höglund 1964, Storer 1966, Mueller & Berger 1970, van Beusekom 1972, Opdam 1975, Snyder & Wiley 1976, Schipper 1973, Cade 1960). It reduces competition for food between the sexes, and perhaps allows each pair to live off a smaller area than they would need if both mates took prey of the same size. Some authors have proposed that the benefit of a diet difference has been the main selective factor behind the evolution of the size dimorphism (Selander 1966). But this view on its own does not explain why the female is the larger sex and not the male.

Explanations for reversed size dimorphism

The female's larger size would seem to be connected with the raptorial lifestyle, as already mentioned; and, although various ecological and behavioural explanations have been put forward, this has proved one of the most intractable problems in raptor biology. One might regard the link with feeding habits as the 'ultimate' cause, and the ones discussed below as 'proximate' causes. To begin with an ecological view: since the male does the hunting for most of the breeding period, it is best for him to be the smaller sex because he thereby specialises on smaller prey which are generally more numerous in the environment, as species and as individuals, than are large prey (Storer 1966). The male can then supply frequent meals for the female and young. The larger size of the female has supposedly been retained, because she must supplement the male's foraging when the young are in greatest need, and her larger size enables her to catch from a greater size spectrum when confined to the immediate nest vicinity (Reynolds 1972). These points may be valid for most environments, but they apply only to species in which dimorphism is great enough to bring about a big difference in diet. And even in these species it is arguable whether a frequent supply of small prey items has much advantage over an infrequent supply of large prey items, especially when the latter can be cached and eaten at will. In other words, I do not think the theory has much to commend it for general application. Another ecological view is that the male is smaller in order to reduce the total food needs of the pair (in one species the male's daily energy needs have been measured to be smaller than the female's). More saving is achieved by the male rather than the female being smaller, because for most of the breeding cycle he is the active member (Mosher & Matray 1974, Balgooyen 1976). The reason why the male is not even smaller, or the female is not small as well, was attributed to supposedly conflicting selection pressures. This kind of argument would apply equally to many other birds, and is obviously unsatisfactory.

Turning now to the explanations for reversed dimorphism that have a behavioural basis, one feature that larger female size confers is female dominance. In conflicts over food, perches, roosts or territories, females can if they wish displace males (Cade 1955, 1960). This seems to be true of all raptors with reversed dimorphism, and in the extreme examples among the accipiters the males seem generally afraid of the females, and are reluctant to approach them closely (Snyder & Wiley 1976). There is thus no doubt about the existence of female dominance, but whether it is the incidental result or the underlying cause of reversed dimorphism is an open question. Nonetheless, hypotheses have proposed that larger female size and dominance in raptors is necessary: (a) to promote successful pair formation and to prevent the male from harming the female; (b) to make the male surrender food for the female and young; and (c) to prevent the male, accustomed to killing small animals, from killing his own young. In my view, none of these ideas stands up to critical examination. Any problem of pair formation is not solved merely by reversed size dimorphism, because it then gives the possibility of females harming males (for records of female accipiters eating males in the breeding season, see Chapter 12). Larger female size is not necessary in other birds to make the male surrender food for his own offspring and, in some highly dimorphic raptors, the male regularly broods and feeds the young (for Peregrine, see Enderson *et al* 1973), or can do so if the female dies (for Cooper's Hawk *A. cooperii*, see Schriver 1969). Likewise, the males of many other birds that prey on small animals do not harm their young, and any such trait in either sex would be swiftly eliminated by natural selection. Another suggestion is that the female is the larger sex because, being responsible for most nest-duties, it is she who has to defend the nest against predators (Storer 1966, Snyder & Wiley 1976). Some raptors do defend their nests vigorously, and it is true that the task usually falls to the female, but the extra advantage that larger size would confer would presumably only count in the species in which dimorphism was great. Nonetheless raptors are some of the few birds which stand much chance of success in nest defence, so this view has more to commend it than the others.

The idea that larger female size is necessary for successful breeding was somewhat weakened by experiments with American Kestrels. By selecting birds from different races, Willoughby & Cade (1964) were able to form pairs in captivity in which females were smaller than males, the same size as males, and larger than males respectively. No differences in behaviour or production of young were noted between the three groups and all groups were highly successful. So in this slightly dimorphic species, greater female size was not important to successful breeding.

Apart from the owls and skuas mentioned earlier, other birds give little clue to the significance of sexual size reversal in raptors. In the waders (Limicolae) that show it, it is associated with unusual breeding behaviour, for the females are more brightly coloured and dominant in courtship, while the males incubate and raise the young unaided. So in these birds, different factors are presumably involved. The frigate birds (Fregatidae) and boobies (Sulidae) also show slight reversed dimorphism, but whether from the same or different selection pressures as the raptors is anyone's guess (Amadon 1975).

In summary, it can be concluded that: (a) reversed dimorphism, and its extent, are closely related with diet, being most marked in species which eat large, fast and agile prey; (b) at least in the moderately and highly dimorphic species it is associated with a measurable food difference between the sexes; (c) irrespective of extent, it also confers female dominance; but (d) this is probably an incidental result of reversed dimorphism, rather than the reason for it. Some existing views on why the female rather than the male is larger in raptorial birds are improbable, while others would apply only to species in which dimorphism is great. In this account, I have discussed only the more important or frequent ideas on size reversal but, as expected with so intractable a subject, some even less likely ideas have been proposed.

SEX RATIOS

In view of size dimorphism, sex ratios in raptors are of special interest. The sex of any individual is determined genetically, and in most animals the ratio of males to females among the offspring is about 50:50. The explanation for equality was first offered by Fisher (1930), based on the action of natural selection on the parents and the sexes of offspring they produce. Suppose that the majority of offspring produced at any time are females. Then a parent who produces only sons is likely, because males are scarce in the population, to leave more descendants than one who produces only daughters or a mixture of sons and daughters. Any inherited trait to produce sons will be perpetuated by selection until there is an excess of males in the population, at which point the tendency to produce daughters will again be favoured. Thus, whenever the sex ratio deviates from unity, selection will tend to restore it, and 50:50 is the only ratio that is stable in the long term. This does not of course mean that all parents will have equal numbers of sons and daughters, only that the chances of any one offspring being male or female are equal, so that in the population as a whole the two sexes will be produced in similar numbers.

In fact Fisher argued that it was not the number of males and females which would be equal, but the parental expenditure (in terms of food and care) on males and females. Where the sexes are the same size at the end of parental care, this generally means that the parent generation will have equal numbers of sons and daughters, as described above. But where the sexes differ greatly in size, the same investment could go towards fewer large offspring (females in raptors) and more small ones (males in raptors). In other words, if it takes less food to raise a male than to raise a female, the sex ratio at conception should be tipped in favour of males, so that in the end the same is spent on raising both sexes. The position is further altered if mortality during the period of parental care falls more heavily on one sex than on the other. Several writers have suggested that male raptors might survive less well in the nest than females because they are smaller and lose in fights for food. If this is true, the cost of raising the surviving males is increased, to allow for those that die before reaching independence. The sex ratio at conception should then biased even further in favour of males, whose proportion will decline during the period of parental care, so that by the end of this period the total expenditure on each sex has been equalised. Any differential mortality which might occur after the young become independent has no relevance to the primary sex ratio. In view of these ideas, it is of special interest to examine the data available for raptors to find what holds in practice.

Sex ratio in nestlings

In species with marked size dimorphism, the young can be sexed at a glance from the time they are about half grown, but in other species, sexing depends on careful measurements, or on plumage differences (eg kestrels) or eye colour (eg harriers). Only for the European Sparrowhawk, the American Kestrel and the Peregrine are figures available for the sex ratio in the egg, and in all three it was effectively 50:50 (Table 2). The Sparrowhawk and Peregrine are highly dimorphic and the Kestrel slightly so. The egg ratios were deduced retrospectively from nests in which all eggs laid gave rise to young which survived to an age at which they could be sexed. Such a sample might not be representative, however, because it takes no account of nests in which mortality of eggs and young occurred.

For well grown nestlings, much larger samples are available for a greater range of species, but by this stage mortality has occurred in some broods, both at egg and nestling phases. To begin with the highly dimorphic Sparrowhawk, the sexes hatch from eggs of similar weight, but they diverge so rapidly that by the time they are 24 hours old, females are already heavier than males to an extent that is significant statistically ($P<0\cdot05$). As they grow, the sexes diverge further, so that by the twelfth day they can be easily distinguished by eye, and by the time they leave the nest at 26–30 days, they are as dimorphic as adults. Nevertheless, 651 broods examined when near fledging contained about equal numbers of each sex, a total of 1,102 (50·95%) males and 1,061 (49·05%) females (Newton & Marquiss 1979).

Individual Sparrowhawk broods contained from one to six nestlings. In each size of brood a number of sex ratios was possible. A family of five, for example, might contain five males, or one male and four females, two males and three females, and so on through to five females. Whatever the brood size, however, the frequency with which broods of different sex composition were encountered in the wild did not deviate significantly from the frequencies expected on an overall 50:50 ratio (Figure 4). In a further analysis, no significant variation in the overall sex ratio at fledging occurred between different areas, nor between years in the same areas. Hence, in this species, no evidence was found for differential mortality between the sexes at any stage from egg to fledging, nor that the overall sex ratio at fledging was other than equal.

Other data on the sex ratio near fledging were obtained in at least 17 studies of nine species varying from slightly to highly dimorphic (Table 2). In most of these the sex ratio did not differ significantly from unity; but in one out of four studies of Peregrines and one out of two of Hen Harriers, the surplus of females was statistically significant ($P<0\cdot05$). Differential mortality of the sexes in the nest may have caused this imbalance.

Among nearly fledged Goshawks *A. gentilis* in Finland, males were 57% in broods of four young, 51% in broods of three, and 47% in broods of two (overall in these broods 52%). The smaller broods were

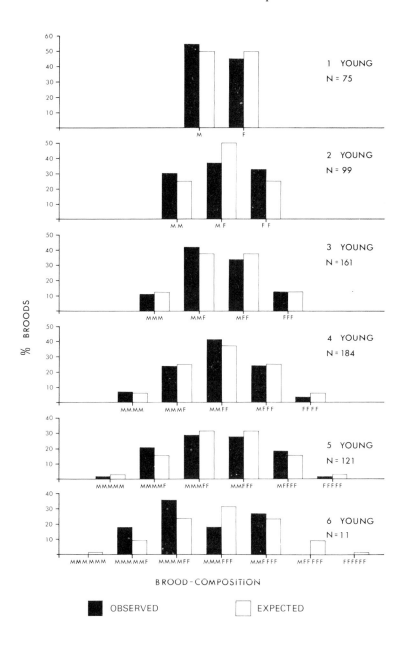

Figure 4. Sexes of young in Sparrowhawk broods. The frequencies with which broods of different sex composition were observed in the wild did not differ from the frequencies expected on the assumption of an overall 50:50 ratio, M – male; F – female. Drawn from data in Newton & Marquiss 1979.

derived mainly from large ones after mortality, so Wikman (1976) took this trend to indicate an initial surplus of males and a greater loss of males than of females at the nestling stage. He excluded broods of one young from the calculation because he thought that observers were less likely to sex such young correctly than young in larger broods, in which sexes could be compared. If such single-young broods are included as recorded, however, then small broods no longer contain fewer males than those of 4–5 young. So further data are needed for this species to check the possibilities of greater numbers and subsequent mortality of male young.

Summarising, no significant departure from a 50:50 ratio at hatch was apparent in the three species for which data were available, nor was it usual in the nine species examined near fledging. An apparent predominance of females occurred in particular studies of two species, but not in other studies of these same species. As yet there is no firm evidence for differential mortality of either sex in the nest, but the Goshawk is a promising subject for further investigation.

At first sight, these observations appear to contradict Fisher's prediction that the parents should equalise investment rather than numbers. However, in the highly dimorphic Sparrowhawk, male and female nestlings were found to eat the same amounts, not different amounts as might be expected from their size (Newton 1978). Females put on weight faster, but males feathered sooner and left the nest 3–4 days earlier. Hence, parental expenditure on the two sexes was equal, at least until fledging (the young were fed for 2–3 weeks after leaving the nest, but food intake could not be measured then). As the Sparrowhawk is one of the most dimorphic of raptors, it would clearly be unwise to assume that parents of other species invest differently in individual male and female offspring merely because these offspring differ in size. If, in other raptors, food consumption of males and females is the same, and there is no sex difference in mortality, then the data are consistent with sex ratio theory. It would be useful to have more information on sex ratios, and on the comparative food intakes of male and female nestlings, from a greater range of species.

Sex ratio in adults

Only in those species in which males and females differ strikingly in size or colour can the sexes be distinguished in the field, but species with lesser differences can often be sexed in the hand by taking weights and measurements. The overall sex ratio is almost unattainable, however. In the breeding season, one sees mainly males because females are on their nests, while in the rest of the year the sexes behave in ways that make it hard to get unbiased counts. They may differ in migratory habits, so that one sex leaves in greater proportion or migrates earlier or further than the other (Chapter 11). This leads to unequal sex ratios among birds seen on migration or in particular wintering areas. The sexes may also differ in habitat preference; or in the ease with which they can be seen or trapped. Hence, most published information on the sex ratios of full grown raptors shows a strong preponderance of one or other sex that reflects the ecological differences between males and females rather than their overall numbers (Table 3).

Where the sexes were found to differ in wintering habitat, the females generally predominated in the more open of the various places used. Among Sparrowhawks in Scotland, males predominated in forest and females in farmland. Among Kestrels in California, males predominated in orchards and well timbered farmland, and females in open pastures and fields. Among Peregrines in Texas, males were commoner relative to females inland than on open sea coasts (Hunt *et al* 1975). All these differences may have been due to prey selection, or they may have been linked with the larger size and lesser manoeuvrability of the females, which might have led them to prefer more open situations than males. Among Goshawks in northern Europe, males predominated in trapped samples but only after the first year of life. This was because in the first year both sexes moved around equally (and were thus equally susceptible to capture), but thereafter males moved more than females (and were thus more susceptible to capture) (Haukioja & Haukioja 1970).

The tendency for the sexes to winter in partly different areas or habitats, together with a difference in diet, presumably reduces competition between them. It also makes it unlikely that the two sexes would survive equally, since they occupy partly different niches, and are dependent on different resources and exposed to different risks. There are thus strong grounds for expecting that the overall sex ratio among adult raptors would not be equal, especially in the most dimorphic species. So far, the only clue that this is so comes from breeding populations, in which one sex bred at a younger age, on average, than the other. Females bred more often as yearlings than did males in populations of Goshawks, Red-shouldered Hawks *Buteo lineatus*, Hen Harriers *Circus cyaneus* and Peregrines; and males more often than females in European Sparrowhawks (Table 21). One interpretation of these findings supposes a shortage of older females in the first four species, and a shortage of older males in the fifth. Most such species are strongly dimorphic. A bias towards more males might also be expected in populations which are heavily persecuted, because females are more easily shot at the nest, but for this view there is no evidence one way or the other. In time, enough ring recoveries for some species should accumulate to assess the survival of each sex separately.

Summarising, the overall sex ratio in full grown raptors is hard to get because of ecological differences between the sexes, but in some monogamous species the age structure of breeding populations suggests an unequal ratio.

MATING SYSTEMS

Most raptors are monogamous but, whereas in some populations the individuals have a different mate each year, in others they keep the same mate for several years (Chapter 10). Either way, the sexes have different roles: the male provides the food and hunts mainly away from the nest; while the female stays near the nest, and is responsible for incubating the eggs and for looking after the young. This difference between the parents persists at least until the young are about half grown, after which the female may also begin to leave the nest area to hunt.

Since the division of duties is more marked in raptors than in many other birds, this system presumably has advantages for raptors over one in which the sexes share equally in incubation and foraging. Several possibilities come to mind. First, the total food needs of the pair are reduced if the larger sex can become inactive; second, the female can afford the luxury of accumulating larger body reserves for breeding if she becomes inactive at that time rather than continuing to hunt for herself (Chapter 6); and third, as already mentioned, it is better that the larger female be present at all times to defend the nest, rather than the male at some times and the female at others. Moreover, in the most dimorphic species, it may be hard for the small male to incubate the eggs effectively. Some of these points apply only because the female is the larger sex, and would not apply if the reverse were true. The vultures are the only raptors in which male and female are known to share equally in incubation and foraging; this may be connected with the lack of an appreciable size difference between the sexes, or with their particular feeding habits (Chapter 9).

Alternatives to monogamy

I shall use the term 'polygyny' when the male has more than one female at a time, and 'polyandry' when the female has more than one male. Polygyny has been recorded in at least eleven species in five genera (Table 4). In type A, the females used the same nest, in type B they used separate nests close together in an area that would normally hold one pair, and in type C they used separate nests in areas that would normally hold separate pairs. More than one type has been found in the same species. Non-monogamous relationships are easily overlooked, however, and are almost certainly more widespread than indicated in the table. They do not entail any marked change in the usual duties of male and female, except that in polygyny the female has to fend for herself more than usual. On present data there is no reason to suppose that any of the individuals within a mating group are closely related to one another, but more studies with marked birds are needed.

The harriers are the only species in which polygyny is known to be frequent, the females using different nests. In the Hen Harrier (or Marsh Hawk) *C. cyaneus*, it has been noted in widely separated parts of Europe and North America, but at different frequencies (Table 5). In some areas all birds were monogamous, while in others a small or large proportion were polygynous. The most extreme was on the Orkney

1   (Upper) Ospreys at nest in Sweden. During the nineteen-fifties and sixties this species declined in parts of its range owing to contamination with organo-chlorine pesticides and other pollutants. Photo: M. D. England. (Lower) Martial Eagles at their nest in Zimbabwe Rhodesia. One of the largest of the eagles, this species lives at some of the lowest natural densities recorded for birds, with one pair per 125-300 km^2 of suitable habitat, depending on region. Photo: Peter Steyn.

2 (Upper) Black-shouldered Kite at nest. This widely-distributed species of the Old World lives primarily on small rodents and, when these animals are plentiful, some kites may raise two or three broods in a year. Photo: M D. England. (Lower) Snail Kites at their nest in a Florida marsh. This New World species breeds in colonies in bushes or reeds and feeds on aquatic snails. Photo: H. A. Snyder.

Islands, off northern Scotland, where the overall sex ratio in the breeding population over several years was two females to every male (Balfour & Cadbury 1974, 1979). Up to 55% of breeding males and 75% of females were involved in polygynous matings, depending on their age. Yearling males usually had only one female, but older males had up to three females at a time and occasionally up to six. In the Netherlands, a single male was thought to hold an area containing seven different females (Kraan & Strien 1969). When the number of females gets so large, however, it becomes hard, even with marked birds, to be sure who is mated to whom, and the records of such large groups should be treated with caution. Polygyny has also been recorded commonly in the Montagu's Harrier *C. pygargus* and Marsh Harrier *C. aeruginosus*, with 2–3 females per male (Table 5), and may well occur in other harrier species not yet investigated.*

In each polygynous harrier group studied, one female was favoured by the male and was apparently dominant over the other female(s). This alpha-female received most food from the male at the start of the season, and was first to lay (Hecht 1951, Dent 1939). Once she was incubating, the second female began to get more food from the male and laid her eggs soon after; any other females followed in turn (N. Picozzi). During the incubation and nestling periods, the male often began to favour the alpha-female again, and since this female was first to hatch her eggs, she was in a strong position to reclaim the male's attention. In consequence, the other female(s) more often failed in their breeding, or at least had to begin hunting for their young at a much earlier stage. This was the general picture, but there was some variation in behaviour from group to group. In some groups the female hierarchy appeared well defined, in others less so.

In the Common Buzzard *Buteo buteo*, males mated to two females were noted in four out of 33 territories studied in northeast Scotland (Picozzi & Weir 1974). Over the years, a total of seven instances was noted, and almost always both females laid. In those territories in which bigamy was recorded in more than one year, one or both females were different in the second year. This implied that the bigamy was regulated, depending on some aspect of territory or male. Another instance of bigamy was recorded among Buzzards elsewhere in Scotland. It occurred in a vole plague year and both females raised young (Table 4). The Buzzard has been studied many times in other countries, but not found to be other than monogamous. However, males with two females have been found in the closely-related North American Red-tailed Hawk *B. jamaicensis*, and also in the Osprey *Pandion haliaetus*, the European Sparrowhawk and in three falcon species (Table 4).

Polyandry, involving one female and more than one male, has been noted as frequent in at least two species, in both of which the males in a

* The harriers in general show much greater colour difference between the sexes than is usual in raptors, and the males are of particularly striking appearance. This could result from more intense sexual selection which is likely in polygynous compared with monogamous species.

group all copulated with the female, participated in incubation and provided food for the female and young. In Harris' Hawk *Parabuteo unicinctus* (two males per female), it apparently involved 23 out of 47 females in one Arizona area (though the sexes were confirmed at only three nests), and one out of 19 in a Texas area; in the Galapagos Hawk *B. galapagoensis* (two or three males per female), it involved from 0 to 73 % of females on different islands (Mader 1975, 1975a, Griffin 1976, de Vries 1975).* In other raptor species, three adults have been seen to occupy a territory together, but the third bird was neither sexed nor known to participate in breeding. Such threesomes have been observed several times in Bateleurs *Terathopius ecaudatus* (once the surplus bird was a male) and at least once in Bald Eagles *Haliaeetus leucocephalus* in which the trio was seen together for four years (Brown 1952–53, Sherrod et al 1977).

Conditions favouring non-monogamous mating systems

To what extent do these various mating systems depend on uneven sex ratios? Only for two relevant species have good data been published on the sexes of nestlings, the polygynous Hen Harrier and the polyandrous Harris' Hawk, and only in one study of the harrier was there a significant predominance of females (ratio 1:1·16), though not as great as in the breeding population (Table 2). Again there were problems with the overall adult sex ratio because counts were restricted to breeding areas, where the ratio might have been set by the mating system (rather than the other way round), and non-breeders of one or other sex might have been excluded from such areas. Among Hen Harriers, more yearling females than yearling males were found breeding, which is one way in which some extra females for polygyny might have been found (Table 21). Unfortunately, female ages can only be found from marked birds because, while brown yearling males can easily be distinguished from grey older ones, both yearling and older females are brown and look the same. This difficulty applies to most harrier species.

In birds other than raptors, neither polygyny nor polyandry is a mere byproduct of an uneven sex ratio. Some species have one or other mating system despite an equal sex ratio among adults, whereas others are monogamous despite an unequal ratio (Verner 1964, Zimmerman 1966, Lack 1954). As occurs to some extent in the Hen Harrier, non-monogamous mating systems often entail one sex starting to breed at an earlier age than the other (Wiley 1974). So although a surplus of one sex might in some species contribute to the development of non-monogamous mating systems, other circumstances must presumably be involved.

What are these other circumstances likely to be? They must be

* The term 'nest-helper' has sometimes been used for the extra males, but this is inappropriate for birds which also copulate; moreover, they are not analogous to the nest-helpers found in other birds.

different from those that would merely lead to a higher or lower density of breeding pairs. In polygyny, the mated males must be capable of excluding from the breeding areas any other males that might wish to breed, for only then could they keep the extra females for themselves (Orians 1961); in polyandry the mated females must be capable of excluding other females. Since the male in breeding raptors is responsible for most hunting, one might expect that polygyny would arise in good food conditions, when a male could feed more than one female at the time of settling (or alternatively when the females could start to breed on a minimum of food from the male), and that polyandry would arise in poor food conditions, when one male alone was unable to feed a female properly (Newton 1976a). Three lines of evidence suggest a link between polygyny and ease of foraging: (1) polygyny occurs mainly among raptors found in rich treeless habitats, such as marsh or prairie, where nearly all animal life is concentrated near the ground (this is also true in other birds, see Lack 1968, Orians 1971); (2) in any one species polygyny is most prevalent in years or in areas which on other evidence seem rich in food (see Balfour 1962 for Hen Harrier, Picozzi & Weir 1974 for Buzzard); and (3) in the Hen Harrier it is more prevalent among experienced, older males than among inexperienced young ones (see above). Polygyny controlled by food could of course develop only where the number of females was not limited by some other factor, such as scarce nest-sites.

The proposed link between polyandry and difficulty of foraging is less firm, but Harris' Hawk practises group hunting, so two males working together may be disproportionately more successful than two singles (Mader 1975, 1975a). The polyandrous Galapagos Hawk apparently does not practise group hunting, but de Vries (1975) thought that polyandry was most frequent on islands where competition for food was unusually strong. In both species, the individual groups persisted for more than one year, which suggests more than a casual bond between the members.

Given appropriate food conditions, such mating systems are also likely to arise only when they offer advantages to each participant over the alternatives of monogamy or non-breeding. Only in this way could they be favoured by natural selection. In polygyny, the advantage to the male in having more than one female was borne out by results from the Hen Harriers in Orkney, where the average number of young produced by each male rose with the increase in the number of females in the group (Table 6). The average adult male raised 30% more young with two females than with one, and 74% more young with four females. The advantage to the individual females in these groups was less obvious, because they lost the undivided attention of the male and, on average, they each produced fewer young than did the females in monogamous matings (Table 6). It seems, then, that the average male benefits from polygyny, but the average female does not. Most failures occurred at the egg stage, apparently because the males were unable to provide all their females with enough food to enable them to continue with incubation. These females therefore hunted for themselves and allowed their eggs to chill, or deserted them.

So, with these disadvantages, why did any females enter into polygynous groups? The answer probably differed for different females, according to their relationship with the male. Within groups, the alpha-females raised young more often than the others. The others were successful either when the male was able to provision more than one brood, or when the alpha-female failed, releasing the male to give more attention to other broods, or when the weather was warm, so that the females could leave their young unbrooded and hunt themselves. The alpha-female may have accepted other females into the group because she was unable to keep them out, especially when she had eggs to incubate. Also they posed little threat to her, and probably made little difference to her breeding success. The less favoured females, on the other hand, may have accepted polygyny because it was the only alternative for them to non-breeding. These views are tentative, however, for none of the people who studied polygyny in raptors considered these questions in their papers. More information is especially needed on the success of individual females in a group in relation to their position in the hierarchy, and also on their fortunes from year to year.

In polyandry, the obvious advantage to the single female is evident from results on Harris' Hawk, in which trios raised more young than pairs, on average 2·0 and 1·3 respectively (Table 7). The advantage to the males is not so obvious, because the number of young raised per male was slightly less in trios than in pairs, at 1·0 and 1·3 respectively (though samples for this comparison were small). More information is also required on the behaviour of the males in a trio, and particularly on whether one copulates with the female much more than the other does. With their rigid dominance hierarchies, such trios seem models of peaceful co-existence, and show various intimate behavioural interactions and a lack of aggression that would be hard to imagine in other raptors (including perching on one another's back, Mader 1975a). The same behaviour may also help to strength social bonds for joint hunting, in which case Harris' Hawk resembles some of the pack-hunting mammals.

SUMMARY

Reversed size dimorphism, with the female larger than the male, is an important feature of raptors and its extent is correlated with diet. It is least pronounced in species which eat immobile or slow-moving prey, and most pronounced in species which catch large fast-moving prey. It means that females generally dominate males. Sex ratios in nestling raptors tend to be equal, but in adults they are hard to evaluate because of sex differences in migration or habitat. In breeding, there is a marked division of duties between male and female (except in vultures). Most raptor species are monogamous, but others are some-times polygynous or polyandrous. I suggest that mating systems involving more than two birds arise in specific environmental condi-tions, and when the advantage to each participant (in terms of offspring raised) is greater, on average, than in the alternatives of monogamy or non-breeding.

Buzzards

CHAPTER 2

Dispersion

By dispersion, I mean the pattern in which birds distribute themselves over the landscape in relation to food and other resources. As a group, raptors vary in this respect almost as much as do birds as a whole: they nest sparsely in individual large areas or densely in closely packed colonies, and they feed singly or in flocks, according to conditions. Dispersion is thus concerned with the spacing between nests and colonies, with the size and density of colonies, and with the day-to-day movements of individual birds. Its study tells us about the social organisation of populations, and about the behaviour of individuals towards one another.

An understanding of dispersion depends partly on being able to track particular birds in the field. This is easier if the birds have been wing-tagged or marked in some way so that they can be identified, for only then can we be sure that we are watching the same individuals. In recent years, radio-telemetry has often been used for this purpose. A small radio transmitter is fixed to the bird and, with the help of a receiver, the bird can usually be located, and can then be tracked on foot or from a car or aircraft, as the need may be. By using different wave frequencies, several birds can be studied in the same area. The method is especially useful with species which range over large areas

Illustration: Peregrine on nest-cliff

or which live in dense cover, where they cannot easily by followed by eye. If the points where an individual has been found over a given time period are plotted on a map, they can be used to provide an estimate of the home range of the individual.

In addition to watching the birds themselves, much can be learned about breeding dispersion from the locations of nests. In many raptors these are conspicuous and long-lasting structures, which can be readily identified, even outside the breeding season. With the help of maps, the dispersion of birds or nests can then be examined in relation to habitat and other landscape features.

Nesting places
Whatever their dispersion, most birds-of-prey choose special places for their nests. Such places may be cliffs, isolated trees, groves of trees, or patches of forest or ground cover, depending on species. Peregrines use cliffs, Goshawks use woods, Marsh Harriers use reedbeds, and so on. Such features are important because in any landscape, they form the basis of distribution for any breeding population. The locations of other pairs are also important, because in solitary species the pairs tend to space themselves out, while in colonial ones they clump together.

A striking feature of raptor nesting stations is that they are often used over long periods of years. Particular cliffs are known to have been used by successive pairs of Golden Eagles *Aquila chrysaetos*, or of White-tailed Eagles *Haliaeetus albicilla*, Peregrines or Gyr Falcons *F. rusticolus* for periods of 70–100 years (Newton 1976a). Among 49 British Peregrine cliffs known to falconers between the 16th, and 19th centuries, at least 42 were in use during 1930–39 (Ferguson-Lees 1951). Continued occupancy may thus have held at many cliffs for centuries, long before there were ornithologists to record it. In trees, too, certain eagle nests have been used for longer than a man's lifetime and, added to year after year, have often reached enormous size. One historic Bald Eagle nest in America spanned 8 m² on top and contained 'two waggon-loads' of material, while another was 3 m across and 5 m high (Bent 1938). Such nests sometimes become so heavy from the continued addition of material over the years that the branch supporting them breaks off, and the birds are forced to start anew. Some Osprey nests were in continued use for periods of 45, 44 and 41 years, and Red-shouldered Hawk nests for 47 and 37 years (Bent 1938). Certain patches of forest (though not the same nest) have been used for long periods by other species, and even patches of ground cover were used by Hen Harriers for more than 50 years (Balfour 1957). In general, of course, sites on rock must be more permanent than those in trees, and sites in trees more permanent than those in herbaceous cover.

Several authors have felt that the continued use of particular places by solitary species depended on each bird, whenever it lost a mate, attracting another partner to the same site. This may be so to some extent, but from experience in Britain, even when both occupants were shot every year, places still remained in use (Chapter 3). So perhaps continued occupancy is best explained by the superiority of particular places over local alternatives, together with the need for any incoming birds to fit within an existing territorial framework. The tendency of young birds to return to breed in the general area where they were born normally helps to ensure a continuing supply of new recruits to fill the gaps (Chapter 10).

Colonial raptors also tend to nest in the same places year after year, and in southern Africa many cliffs whose names indicate that they were used by vultures in previous centuries are still used by these birds today. As in other colonial birds, the individuals defend only a small area around their nests, so that, given enough ledges, many pairs can crowd onto the same cliff, leaving other apparently suitable cliffs vacant. In birds such as these, continuity of use can be explained partly by the obvious attraction that existing colonies have for newcomers, as well as by the superiority of particular cliffs and their spacing in relation to other colonies.

At regular nesting stations raptor pairs usually have favoured perches, which they use when eating and roosting, and which become marked by an accumulation of white-wash, pellets and prey remains. In species which build their own nests, nesting stations also contain old structures from previous years, and are thus easily recognised. Typically, you find several nests, at various stages of dilapidation, within a short distance of one another. When occupied, nesting stations could equally well be called 'nesting territories', because normally each is held by a single pair which keep away others of their species. I have therefore used the term 'nest-site' for the situation of the nest, 'nesting territory' for the area around the nest that is defended, and 'home range' for the area that embraces all the activities of a bird or pair over a given time period. In the case of a breeding pair, the home range would include the nesting territory and any hunting areas, whether defended or not. In the polygynous harriers discussed in Chapter 1, the males were evenly spaced through the breeding habitat, and females were clustered around each male, on their individual nesting territories (Balfour & Cadbury 1978).

Sometimes the Peregrines associated with a particular home range alternate between two or more cliffs for different nesting attempts (Figure 5). Such cliffs may be more than a kilometre apart, but are judged to be alternatives in the same home range because only one is occupied at a time, and the same ringed birds have been trapped at the different cliffs (R. Mearns). Golden Eagles and other cliff-nesting species also use alternative nesting stations, as do Goshawks and other forest species, and Merlins and other ground nesters. It is not always obvious what influences the choice in particular years. The contemporary distribution of neighbours or competing species may be important, as may prey populations, disturbance, or previous nest success. In some species, pairs more often return to the same site if they were successful there the year before than if they failed (Chapter 8). Also, within a season, Peregrines usually shift to an alternative cliff for a repeat clutch if the first fails.

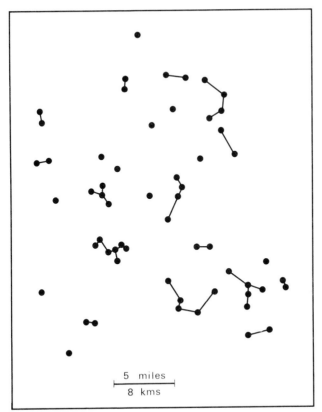

Figure 5. Nest-spacing among Peregrines in part of Britain. Each point
marks a nesting place, and lines join the alternative nesting places of the
same pair. The different pairs are spaced fairly regularly in this area
where nest-sites are not limiting. Re-drawn from Ratcliffe 1962.

Hunting places

How far from the nest raptors hunt seems to depend largely on food
availability. In general, birds range over small areas where prey is
abundant near the nest, and over large areas where prey is either
generally scarce or far distant. Home ranges are thus determined
largely by the number of feeding places (or hunting perches) used, and
by the distances between them. To give some examples of long-
distance hunting: some European Sparrowhawks that nested one
kilometre apart in a forest flew up to nine kilometres to hunt outside
the forest; Montagu's Harriers flew up to 12 km from the nest to hunt,
Red Kites flew up to 15 km, Sakar Falcons *F. cherrug* up to 20 km,
Peregrines up to 20 and 27 km in different areas, and Black Eagles *A.
verreauxi* up to 27 km (refs. in Newton 1976a). In the early studies,
such extreme distances were estimated from observation or from the
regular appearance at the nest of prey-species which could not have
been obtained nearby; but in recent studies, such distances have been

checked using radio-telemetry. Birds that hunt far away spend much time in travelling back and forth from the nest. They may often have to cross the nesting territories of other pairs, or fly long distances over terrain in which hunting is impossible. This is true for example of Ospreys which nest a long way from water, or of European Kestrels which nest in towns a long way from open land. Thus, long-distance hunting, which occurs when good nest-sites are well separated from good hunting places, tends to be inefficient and often results in poor breeding success (Chapter 8).

Individuals may change their hunting areas during the course of a breeding cycle. They do so partly in response to seasonal changes in prey-distribution or to changes in their prey needs, and partly because they are freer to range long distances in late season, when they no longer need to guard the nesting territory or the young so closely. Successful pairs can also exploit areas in late season which had previously been defended by other pairs that have failed and moved away, as noted in Marsh Harriers and other raptors (Craighead & Craighead 1956, Sondell 1970, Johannesson 1975). The net result is that the home ranges of particular pairs often change in shape and size during the season, and get larger towards its end. The female, in particular, remains very close to the nest until the young are feathered, but may then begin to range over an even larger area than the male. In addition, pairs that are breeding tend to range over larger areas than those that are not, while single birds use even smaller areas, corresponding with their differing food needs (Smith & Murphy 1973).

Some of these points are illustrated for the Sparrowhawk in Figure 6, based on radio-telemetry studies in south Scotland (M. Marquiss & I. Newton). Males had fairly small ranges around the time of laying, then hunted over an increasingly larger area in the nestling period, contracting slightly again in the post-fledging period. Females stayed at the nest from before laying until the young were about half grown, after which they ranged further and further afield, eventually hunting at greater distances from the nest than the males. Moreover, both sexes ranged over larger areas in habitat that was poor in prey than in habitat rich in prey. Neighbouring males held mutually exclusive ranges around the time of laying and early incubation, whereas females were exclusive whenever they were restricted to the nest vicinity; at other times neighbouring birds of both sexes overlapped widely in their ranges, with no great correspondence between the members of a pair. In winter, all birds had large and overlapping ranges, and juveniles hunted over larger areas than adults (Figure 7). In general, home range sizes seemed to depend on habitat and local food availability, on the age and competence of the birds concerned, and on their immediate food-needs, which were greatest when there were large young to feed. Thus, range-size at all times of year seemed to represent the area that an individual had to cover to obtain enough food.

Some radio-marked Goshawks, which were studied during autumn in Sweden, also had widely overlapping ranges (Kenward 1977a).

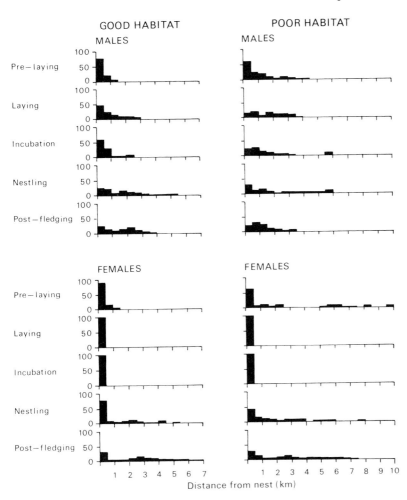

Figure 6. Occurrence of radio-tagged Sparrowhawks at different distances from the nest, according to stage of breeding and habitat. Each histogram shows the mean values obtained from several individuals studied at each stage. Differences in foraging behaviour are evident between the sexes, between different stages of breeding and between habitats rich in prey and habitats poor in prey. From M. Marquiss & I. Newton.

From 14 ranges defined, the two smallest (both about 7 km²) belonged to adults, and the two largest (46 and 49 km²) belonged to juveniles. However, the ranges of juveniles were large partly because these birds occasionally flew up to ten km outside their regular areas. Such flights often involved soaring, and may have served for exploration, with the hawks hunting primarily in one area, but checking other areas from time to time, in case they offered better feeding.

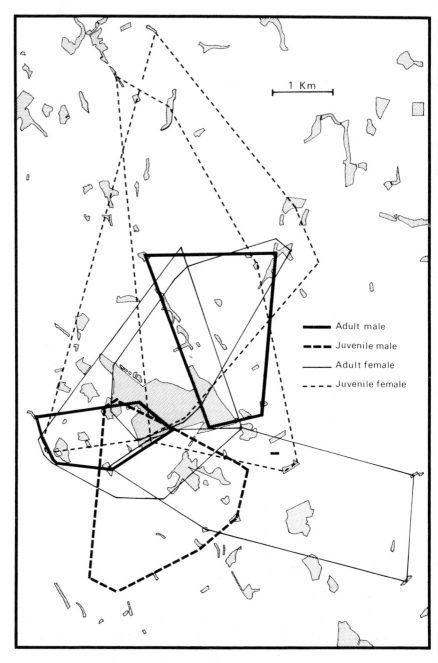

1 Km

—— Adult male

▬ ▬ ▬ Juvenile male

—— Adult female

- - - Juvenile female

Figure 7. Overlapping home ranges of wintering Sparrowhawks in southern Scotland. Stippled areas show main woodlands. Females extended more into open areas than males, and juveniles had larger ranges than older birds of the same sex.

Defence of hunting areas

Apart from an area around the nest itself, the amount of ground that raptors defend in the breeding season seems to depend largely on what is feasible, which in turn depends on such factors as food and habitat and on the number of intruders they have to keep out. Where birds can get all they need in a small area they often defend it all, but where they range over a larger area, this tends to be less exclusive, with poorly defined boundaries. Exclusive ranges containing nesting and feeding areas have been noted in various *Buteo* species, including the Common Buzzard in different parts of Europe, and the Red-tailed Hawk in North America (Dare 1961, Mebs 1964, Fitch *et al* 1946). Such ranges could equally well be called territories because they were defended in their entirety. Moreover, the boundaries between them were often stable from year to year, corresponding with landscape features. Similarly exclusive ranges were held year-round by the Black Eagles which nested at high density in the Matopos Hills in Rhodesia, and by Greater Kestrels *F. rupicoloides* and Black-shouldered Kites *Elanus caeruleus* in South Africa (Figure 8, Gargett 1975; Kemp 1978; J. Mendelsohn). On the other hand, widely overlapping ranges were found in Red Kites, Hen Harriers and Prairie Falcons *F. mexicanus*, and inevitably in various colonial and flock-feeding raptors (Davies & Davis 1973, Balfour 1962, R. Fyfe). Some species, such as the European Kestrel (and the Sparrowhawk just mentioned), were found to have almost exclusive ranges near the start of the breeding season but much larger, overlapping ones by the end (Village 1979). Hence, the amount of overlap in the home ranges of raptors varies from almost nil to almost complete, apart from the nesting territory. Furthermore, in places the overlap may involve many different individuals, and not only those from adjacent pairs. The spatial overlap may be greater than the temporal, however, for in solitary species the different individuals often hunt the same areas at different times. This is a consequence partly of chance, and partly of a tendency the birds have to avoid one another, and separate if they meet.

Behaviour towards other species

Most birds defend territories only against their own kind, but some also exclude other species with similar ecology. Red-tailed and Red-shouldered Hawks hold mutually exclusive feeding territories in North America, as do Common and Rough-legged Buzzards *B. lagopus* in northern Europe, and Golden and Bonelli's Eagles *Hieraaetus fasciatus* in southern Europe (Bent 1938, Craighead & Craighead 1956, Sylvén 1978, Cheylan 1973). The species in each pair occupy mainly different regions or habitats, and overlap only in restricted places. Here they eat similar foods, so the advantage in mutually exclusive territories is clear, at least to the larger species which is dominant.

Other raptors defend only their nesting places against other species, so that no more than one species is found on the same section of cliff or in the same patch of forest. This may serve to reduce mutual interfer-

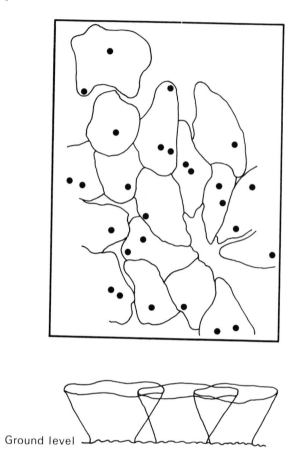

Figure 8. Mutually exclusive nesting/feeding territories of Black Eagles in the Matopos Hills, Rhodesia. Lines show territory boundaries and spots show nest-sites. In contrast to the well defined boundaries on the ground (upper), there was considerable overlap between pairs in the air, and as birds soared higher they often drifted further beyond their ground boundaries (lower). Re-drawn from Gargett 1975.

ence with breeding, but the interactions sometimes lead one species to affect the numbers, distribution or breeding of another (Chapter 3). In North America, Bald Eagles behave aggressively to Ospreys nesting nearby and occasionally rob them of fish. On one islet in Florida Bay, the establishment of an eagle pair was followed by the relocation of nest-sites by eight out of 13 Osprey pairs and a reduction of Osprey breeding success that year. Aerial surveys of the entire bay over several years showed increases in the number of Osprey nests on islets which had lost an eagle pair, and reductions in Osprey nests on islets which had gained an eagle pair (Ogden 1975). Interactions between other

Haliaeetus eagles and Ospreys have been noted elsewhere in the world.

In their dispersion patterns, raptors show the same broad relation-ships with food as do other birds, nesting solitarily when food is fairly evenly distributed and predictable, and colonially when food is clumped and unpredictable (Crook 1965, Lack 1968, Ward & Zahavi 1973). Given widespread nest-sites, the three main patterns are as follows.

(1) *Pairs spaced out in individual home ranges.* This seems to be usual in about 75% of the 81 raptor genera, including some of the largest, such as *Accipiter, Buteo, Aquila* and *Falco.* Each pair defends the vicinity of the nest and a variable amount of surrounding terrain, so that home ranges may be exclusive or overlapping, as described above. Through suitable habitat, the nests of different pairs tend to be spaced fairly regularly, at distances from less than 200 m apart in some small raptors to more than 30 km apart in some large ones (Chapter 3). Most species that show this dispersion system feed on live vertebrate prey and show considerable stability in numbers and distribution from year to year (exceptions discussed later). Individuals usually hunt and roost solitarily.

Outside the breeding season, the birds may occupy much the same range as within it; they may stay together in pairs or one member may stay and the other may leave. Adults that have left their breeding areas occupy individual ranges, as do immatures.

(2) *Birds nest in loose colonies and hunt solitarily.* This system is shown, among others, by the kites *Milvus migrans* and *M. milvus, Elanus scriptus* and *E. leucurus,* and by the harriers *Circus aeruginosus, C. cyaneus* and *C. pygargus.* Groups of pairs nest close together in 'neighbourhoods', and range out to forage in the surround-ing area. The different pairs may hunt in different directions from one another, or several may hunt the same area independently, from time to time shifting from one area to another (Sondell 1970, Dickson 1974). The breeding groups usually contain less than ten pairs, with nests spaced at 70–200 m apart. Larger groups have sometimes been found, including up to 30 pairs of Montagu's Harriers and up to 54 pairs of kites (49 Black Kites and five Red Kites) (Laszlo 1941, Meyburg 1969). In harriers, the tendency to coloniality is sometimes accentuated by polygyny, and in harriers and kites by the frequent need to concentrate in patches of restricted nesting habitat. Even where nesting cover is widespread, however, the colonial habit is still apparent. Such species often exploit sporadic food sources, such as local grasshopper or rodent plagues. They are nomadic to some extent, concentrating to breed wherever food is temporarily plentiful, so that local populations can

fluctuate substantially from year to year (Galushin 1974). Not all pairs of such species nest in groups, however.

Outside the breeding season, kites and harriers tend to base themselves in communal roosts, from which they spread out to hunting areas during the day (Weiss 1923, Davies & Davis 1973, Renssen 1973, Watson 1977). Kites roost in trees and harriers in reeds or long grass, in which each bird tramples the vegetation to form a platform. The roosts usually contain up to 20 individuals, occasionally more, and up to 300 harriers of several species were counted at one place in Africa. From a roost in Michigan, Hen Harriers flew up to 36 km to hunt, while in Germany Red Kites flew up to 34 km (Craighead & Craighead 1956, Hölzinger et al 1973). Outside the breeding season, some species which nest solitarily also form communal roosts when exploiting sporadic food sources, as do Rough-legged Buzzards in North America, Common Buzzards in central Europe and Black-shouldered Kites in Africa; the birds disperse to individual hunting ranges during the day (Schnell 1969, Glutz et al 1971, Brooke 1965, Tarboton 1978). In all these species, the same roosts may be used year after year, but by greatly varying numbers of birds.

(3) Pairs nest in dense colonies and forage gregariously. This system is shown by the small snail-eating Everglade Kite *Rostrhamus sociabilis*; by the insect-eating kites of the genera *Elanoides*, *Gampsonix*, *Chelictinia* and *Ictinia*; by the insect-eating falcons, *F. naumanni*, *F. vespertinus*, *F. amurensis*, and *F. eleonorae*; and by the large griffon vultures (*Gyps*). In these species, the pairs typically nest closer together (often less than 70 m apart) and in larger aggregations than do those mentioned above. They also feed communally in scattered flocks or, in the case of vultures, spread out in the air, but crowd around carcasses. The feeding flocks are not stable, but change continually in size and composition, as birds join or leave. Colonies usually contain up to 20 or 30 pairs, but those of Everglade Kites sometimes reach about 100 pairs and those of some *Gyps* vultures more than 250 (Haverschmidt 1970, Ledger & Mundy 1975). The food sources of these various species are even more sporadic and fast-changing than those of the previous group. Food may be plentiful at one place on one day and at another place on the next. Such species roost communally at all times and, when not breeding, may gather in enormous numbers. The insect-eating falcons in their African winter quarters use the same roosts year after year, which often contain thousands of individuals of several species. One was found to hold 50,000–100,000 birds, mainly Eastern Red-footed Falcons *F. amurensis* (Benson 1951, Moreau 1972). Such birds exploit local flushes of food, such as termites and locusts, and move around over long distances in response to changes in prey availability.

At times various other raptors, including Black Kites (*Milvus migrans*), harriers, buzzards, Lanner Falcons (*F. biarmicus*) and eagles, congregate to feed at *Quelea* colonies, termite and locust swarms, or on the

(Upper) Red Kite with wing tags. The colour of the tag, which can be seen at a distance, gives the year of birth; the number must be read at shorter range and identifies the individual. Such tags are helpful in studies of movements and age; they do not interfere with the birds' flight. Photo: R. J. C. Blewitt. (Lower) Black Kites on a pylon at a rubbish dump in India. In parts of SE Asia these scavengers are especially numerous around towns, living off masses of human waste, garbage and carrion. More than 2,400 pairs breed within the city of Delhi, at 16 pairs per km². Photo: B. Little.

4 Bald Eagles usually nest in trees (upper
but on the Aleutian Islands, where there ar
mammalian predators, they often nest on
ground (upper right). On these islands they h
not been persecuted by man and will put (
vigorous nest defence (lower left) and may (
stoop at anglers carrying fish. Photos: H
Snyder (upper left); C. M. White (upper right
lower left).

insects and other small animals flushed by bush fires (Brooke *et al* 1972, Smeenk 1974). In southern Africa this applies to the Steppe Eagle *A. rapax* and the Lesser Spotted Eagle *A. pomarina*, of which more than 100 individuals have been seen together on the ground picking up winged termites as they emerged from their nests (Steyn 1973). On the South American pampas, Swainson's Hawks (*B. swainsonii*) feed in flocks on insect concentrations, as to a lesser extent do vultures and harriers (Bent 1938). Moreover, these various birds may be in single species or mixed flocks.

The above division of dispersion patterns into three categories is arbitrary, and in practice gradations exist between, on the one hand, exclusive highly-defended home ranges spread evenly through the habitat and, on the other, dense breeding colonies and feeding flocks occupying only a fraction of the habitat at any one time. The former is associated with fairly uniform, stable, and predictable food supplies and the latter with unpredictable and sporadic super-abundances. Rodent plagues, insect swarms, or large carcasses, are all food sources which are irregular and continually changing in distribution, but on which, once located, many birds can feed together. Particular dispersion patterns are not necessarily characteristic of species, and the same species of eagle, buzzard, harrier, kite or falcon may adopt different systems according to how its food is distributed. Nonetheless, many species show only one dispersion pattern over most of their range, or for most of the time.

In an exclusive range, the individual (or pair) has sole use of food sources; in an overlapping range, each pair holds an exclusive area round the nest, but has shared access to the food sources of a wider area; and in a nesting colony, each pair has exclusive use of a small nesting area, but can forage with others over a very wide area, wherever food happens to be plentiful at the time. The three systems may thus be regarded as progressive adjustments to increasingly widely spaced and sporadic food sources. An even more flexible system is possible outside the breeding season, when the birds no longer need to return frequently to a fixed point (the nest), and so can accumulate in bigger numbers and become nomadic over larger areas. In any one place, moreover, different segments of a population may forage differently, non-breeders having greater freedom of movement than resident pairs. Among *Haliaeetus* fish eagles, it is common to find the breeders in widely spaced nesting territories, and the non-breeders grouped around rubbish dumps, slaughter houses or other local abundances of food. In these respects, such raptors parallel the more familiar crows and ravens.

Dispersion patterns have been explained mathematically in terms of energy budgets. If food is uniformly distributed, it is more economical (requires least travel time) for individuals (or pairs) to space themselves out and for each to forage within a small radius of its nest; but if food is concentrated unpredictably at various places and times, it is more efficient for individuals to clump together in a central colony and forage outwards over a wider radius (Smith 1968, Horn 1968). Nesting in a group may also aid in food detection, the individuals communicating with one another mainly through mutual observation. Ward & Zahavi (1973) gave the name 'information centres' to the communal roosts of birds, on the assumption that individuals learnt about food-sources from one another. The individual that knows where food can be found might 'lose out' at the time by having to share the feast, but it gains from other birds when it is hungry itself. No one who has seen vultures home in to a carcass can doubt the effectiveness of visual communication.

It has also been suggested that some bird species breed colonially in order to improve their defence system, with several individuals joining together to drive away predators. Communal nest defence has been seen among Marsh Harriers, Swallow-tailed Kites *Elanoides forficatus* and other gregarious raptors, but in my view it is more likely a secondary development than the primary reason for colonial nesting (Axell 1973, Snyder 1975).

DISPERSION IN RELATION TO NEST-SITES

Food is not the only factor influencing dispersion. Nesting places are also involved. Where suitable places are widespread, many species nest solitarily in contiguous or overlapping home ranges, as described; but where suitable nesting places are concentrated, pairs of these same species may have no choice but to nest close together, and range over surrounding land to feed. The Common Buzzard is instructive in this respect. Studied at several localities in Europe, it illustrated a continuum from one extreme in which territories were spread evenly over the study area and provided all the food for their occupants (Dare 1961, N. Picozzi & D. N. Weir); through intermediate situations in which territories were spread over part of the study area and provided a large part of the food for their occupants (Tubbs 1974); to another extreme in which territories were clumped into restricted nesting habitat, and provided only a small part of the food, the rest being obtained from undefended ground nearby (Joensen 1968). The system adopted varied with the local landscape, the first extreme occurring where nesting places (= woodlands) were freely available through the foraging area, and the latter where nesting places were concentrated in part of the foraging area.

Some of the greatest concentrations of nesting raptors are found where cliff escarpments or isolated woods occur in otherwise open terrain rich in food. One of the most remarkable is in the Snake River Canyon in Idaho, where 211 pairs of 11 species, none of which are normally colonial, were counted on a 60 km stretch of cliffs (Olendorff & Kochert 1977). These birds included 117 pairs of Prairie Falcons, and many Kestrels, Red-tailed Hawks, Golden Eagles and others, all of which foraged in the surrounding sage brush flats. In places, the Prairie Falcons nested one above another, forming up to three tiers on the low, medium and high cliffs respectively (M. Nelson). The use of radio-telemetry showed that individual falcons flew up to 27 km from the canyon to hunt, and overlapped in range with many others of their species.

Another concentration, consisting of 73 pairs of four species, was found in some small stands of timber, only 2 km² in area, which were scattered like islands in an open area of rich feeding in Germany. The pairs included five of Red Kites, 49 of Black Kites, 18 of Buzzards and one of Honey Buzzards *Pernis apivorus* (Meyburg 1969). To mention some more extreme examples: several pairs of Black Kites nested in the same isolated tree, as did a single pair each of Black Kite, Kestrel and Hobby *F. subbuteo* and of Peregrine and White-tailed Eagle (Dementiev & Gladkov 1954, Deppe 1972). Close nesting Kestrels were found many times, as many as 50 pairs on a coastal cliff in Japan, and more than 100 pairs in nest-boxes in the Netherlands (Fennell 1954, Cavé 1968). In all these instances, comprising pairs of the same or different species, the birds spread out to forage over a wide area devoid, or almost devoid, of nest-sites.

Hence some raptors may be concentrated in the breeding season for reasons of food, and others for reasons of nest-sites. In the latter, dispersed nesting occurs where sites are widespread through the feeding area, and clumped nesting where sites are abundant yet concentrated into small parts of the feeding area. This re-affirms the importance of landscape structure as a factor in nesting dispersion. The remaining parts of this chapter deal in more detail with some raptors whose dispersion is of special interest.

FISH-EATING SPECIES

Ospreys and various fish-eagles (*Haliaeetus*) occasionally nest in loose colonies. In some places, this results from nesting sites being concentrated among widespread feeding areas, and in others from individual pairs holding extremely small home ranges. The first system occurs, for example, where more than one hundred Osprey pairs nest within a few square kilometres on the same island, with nests as close as 50 m, but hunt over communal feeding areas in nearby coastal bays and marshes (Abbott 1911, Tyrell 1936, Bent 1938). Colonies of Ospreys occur on islands off eastern North America, the keys of Florida, and on islands in the Red Sea. The second system is shown by the African Fish Eagles (*H. vocifer*) found by Brown (1960) to have small but overlapping home ranges, each containing a nest, and good hunting perches near fish in shallow water. The ranges of these birds were smaller than those of any other eagle so far studied. This was possible in such a situation (shallow water around islands) presumably because the birds did not rely on fish produced within their ranges, but on a constant through-movement of fish from a larger area.

The tendency of *Haliaeetus* eagles to flock around abundant food supplies was mentioned above. This can apply to non-breeders at any time of year, and to the whole population in winter. The most remarkable concentration known is on one 15 km stretch of an Alaskan river which remains unfrozen in winter and packed with thousands of spent salmon (Gregg 1961). Some 3,000–4,000 adult and immature Bald Eagles have wintered in this one area, standing within a few metres of one another to feed and roost. Concentrations of a few hundred birds have been found on the Mississippi River (Illinois) and on McDonald Creek (Montana), feeding on fish or on water-fowl wounded by hunters, while smaller numbers have been noted on the Aleutian Islands, feeding from garbage dumps or stranded whales, and at several other localities further south (Southern 1964, McClelland 1973, Sherrod et al 1977). The birds roost communally, favouring large bare trees, and use the same perches night after night.

ELEONORA'S AND SOOTY FALCONS

These two falcons nest on offshore islands, the former in the Mediterranean and off the Atlantic coast of Morocco, and the latter in the Red Sea and the Gulf of Aden (Vaughan 1961, Walter 1968, Clapham 1964). In the early part of the season, they live largely on insects and, in their colonial nesting, they parallel the other insectivorous falcons. Later in the season, when the song-bird migration begins, they live on such birds and, like the fish-eagles just discussed, they can nest close together because of the continual movement of prey past the nests, which makes them independent of the sparse local bird life. Colonies of Eleonora's Falcon *F. eleonorae* usually contain 5–20 pairs, but some hold up to 100 or more; they may be quite dense, and one 2·5 ha islet off Mogador held 80 pairs, with some nests only a few metres apart (Vaughan 1961). Sooty Falcons *F. concolor* are less strictly colonial; more than 100 pairs were found on one desolate island of 40 km², but their nests were seldom closer than 50 m (Clapham 1964). This species also nests solitarily in the Sahara desert, again exploiting migrants (Moreau 1966).

VULTURES

Although they all eat carrion, the vultures vary considerably in dispersion and foraging behaviour. The most strongly gregarious are the large griffons *Gyps fulvus*, *G. rueppellii* and *G. coprotheres*, which nest on cliffs in big colonies numbering up to 100 pairs or more, with some nests only a few metres apart. One of the largest concentrations known in East Africa is on the Gol escarpment and neighbouring cliffs; it contains more than 1,000 nests of Rüppell's Vulture, distributed in several colonies, and is supported largely by the huge ungulate populations of the Serengeti Plain. Such birds feed entirely from large carcasses and, being dependent on migrant game, they often have to fly great distances for food, taking more than one day over each trip. Their nesting on cliffs enables them to gain good lift in the updraughts when setting off on a flight, which in the Serengeti may take them up to 150 km from the colony (Houston 1976).

The food searching of griffons is proverbially efficient. Following the Charge of the Light Brigade in the Crimean War, so many birds gathered on the battlefields that shooting squads had to be posted to protect the injured (Meinertzhagen 1959). Abilities to find isolated carcasses and to gather quickly in large numbers in areas where they had apparently been scarce have caused some writers to suspect smell or telepathy, while some native Africans think that vultures dream the locations of food. In fact the birds rely on vision, but most find food indirectly by watching the activities of neighbouring birds in the air. Together they form a communications network over the landscape, spread out and at higher altitudes over areas of low game density but closer together at lower altitudes over areas of high game density. It is in the latter situation that carcasses are located and consumed most quickly (Houston 1974).*

Other similar species, including the White-backed Vultures (*Gyps africanus* and *G. bengalensis*) of Africa and Asia, also feed entirely on carcasses, but they depend more on resident and less on migrant game, and travel less far than the cliff-nesting *Gyps*. They tend to nest in smaller, more scattered colonies, and occasionally as individual pairs, but on trees rather than cliffs (Brown & Amadon 1968, Kemp and Kemp 1975). They also weigh less, so can take wing earlier in the day.

Another category of Old World species includes the Lappet-faced Vulture *Torgos*, the European Black Vulture *Aegypius*, the Indian Black Vulture *Sarcogyps*, and the White-headed Vulture *Trigonoceps*, all of which nest near the tops of trees. Then there is the Lammergeier *Gypaetus* (cliffs), the Egyptian Vulture *Neophron* (cliffs or trees) and the Hooded Vulture *Necrosyrtes* (trees). All these species behave in some respects like eagles. They nest much farther apart than *Gyps* vultures and hold large ranges round their nests. They feed partly from large carcasses, but also take smaller items, including living prey; they do not fly long distances to forage, so it is rare to find more than two or three pairs at the same corpse (Pennycuick 1976, Anthony 1976, P. Mundy). However, bigger concentrations occur at watering places, and in some regions numbers of non-breeding Lammergeiers and Egyptian Vultures occur at rubbish dumps, and other places where food is

---

* In the New World vultures, smell may also play a part in food detection, as Turkey Vultures have been shown to have this sense unusually well developed compared to other birds, and to be capable of finding covered carcasses (Stager 1964). There are no indications that Old World vultures use smell, and plenty to the contrary.

concentrated (Brown 1977, Sauer 1973). The various New World vultures and condors resemble this last category of Old World species, in that they usually nest well apart from one another, but they all roost communally and feed in groups, especially the non-breeders (Bent 1938, Koford 1953). Pairs of Turkey Vultures *Cathartes aura* or Black Vultures *Coragyps atratus* occasionally breed in loose aggregations where large cliffs hold several suitable caves, and the two species often roost together.

In conclusion, among the vultures as a whole, nesting dispersion seems to depend mainly on the proportion of large-carcass carrion in the diet, and on the distances flown from the nest.

SUMMARY

Both solitary and colonial raptors tend to nest in the same restricted places year after year. The term 'nesting territory' is appropriate for the area around the nest that is defended, and home range for the area containing the nesting territory and hunting places of a pair. Except in some *Buteo, Aquila* and *Elanus* populations, home ranges usually overlap widely among pairs. Normally, nest-sites and food govern the distribution of breeding raptors. Where nest-sites are widespread, many species nest solitarily in contiguous or overlapping home ranges; but where sites are concentrated, these same species nest close together and hunt over communal land nearby. Given widespread nest-sites, exclusive or partly exclusive home ranges are also associated with fairly even and stable food sources, whereas gregariousness (communal breeding, roosting or hunting) is associated with abundant, yet widely spaced and transient food-sources. The latter include rodent plagues (exploited by kites and harriers), insect swarms (utilised by some small falcons and small kites), various aquatic foods (snails utilised by the kite *Rostrhamus*) and large carcasses (utilised by some vultures). Individual species show different dispersion patterns in different areas depending on the distribution of nest sites and food.

CHAPTER 3

Breeding density

Population studies usually entail finding all the pairs of a species in a given area over several years. Up to a point, the value of such studies is increased the longer they are continued and if information on other aspects, such as nest success, is obtained at the same time. To be sure of finding all the pairs, the study area must usually be checked at the start of each season because, in many raptors, pairs which fail in their breeding often move away soon afterwards, and so cannot be detected in checks later in the breeding season. Many species use the same nesting places in different years, so to some extent monitoring becomes easier the longer it is continued. Some raptor populations have now been studied for more than ten years, and many others for periods of 3–5 years, so that comparisons between areas are possible. Within each area, the spacing of established pairs is usually expressed as a density in pairs per unit area, or in terms of the distances between pairs. The latter is an especially useful measure where pairs are distributed

Illustration: Martial Eagle at nest

linearly, as along rivers or cliff escarpments. This chapter is concerned with a restricted aspect of population studies, namely with the natural limitation of breeding density. Other factors that influence density, such as human persecution and pesticide use, are discussed later.

What, then, is the evidence that raptor breeding populations are normally regulated or limited in some way, rather than fluctuating at random? When applied to many species, the concept of limitation is based on four main findings: (1) the stability of breeding population, in both size and distribution, over periods of many years; (2) the existence of 'surplus' adults, capable of breeding but attempting to do so only when a territory is made available through the death or removal of a previous occupant; (3) the re-establishment, after removal by man, of a population showing roughly the same size and distribution as the previous one on the same ground; and (4) in areas where nest-sites are not restricted, a regular spacing of breeding pairs. In the sections below, I shall discuss these findings in detail. Stability would of course be expected only from populations in stable environments (including food), and not from populations that were changing in response to some human pressure.

1. Stability of breeding population

Often taken for granted, stability of breeding numbers can be documented to some extent from long-term studies (Table 8). The best studies for this purpose were those on Peregrines, Kestrels, Buzzards, Golden Eagles and Black Eagles, all involving populations of more than ten pairs observed over periods of ten or more years. In all these cases, breeding numbers remained either absolutely constant, or changed by less than 15% of the mean over the period concerned. Compared to findings on some other birds, and to what is theoretically possible, this represents a remarkable degree of stability, unexpected unless numbers were controlled in some way. Evidence for stability involving fewer pairs or shorter time periods is available for other populations of these same species, and also for at least 13 other species (Table 8).

Population stability was also found in studies of the total raptor fauna of particular areas, in which the pairs of all the species present were counted year after year. Near Berlin, Wendland (1952–53) found constancy in the numbers of individual species, and of the raptor population as a whole, which varied over eleven years between 61 and 65 pairs only (Table 9). Likewise in a Michigan area, the total number of raptor pairs was 36 in 1942, 37 in 1948 and 35 in 1949, even though in this period one *Buteo* species partly replaced another through habitat change (Table 9).

The degree of stability recorded in any population presumably depends partly on the duration of study, as the environment itself is unlikely to remain constant over periods of many decades. Hence, populations which appear stable over the usual study periods may in fact be changing slowly in the longer term. Also, populations of some non-raptorial birds tend to fluctuate more in poor habitats than in good ones; so the stability recorded in some raptor populations might not be wholly typical of the species concerned if observers had worked chiefly in the better habitats.

2. Surplus birds

Non-breeding, non-territorial adults were seen to be present in some populations, and were occasionally observed fighting with breeders over nesting territories (Gargett 1975). But the chief evidence for the existence of surplus birds, which breed only when a place becomes available, is that lost mates are sometimes replaced in the same season by other birds, which then proceed to nest. Scottish gamekeeper Dougal Macintyre (1960) wrote that whenever a female Peregrine was shot from the eyrie the male generally re-mated within 24 hours. He several times saw a bereaved male return with a new mate, and knew one male to acquire four adult mates in quick succession. Only once in many years of Peregrine shooting was the new female not in adult plumage. Macintyre suggested that this indicated the existence of a considerable reserve of non-breeding adults, as he found no indication that the new mates moved in from neighbouring eyries. Many other instances of swift replacement have been recorded in other Peregrine populations, in both Europe and in North America, and the phenomenon is also widespread in other raptors. From the literature available to me, I have come across records in 26 species, from small falcons to large vultures (Table 10).

Sometimes two hens appeared as replacements at the same site in a short time, as in White-tailed Eagle, Sparrowhawk, Cooper's Hawk and Merlin (Saxby 1874, Owen 1936–37, Bent 1938, Seebohm 1883), while two males appeared as replacements at nests of Merlin and Peregrine (Brooks 1927, Berg 1962 in Lindberg 1975). In all these instances only one partner was shot at a time and the other was left. In the Lesser Kestrel no less than eight females were shot in quick succession at one nest box, until shooting the male as well as the eighth female ended the series for that year (Lucanus 1936). On the other hand, when a pair of Ospreys was shot on the same day, they were soon replaced by a fresh pair (Bent 1938). Overall, most replacements referred to hens. This could have been because hens were most easily shot, or because hens were more numerous among surplus birds, or for both reasons, but on the information available it is not possible to tell. Also, although some replacement birds were in adult plumage, others were in immature plumage. The adults provide the best evidence for the existence of surplus birds capable of breeding, because immature-plumaged raptors often do not produce eggs or, if they do, the eggs are fewer and less successful than those of older birds (Chapter 8).

Most of the instances in Table 10 were from the early literature and, as evidence for the existence of surplus birds, they fall short of the standard required today. For example, near the start of a season, replacement might have been by birds that would otherwise have bred elsewhere, whereas near the end, it might have been by birds that had failed or lost a mate at another site. And how often, I wonder, were the possibilities of movements between territories checked, or of new pairs taking over, rather than the widowed bird re-mating? Losses followed by replacement were more likely to have been documented than losses followed by no replacement, so more information is needed on how often replacement occurs. In the Buzzards studied by Dare (1961), which resided all year on their territories, most vacancies were filled at periods of local movement (autumn and late winter) rather than in the breeding season, and six weeks was the shortest period between a loss and its replacement. The inference of this study was that replacement depended not only on the existence, but also on the proximity, of suitable recruits. This whole field is in need of controlled removal experiments of the kind done on other birds, for only then can the early anecdotal evidence be superseded by something more reliable.

The early records repeatedly show another important point, however, namely that the newcomer often carried on with a nesting attempt where the previous bird left off. One female Cooper's Hawk *Accipiter cooperii* that was shot at the nest was soon replaced by another which incubated the same eggs; this female was in turn shot and was then replaced by a third female, which took over and subsequently hatched and raised the brood (Bent 1938). Similarly, a male Goshawk, which took over in the incubation period, assumed all normal duties and helped to rear the young (Holstein 1942). At one Peregrine eyrie, both adults were replaced so quickly that two birds eventually raised a brood that neither had parented (Taverner, in Hickey 1942). Both birds of the pair that had produced the eggs were killed a few days apart. The bird that was first to be widowed acquired a new mate and was subsequently killed. The new mate in turn acquired a new mate, and together they raised their fostered brood. Also in the Peregrine, Macintyre (1960) found it usual for new females to take over 'at whatever stage of hatching or rearing' the nest had reached. And there is a recent record from California of a new male joining a female in the post-fledging period and helping to rear the flying young (Cade 1977). In all these cases, the behaviour of the incomer was apparently influenced by what the nest contained, and females laid in empty nests, incubated an existing clutch, or brooded and fed young, as the situation required. The evolution of this behaviour needs explanation, because any birds that spend energy raising young that are not their own could hardly be expected to be favoured by natural selection unless there were some counter advantage. This advantage is presumably a place in the breeding population, the initial cost of which is helping to rear the brood of the widowed bird. The selection pressure could be applied by the widowed bird which, given a choice between a potential mate that would or would not help to rear existing young, pairs with the former, so that in the long run the trait to help with a previous attempt is perpetuated in the population. The trait may not be invariable, however, for one replacement Peregrine laid a new clutch near an existing one (Nethersole-Thompson 1943); and one replacement Cooper's Hawk laid eggs beside a two-week old chick, but it raised the chick rather than incubating the eggs (Schriver 1969).

3. Re-establishment of populations
In Britain, cases are known of raptor populations being removed or

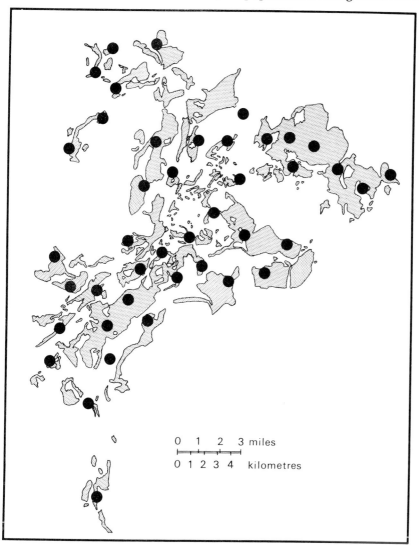

Figure 9. *Regular spacing of Sparrowhawk nesting territories in an area in northern Scotland. The birds nested only in woodland (stippled), and territories are shown by spots. In the field, each nesting territory was recognised by an accumulation of different years' nests, which usually fell within a circle of 50 m radius. Re-drawn from Newton et al 1977.*

depleted by human action and then recolonising or recovering to about the same level as previously, with pairs in the same places. On parts of the south coast, Peregrines were deliberately shot out during the 1939–45 war, because they attacked the pigeons used to carry military messages. After the war, the birds recolonised some areas to their former level within a few years, but in other areas the birds were still moving back when their numbers were again reduced by organo-chlorine pesticides (Ratcliffe 1969, 1972). At the time of writing, Peregrines in certain other areas of Britain have recovered from the low level resulting from organo-chlorine use, and the numbers of breeding pairs in these areas are remarkably similar to those in pre-pesticide days (D. A. Ratcliffe). In each area, not only did the newcomers use the same cliff-sections as their predecessors, but also the same nest ledges. Likewise, in many areas the Sparrowhawk has reoccupied former nesting places, so that the latest densities are the same as those found 15–20 years earlier (I. Newton). Rowan (1921–22) wrote that the Merlins and their successors in two neighbouring territories were shot every year for 19 years, and produced not a single young, yet every year without fail two new pairs settled in the same places. In one place the birds used the same heather patch for all 19 years and in the other for the first 12 years until it was destroyed by fire. All this implies some constancy in the carrying capacity of the environment over the years, with consistent limitations on raptor breeding numbers.

4. Regular spacing of territorial pairs

In continuously suitable habitat, the nests of pairs of a species are often separated from one another by roughly equal distances (Figure 9). This has been noted in many species, including Sparrowhawks, Red-shouldered Hawks, Red-tailed Hawks, Buzzards, Wahlberg's Eagles *A. wahlbergi*, Golden Eagles, Wedge-tailed Eagles *Aquila audax*, Martial Eagles *Polemaetus bellicosus*, Kestrels and Peregrines (Newton *et al* 1977, Henny *et al* 1973, Hagar 1957, Tubbs 1974, Brown 1970, Lockie 1964, Leopold & Wolfe 1970, Smeenk 1974, Taylor, 1967, Ratcliffe 1962). Such regular spacing is consistent with the idea that breeding density is limited by the territorial behaviour of the birds concerned; but it does not alone prove the idea because of the alternative possibility, that pairs already limited by some other factor then divide the area equally between them. Either way, one advantage of uniform spacing is that, under any given density, each pair is as far from its neighbours as possible, thus reducing interference in breeding and hunting. Regular spacing would not, of course, be expected where habitats were patchy or where nest-sites were restricted.*

When taken together, these four arguments provide strong circum-

---

* Pennycuick (1976) concluded that the spacing of Lappet-faced and White-headed Vultures in Serengeti, Tanzania, was more or less random, but he did not take account of local variations in habitat. His conclusion thus contrasted with that on Lappet-faced Vultures in Rhodesia, where Anthony (1976) did take account of habitat and found regularity of spacing within the suitable patches.

stantial evidence: (a) that breeding density is limited, (b) that the limitation is through competition for nesting territories, and (c) that long-term stability of breeding numbers is helped by the existence of surplus birds, able to breed only when a territory becomes vacant, but otherwise excluded by the birds in occupation. This is not to imply that surplus birds are available at all times in all populations and, as mentioned, the phenomena discussed here would not apply to many modern populations reduced by pesticide poisoning or other human action. Moreover, the evidence as presented has been pieced together from many different studies, a procedure necessitated by the present poor state of knowledge; only in a few populations have all aspects been properly studied.

A curious feature in many studies was that not all territories were occupied in any one year, despite the relative stability of breeding numbers and the proven existence of surplus birds. In Britain, the usual occupancy of Peregrine territories until 1939 was about 85%, and in some Buzzard territories during ten years it was 77–83% (Ratcliffe 1972, Tubbs 1974). In each case certain territories were vacant much oftener than others, so possibly these territories were inferior, or suitable for occupation only in certain years, or attractive only to particular pairs. For there is no doubt that territories vary in quality, that is, in the opportunities they offer for existence and for successful breeding (Newton & Marquiss 1976; Chapter 5).

The stability shown by many populations might be maintained in one of two ways: by the annual production of young whose numbers happened to approximate the annual loss of adults; or by a breeding density limited by a fairly constant carrying capacity of the environment in which, whatever the local breeding success, there were always enough new recruits available to fill the gaps. The recruits might come from local or outside areas. Stability of local breeding numbers has often held irrespective of breeding success; in some populations it has held despite prolonged good production of young, in others despite prolonged poor production, and in yet others despite great annual fluctuations in production (Mebs 1964, Ratcliffe 1972, Tubbs 1974). Local populations which consistently produce insufficient young to offset adult mortality could only be maintained by immigration. In fact, limitation of breeding numbers by carrying capacity of the environment is the only tenable view, as indicated by the existence of surplus birds and discussed further below.

BREEDING DENSITY IN RELATION TO FOOD SUPPLIES

In the limitation of breeding density, two resources seem of overriding importance, namely, food and nest-sites. The evidence for a link between density and food in areas where nest-sites are not limiting is circumstantial, and based on the following findings: (1) an overall trend for large raptor-species, which depend on large, sparse prey-

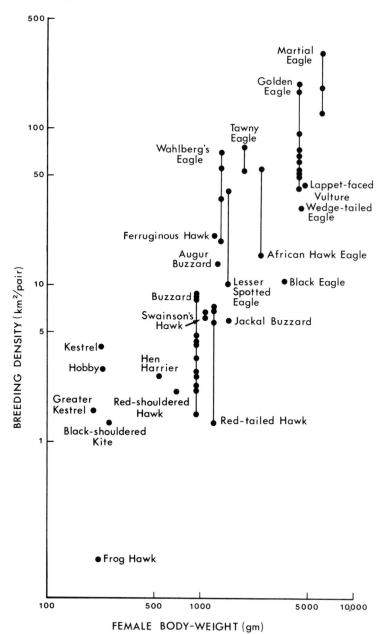

Figure 10. Breeding density and mean range size in relation to body weight. The overall trend is for large raptors to breed at lower density, in more extensive home ranges, than small raptors. This is associated with larger raptors eating larger, sparser and slower-breeding prey than small ones. The graphs are based on populations in which the individual pairs forage in more or less exclusive ranges, and the more numerous data on breeding density are taken from populations not limited by shortage of

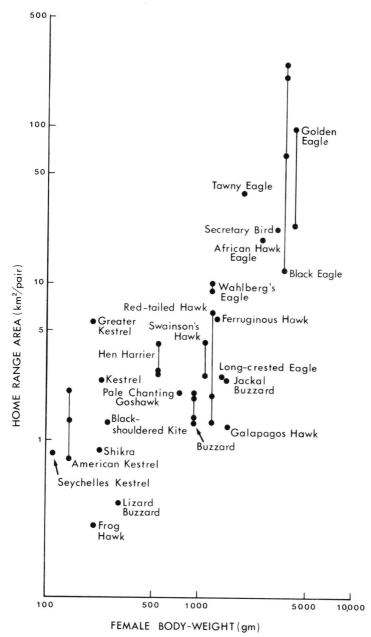

nest-sites. *Different studies on the same species are joined by lines. The scales are logarithmic.*

Sources: *Black-shouldered Kite – W. Tarboton; Lappet-faced Vulture – Pennycuick 1976; Hen Harrier – Breckenridge 1935, Craighead & Craighead 1956, Smith & Murphy 1973; Pale Chanting Goshawk – Smeenk & Smeenk – Enserink 1975; Frog Hawk – Kwon & Won 1975; Shikra – Thiollay 1975; Lizard Buzzard – Thiollay 1975; Red-shouldered Hawk – Stewart 1949; Swainson's*

species, to breed at lower density than small raptor-species, which depend on small, numerous prey-species; (2) area differences in breeding density within species that are associated with area differences in food; (c) annual fluctuations in density within species that are associated with cyclical fluctuations in food; and (4) sudden and long-term changes in breeding density that are associated with sudden and long-term changes in food. In the sections that follow, I have sometimes expressed breeding densities in terms of pairs per unit area, and at other times in terms of the distances between pairs, depending on the information available.

1. Body-size and breeding density

This relationship can best be shown from populations not limited by nest sites and in which pairs forage in more or less exclusive ranges.* For any one population, breeding density is lower than would be expected from home range sizes because study areas generally include some unused habitat. Nonetheless, the overall trend for large species to breed at lower density, in larger ranges, than small species is readily apparent (Figure 10). The huge African Martial Eagle breeds at what must be some of the lowest natural densities for any bird: one pair per 125 km² in the Embu District of Kenya, one pair per 182 km² in Kruger Park in South Africa, and one pair per 300 km² in Tsavo Park in Kenya (Brown 1952, 1955, Snelling 1969, 1970, Smeenk 1974). In other suitable areas, density is even lower, as nests may be separated by 30–40 km. The species lives mainly on game birds and mammals up to several kilograms in weight. Various other eagles usually breed at densities of one pair per 30–190 km², various buteonine hawks at densities of one pair per 1–8 km², and small falcons and kites at one pair per 1–3 km².

* This is because of the practical difficulty of determining, for birds that have widely overlapping ranges, how many individuals were using a given area.

Figure 10: Sources continued from previous page

Hawk – Craighead & Craighead 1956, Smith & Murphy 1973, Dunkle 1977; *Galapagos Hawk* – de Vries 1975; *Red-tailed Hawk* – Fitch et al 1946, Craighead & Craighead 1956, Orians & Kuhlman 1956, Hagar 1957, Luttich et al 1971, Smith & Murphy 1973: *Buzzard* – Wendland 1952–53, Melde 1956, Holstein 1956, Warncke & Wittenberg 1959, Dare 1961, Mebs 1964, Joensen 1968, Holdsworth 1971, Tubbs 1974, Picozzi & Weir 1974, Rockenbauch 1975; *Ferruginous Hawk* – Smith & Murphy 1973; *Augur Buzzard* – Gargett 1978; *Jackal Buzzard* – Harwin 1972; *Lesser Spotted Eagle* – Steinfatt 1938, Goluduschko 1961; *Tawny Eagle* – Snelling 1970, Smeenk 1974; *Wahlberg's Eagle* – Brown 1955a, Snelling 1970, Tarboton 1977; *Golden Eagle* – Dixon 1937, Watson 1957, Lockie 1964, Brown & Watson 1964, McGahan 1968, Smith & Murphy 1973, Beecham & Kochert 1975, Newton 1972; *Wedge-tailed Eagle* – Leopold & Wolfe 1970; *Black Eagle* – Rowe 1947, Brown 1952–53, Siegfried 1968, Gargett 1975; *African Hawk Eagle* – Snelling 1970, Smeenk 1974, Gargett 1978; *Martial Eagle* – Brown 1952—53, 1955a, Snelling 1970, Smeenk 1974; *Secretary Bird* – Kemp 1978; *Greater Kestrel* – A. C. Kemp; *American Kestrel* – Craighead & Craighead 1956, Smith & Murphy 1973, Balgooyen 1976; *Kestrel* – Village 1979; *Seychelles Kestrel* – Feare, Temple & Procter 1974; *Hobby* – Fiuczynski 1978.

(Upper) Fish Eagle at nest in southern Africa. This species has spread in recent years with the building of dams and the resulting expansion of fish populations. Photo: Peter Steyn. (Lower) White-tailed Eagles at a specially provided carcass in Sweden. With this extensive feeding programme it is hoped to improve the survival of young eagles, and to reduce the pesticide loads of older ones, by providing an abundance of less contaminated food than the birds would find for themselves. Photo: B. Helander.

6 *Immature White-tailed Eagles prior to relea on Rhum, W. Scotland. In this re-introducti programme the birds were taken from Norway nestlings and held tethered for a time on Rhu before being released at an appropriate a Photos: J. Love (left) and I. Newton (below).*

The overall trend presumably holds because larger raptors generally eat larger prey species, and large prey species live and breed less abundantly than small prey species. So in each case, range size may be adjusted to food. The evidence is not decisive, however, for it is possible that some unknown factor varies in parallel with food and causes the trend. Moreover, the Peregrine and Merlin are somewhat unusual, because over much of their distribution they live at exceptionally low densities for their size. This may be connected with their hunting methods, which often involve chasing birds for some distance across country. A similar broad relationship between body-size and range-size holds among birds in general, and also among mammals (Schoener 1968, McNab 1963).

The information in Figure 10 was mostly collected near the start of the breeding season, and the home ranges were mean values for the populations studied. In several populations, single birds held smaller ranges than breeding pairs and, where pairs were not constrained by neighbours, they often extended their ranges once they had young to feed (Smith & Murphy 1973, M. Marquiss & I. Newton). This provides circumstantial evidence for a link within species between the food needs and range sizes of birds. The variation within species is discussed further below.

2. Regional variations in breeding density within species

(a) *European Sparrowhawk*. In twelve well-wooded districts in Britain, nesting territories were evenly spaced, but at distances varying from 0·5–2·1 km apart, according to district. These density differences were related to soil productivity and altitude and, in at least three areas where counts were made, to the abundance of small bird prey (Figure 11). In one long valley in northeast Scotland, Sparrowhawks nested most densely on the richest ground near the coast, at an intermediate density part way up the valley, and at lowest density on poor ground near the top (Newton *et al* 1977).

(b) *Peregrine Falcon*. Ratcliffe (1969) found the following mean distances between pairs in different parts of Britain: 2·6 km on sea-cliffs in southeast England and on some Scottish islands with numerous seabirds; 4·8 km over much of England, Wales and southern Scotland; 5·5–6·4 km in the southern Highlands and northern coasts; 7·2 km in the central Highlands; 8·3 km on northeastern coasts and 10·3 km in the western and northwest Highlands. The average area per pair for three inland populations, not limited by shortage of nest sites and covering the full range of variation, was 52 km^2 in Wales, 96 km^2 in the east central Highlands and 220 km^2 in the western Highlands. These differences were broadly correlated with land-productivity and food, though food was not measured directly. The English south coast was backed by rich farmland, offering a variety of abundant prey, whereas in Wales and southern Scotland the country was less rich, but near to fertile lowland. Northwards, the country became progressively

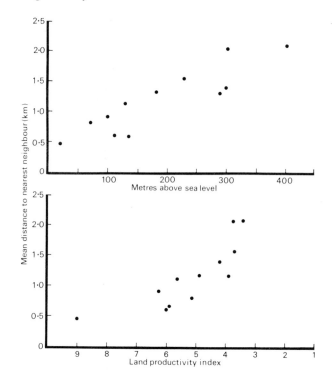

Figure 11. *Mean nearest-neighbour-distances of Sparrowhawk nesting territories in continuous nesting habitat shown in relation to altitude above sea level and land productivity in twelve different areas. Nearest neighbour distances widen with rise in altitude (linear relationship) or fall in land productivity (probably curvilinear relationship). From Newton et al 1977.*

more barren, and high Peregrine densities were found only near those coasts where seabirds were plentiful. The breeding populations of most regions remained stable for as far back as records go, but a one-fifth decline in pair numbers occurred in the western Highlands between 1890 and 1950, which Ratcliffe attributed to a long-term decline in prey caused by extractive land management. He thus argued for a relationship between density and food-supply both on a geographical parallel, and on a local long-term change in density, associated with long-term change in food.

Higher Peregrine densities than any in Britain were found on the Queen Charlotte Islands off western Canada, where the mean distance between about 20 pairs was 1·6 km (Beebe 1960), again linked with massive concentrations of seabirds. This population has recently dropped to about six pairs, following a decline in prey (Nelson & Myres 1975). Lower densities than any in Britain were found along the

Colville River in Alaska, where 8–11 pairs were separated by 11·2–15·4 km (Cade 1960). In other regions, pairs were found at exceedingly low density, say one pair per 5,000 km^2; they were usually isolated near local oases of food in otherwise sterile terrain. Some isolated pairs in the western United States were more than 75 km from their nearest neighbours, and it is remarkable that such territories remained occupied for long periods, requiring, as they presumably did, the continual replacement of birds that died.

(c) *Buzzard.* In general, high densities were again found on fertile soils, and in Britain in areas with most rabbits *Oryctolagus cuniculus*. In the country as a whole, high Buzzard densities occurred more often in farmland, or mixed moorland/farmland, than in moorland or forest alone (Moore 1957). On Skomer Island (Wales), seven pairs nested in 3·1 km^2, feeding from the masses of rabbits and seabirds. Elsewhere in western Britain, densities of 0·4–1·5 pairs per km^2 were found in five good rabbit districts, and of about 0·1 pairs per km^2 in two other districts (Table 11). Buzzard densities were thus 15 times greater in the best rabbit areas than in the worst (and 22 times greater if Skomer was included).

In nine areas on the European Continent, Buzzard densities were generally higher in deciduous woods on rich humus soil than in pinewoods on poor sandy soil, and higher in lowlands than in mountains. The differences were again linked with abundance of prey, in all these areas mainly rodents (Mebs 1964). Home range sizes of pairs also varied within each population and, where studied, were related to local food supply or quality of feeding area, whichever was measured (Dare 1961, Picozzi & Weir 1974).

(d) *Black Eagle.* In some regions, single pairs of this species had home ranges covering about 65, 210 and 260 km^2 respectively (Figure 10). But in the Matopos Hills in Rhodesia, 59 pairs lived in 620 km^2, an average of one pair per 10·5 km^2, associated with an exceptional abundance of two hyrax species, the main prey. The hyraxes occupied rocky outcrops, and where outcrops were close together (i.e. more hyraxes), the eagle territories were smaller than where outcrops were far apart (Gargett 1975).

(e) *Unusual concentrations.* Vultures and kites breed at extremely high density in some Asian cities. In Delhi in 1968–69, in certain sample areas, a mean density of 19·3 raptor territories per km^2 was found. As the city covered 150 km^2, its total raptor population was estimated at 2,900 pairs. Black Kites formed 83%, at 16·1 pairs per km^2, or about 2,400 pairs in all; the rest consisted mainly of large vultures (such as *Gyps bengalensis*), at 2·7 pairs per km^2, or 400 pairs in all, and Egyptian Vultures at 0·5 pairs per km^2, or 100 pairs in all. This remarkable concentration was attributed primarily to the huge amount of food within the city – garbage, carcasses of animals on roads and rubbish dumps, small birds and mammals – but also to an abundance of nest sites and to 'the traditional goodwill of Indians to all living beings' (Galushin 1971). Other concentrations of scavenging raptors occur around towns and villages elsewhere in southeast Asia, as well as in parts of South America and East Africa (Reichholf 1974, Pomeroy 1975).

In all the species discussed in this section, breeding density was broadly related to food supply, and unusual concentrations were associated with unusual food abundance, occurring either naturally (eg seabird colonies) or from human activity. Only for the Sparrowhawk was food actually measured; in the other species density differences were linked with impressions of food differences, often associated with land-productivity; hence, in some such cases, more precise assessments of food are desirable. Golden Eagles in Scotland have been quoted as a possible exception to this general relationship between food and breeding numbers. Brown & Watson (1964) counted carrion (dead sheep and dead deer) and live prey in four areas, and found no consistent relationship between overall food supply and eagle density. Carrion was seasonal, however, and if living prey alone was considered, eagle densities (and brood sizes) were higher in the best (eastern) area than in the three poorer (western) ones. Hence, the overall implication is that raptors respond to the amount of food and become more widely spaced where food is sparse.

It is presumably not just the standing crop of prey which is important in influencing raptor numbers, but also its renewal rate. Any prey species that produces a succession of large broods through the year clearly has the potential to support more predators than a prey species that lives at similar density but produces only one small brood each year. The crucial figure is the number of prey that can be repeatedly removed from an area without causing long-term depletion, and this figure is likely to be higher in more productive environments, and in small fast-breeding prey species than in large slow-breeding ones. It follows that, in comparing areas, one would not necessarily expect a direct relationship between prey numbers and raptor numbers.

The literature contains much fruitless discussion on whether raptor breeding densities are limited by territorial behaviour or by food; in fact both seem to be involved, with the birds using territorial behaviour to adjust their density to the resources available. Spacing behaviour is thus the means whereby food resources and density are attuned. Exactly how this happens is not known, but there are at least two possibilities. First, differences in feeding conditions may lead to differences in the aggressive behaviour of the birds, which in turn may lead to differences in the spacing of breeding pairs; that is to say, the distance at which individuals drive away others of their species, or the area over which they attack others, might increase with deterioration in feeding conditions (Watson & Moss 1970). Thus the further the birds have to range to meet their food needs, the larger the defended areas become. Second, each bird may settle in an area only if it can get enough food there, and move on if it cannot, an outcome influenced partly by food supplies and partly by the birds already present. The latter may or may not behave territorially, but either way they will be utilising and disturbing the prey, making it harder for other birds to get. If enough birds were already present for the food supply, any newcomer would have to move on to survive, or displace another bird which would move on instead; in this way the number of birds would not increase beyond the limit set by resources and would be spaced over the available area in relation to food. Such a reaction would automatically lead to lower densities in the poor environments than in the good ones. It need not entail any change in aggressiveness with feeding conditions, and would apply whether the birds maintained exclusive or overlapping ranges. Either way, differences in dominance between the birds of a population presumably influence which individuals move on and which stay, and which individuals hold nesting territories and which do not.

It seems certain that the tendency to breed at high or low density is not genetically fixed in each bird, because high and low density populations are often adjacent to one another and not reproductively isolated. In the Peregrine and Sparrowhawk, several young ringed in high density areas have subsequently been found breeding in low density areas, and vice versa.

3. Annual variations in breeding density

For some species, the evidence for density limitation in relation to food rests largely on long-term stability of breeding population, but at different levels in different regions; for other species it rests on local fluctuations in population, which follow changes in food. Most such species in the regions concerned have restricted diets based on cyclic prey. Two main cycles are recognised: (a) an approximately 4-year cycle of small rodents on the northern tundras and temperate grasslands; and (b) an approximately 10-year cycle of Snowshoe Hares *Lepus americanus* in the boreal forests of North America (Elton 1942, Lack 1954, Keith 1963). Some grouse-like birds are also involved, but whereas in some regions they follow the 4-year rodent cycle, with peaks in the same years, in others they follow the 10-year hare cycle. Black-tailed Jackrabbits *Lepus californicus* may also follow an approximately 10-year cycle in the dry grasslands of western North America, but the trends are poorly documented. The populations of these various animals do not reach a peak simultaneously over their whole range, but the peak may be synchronised over tens or many thousands of square kilometres.

The main raptor species involved in small-rodent cycles include the Rough-legged Buzzard of northern tundras, various kestrels, harriers and kites of more southern grasslands and semi-deserts, and the Secretary Bird and Black-shouldered Kite on the African plains. All such species tend to breed more densely and more prolifically when rodents are plentiful than when they are scarce (Table 12). The increase in raptor density from one year to the next is often so great that it cannot be due merely to the high survival of adults and young from the previous year, but must be due partly to immigration. Later it is helped by good breeding success, while decline is brought about by a combination of poor breeding, emigration and starvation (Sulkava 1964). Through the worst conditions, birds may remain only in the best habitats, often at reduced density with enlarged home ranges, and vacate the poor habitats completely. In any one locality, breeding numbers are not necessarily limited by food at all stages of the cycle; they may be reduced so markedly in the poor food years that they can then increase again for several years unchecked by further food-shortage, because the prey population is itself rising then.

The main raptor species involved in game bird (or hare) cycles include the Goshawk in the forests, the Gyr Falcon on the tundras, and the Ferruginous Hawk (which eats jackrabbits) in the American grasslands. Goshawk breeding numbers follow the 4-year cycle or the 10-year cycle depending on the region concerned, which is further evidence for a link with food. Moreover, both bird-eating and mammal-eating raptors show more stable breeding numbers in regions where their supply of food is more stable (often through being more varied). Compare, for example, the Goshawk in boreal Europe and temperate Europe, the Kestrel in southern and northern Britain or the Hen Harrier in Wisconsin and eastern Scotland (Tables 8 and 12).

In general, three steps in the response of raptors to annual prey numbers can be recognised. Those birds which are subject to the most marked prey cycles show big local fluctuations in densities and breeding rates (eg Goshawks in boreal regions); those subject to less marked prey cycles show fairly stable densities but big fluctuations in breeding rates (eg Buzzards in temperate regions); while those with stable prey populations show stable densities and fairly stable breeding rates (eg Peregrines in temperate regions) (Chapter 8). Much depends on how broad the diet is, and whether alternative prey are available when favoured prey are scarce. The more varied the diet, the less the chance of all food-species being scarce at the same time.

Controversy exists over whether raptors and other predators actually cause the cycles in their prey. Where the prey breeds faster than the predator, a delayed 'density-dependent' process can be envisaged as follows: an increase in the prey leads to an increase in the predator to the point where the predator begins to over-eat its prey so that the prey declines, then the predator likewise until the point where predation is insufficient to stop the prey increasing again, so the cycle begins anew. If this process occurs, mammal predators may be more important than avian predators in generating the cycle, because the former are less able to move out when prey becomes scarce. The more rapid cycle of rodents than hares could be attributed to the faster breeding rates of rodents and their predators. In each case, the mammals are probably the crucial prey species, but game birds may to some extent be pushed into a similar cycle by the combined predator force turning heavily to them when mammal-prey becomes scarce (Lack 1954). All this is largely speculative, however, and is given to indicate research needs. The most recent work on the 10-year hare cycle indicates that food shortage (not predators) starts the decline in prey, though predators later accentuate it, and the low predator numbers that follow enable the prey to increase again when food-supply improves (Keith 1974).

4. Long-term changes

Long-term changes in raptor densities are often associated with human activities. The development of natural or semi-natural areas for stock-raising or agriculture almost always leads to a drop in the numbers of prey and in turn in the numbers of raptors. The truth of this view is evident from the case-histories of particular areas, and also from a comparison of natural areas with cultivated ones, a difference readily apparent, but seldom documented (Chapter 16). Conversely, the artificial inflation of food-supplies sometimes leads to high raptor breeding densities, as in the city-scavengers discussed above, for it is inconceivable that these birds existed in such numbers in the original habitat (eg 19 pairs/km^2).

To take specific instances of change, in parts of Britain Buzzard breeding densities dropped in 1–2 years, following the sudden and almost total elimination of rabbits (the main prey) by the myxoma virus. In one area the territorial Buzzards dropped from 21 to 14 pairs between one breeding season and the next, and to 12 pairs by the following season (Dare 1961). Twenty years later, neither rabbits nor Buzzards had recovered to their former numbers. Examples of Peregrine populations declining with decline in food supplies were given earlier. Other species have occasionally colonised or increased in areas where extra food was provided by man. Lake Naivasha, in Kenya, originally contained no fish suitable for Fish Eagles, but the lake was stocked artificially, and by 1968 it held more than 56 breeding eagle pairs (Brown & Hopcraft 1972). The Fish Eagle has also greatly increased its range and numbers in southern Africa, following the widespread construction of dams and the resulting expansion of fish populations. All these various instances may seem so unsurprising as not to warrant comment, yet together they form a strong body of circumstantial evidence for the dependence of breeding density on prey populations.

BREEDING DENSITY IN RELATION TO NEST-SITES

Shortage of nest-sites

In some landscapes, shortage of nest-sites holds raptor breeding densities below the level that food would permit. The evidence is of two types: (a) breeding pairs are scarce or absent in areas where nesting sites are scarce or absent, but which seem otherwise suitable (non-breeders may live in such areas); and (b) the provision of artificial nest-sites is sometimes followed by an increase in breeding density.

Kestrels increased in one year from less than 20 to more than 100 pairs when nesting boxes were provided in a Dutch area with few natural sites (Cavé 1968). Similar results were obtained with other populations of European Kestrels, and also with American Kestrels (Hamerstrom et al 1973). An increase in Prairie Falcon pairs from seven to eleven followed the digging of suitable holes in riverine earth banks, and increases in Osprey pairs in several areas followed the erection of nest platforms (Fyfe, in Cade 1974, Reese 1970, Rhodes 1972). Likewise, nesting on buildings and quarries has allowed Peregrines and Lanners to occupy areas devoid of natural cliffs.

All these species breed solitarily, and it is often not shortage of nesting sites as such which limits density, but a shortage of sites far enough from other pairs. Thus a cliff face may have several suitable nest ledges, but support no more than one pair of falcons, for there is a limit to the extent to which different pairs will concentrate to breed (Chapter 2).

Provision of spaced nest-sites has allowed some species greatly to extend their breeding distribution. An example is the Mississippi Kite Ictinia misisippiensis which now breeds on the Great Plains of America 'in hundreds, if not thousands', in places where trees were planted by man (Parker 1974). Several other raptors of American grasslands were found to depend to some extent on planted trees or deserted buildings (Olendorff & Stoddart 1974), while in South Africa, Greater Kestrels and Lanner Falcons have spread by using pylons, as have Black Sparrowhawks Accipiter melanoleucus by using exotic eucalypt or poplar plantations (Kemp & Kemp 1975). In none of these species is it likely that prey populations increased, and the spread could be attributed entirely to the provision of nesting places. On the other hand, the destruction of nesting places has sometimes led to reductions in breeding density: in Peregrines when cliffs were destroyed by mining, and in certain eagles when large free-standing trees were felled (Porter & White 1973, Bijleveld 1974).

Competition for nesting places between species

Where nesting trees or cliffs are scarce, the presence of one species may influence the numbers and distribution of another. In Britain, Golden Eagles take precedence over Peregrines at cliffs and Peregrines over Kestrels. When four eagle pairs occupied an area, two Peregrine pairs immediately moved to alternative crags and two other pairs disappeared altogether (D. A. Ratcliffe). When Peregrines were further reduced by pesticides, Kestrels took over many of their nesting cliffs, only to be displaced when Peregrines returned.

On certain grasslands in western North America, nest-sites were provided mainly by the trees around abandoned farmsteads. At any one farmstead, only one raptor species was usually found, and the early-nesting Great Horned Owl had precedence, followed in order by the Red-tailed Hawk, Swainson's Hawk and Kestrel. Competition for limited sites thus affected the species to varying extents, but the owls could occupy a site only until the nest collapsed, when the site reverted to one of the two large hawk-species, as only they could build a new nest (Olendorff & Stoddart 1974).

The outcome of competition for nest-sites sometimes varies with circumstance. Verner (1909) noted that Golden Eagles and Griffon Vultures in Spain would not tolerate one another on the same cliffs, but they used the same cliffs in different years, and eagles drove away single pairs of Griffons but not groups. The scarcity of eagles in one region was probably because all the best sandstone cliffs were occupied by Griffon colonies. As another example, Prairie Falcons in western Canada could displace Canada Geese *Branta canadensis* from ledges on earth banks but not from holes (R. Fyfe). Another type of displacement sometimes occurs among similar species which, in regions of overlap, hold mutually exclusive feeding (and nesting) territories, as described in Chapter 2.

COLONIAL SPECIES

Although no detailed study has been made, there is no reason to suppose that the numbers of colonial raptors are limited by resources different from those that limit solitary ones. In some regions the spacing of colonies of cliff-nesting vultures is clearly set by the availability of cliffs, and the size of the colonies by the number of suitable ledges. But where cliffs and ledges are surplus to needs, the distribution and size of colonies are presumably influenced by food, which at least provides a ceiling on bird numbers. In Griffon Vultures the colonies fluctuate in size from year to year, sometimes through shifts in individual birds (H. Mendelssohn); in Snail Kites, the colonies also occasionally change in location (Haverschmidt 1970).

The precise behavioural mechanism by which colony sizes are regulated in the face of excess nest-sites is no nearer solution now than when Lack (1954) posed the problem for herons. Because they nest in colonies 'there must be some factor attracting the birds to nest together. But all do not crowd into one large colony, so that there must be some other factor counteracting the tendency to aggregation when each heronry has reached a certain size. . . . Further, the critical size of a heronry varies greatly with the locality.' The same questions apply equally to the *Gyps* vultures, the insectivorous falcons, and the other colonial raptors. In fact, some study of the distribution and size of colonies in such species in relation to food is a major research need.

OTHER POSSIBLE LIMITING FACTORS

Research on solitary species has given a good indication of what sets the upper limits to breeding densities, but it has given no inkling of how many populations are up against these limits at any one time. This is because people have tended to study fairly dense populations, and to avoid sparse or sporadic ones that provide less information. Studied populations tended to be in good habitats, and may therefore have been more stable, with greater occupancy of territories, than many unstudied ones. Friends in various countries have told me of areas in which raptors were unexpectedly scarce and irregular, and in which only small proportions of potential territories were occupied in any one year. They did not study such populations because they did not think it worth the effort. Some such populations are clearly held below the carrying capacity of the habitat by persecution or pesticide use, but others could be kept down by natural factors, not necessarily acting in the breeding areas. Thus some migrant populations might be limited while in winter quarters, so that they could never occupy their breeding habitat fully. To my knowledge, this has not been shown for any raptor (except where pesticides were involved), but such a situation recently became apparent among European Whitethroats *Sylvia communis* and other song-birds that wintered in the drought-stricken Sahel zone of Africa (Winstanley et al 1974). Hence, despite the low yield of data, sparse raptor populations free from obvious human influence would well repay more study. The role of various natural factors in raptor mortality is discussed in Chapter 12.

NON-BREEDERS

So far, I have written chiefly of limitation in the numbers of potential breeding pairs found in association with nesting territories. These birds include pairs which lay eggs and rear young, and others which in certain years do not lay, yet still hold nesting territories. Other (unpaired) non-breeders live away from nesting territories, and may include not only adults unable to acquire a territory, but also individuals too young to breed. Practically nothing is known of how the numbers of such unpaired birds are limited, or even what proportion of the population they form. In general, they go to where they can live free of competition and harassment from breeders, either inside or outside the breeding range. They have been found: (a) unobtrusively within the home ranges of established pairs, at least for short periods, as in Golden Eagle; (b) in spaces between the territories of breeding pairs, as in Buzzard, Red-tailed Hawk and Black Eagle; (c) in those areas in the same geographical region that are unsuitable for breeding (often through lack of nest sites), as in Red-shouldered Hawk, Golden Eagle and certain vultures; or (d) mainly in different geographical areas (eg 'winter quarters') from breeding birds, as in Osprey and Montagu's Harrier (all refs. in Newton 1976a). These are not mutually exclusive alternatives, and more than one system has occurred in the same population.

It depends on food supply whether such birds live solitarily or gregariously. Their numbers have been found to vary between areas and years, and to fluctuate considerably while breeding density remained unchanged. On Lake Naivasha in Kenya, adult Fish Eagles increased by 40% and immatures by 11% between 1968 and 1971, but breeding numbers remained constant at 56 pairs (Brown & Hopcraft 1973). Because they are not tied to a nest, non-breeders have greater freedom of movement than breeders, and more often exploit temporary abundances of food, as is especially true of *Milvus* kites and *Haliaeetus* eagles. Where not restricted by food or habitat, their numbers are presumably influenced by breeding rates and recruitment into breeding populations.

SUMMARY

In any landscape, an upper limit to the number of established raptor pairs is set by food or nest-sites, whichever is in shorter supply. Surplus unpaired adults are present in many populations, but are unable to breed until a gap is made available by the death or removal of a territory owner. In some areas, where populations have been much reduced or eliminated by human action, birds subsequently recolonised to about the same level as previously, with pairs taking up the same places. Where nest-sites are freely available, pairs of many species are evenly spaced, but at different distances apart in different regions, correlated with differences in food supplies.

In general, large raptors live at lower density, and in larger home ranges than small ones; this is linked with large raptors feeding on larger, sparser, slower-breeding prey than small ones. Raptor populations that depend on fairly stable (often varied) food-sources show fairly stable densities over long periods of years, whereas populations that depend on fluctuating (often restricted) food-sources show fluctuating densities, in accordance with prey cycles. Raptors that were subjected to sudden or long-term change in prey numbers, underwent a sudden or long-term change in breeding density. All this evidence suggesting that raptor densities are limited by food-supply is circumstantial and cannot be considered conclusive.

That some populations are limited by shortage of nest-sites at a lesser density than the available food would allow is shown by the absence of breeding pairs in terrain lacking nest-sites but otherwise suitable, and by increases in the numbers of breeding pairs that have often occurred after the provision of artificial nest-sites. Competition between species for limited nest-sites also restricts the local breeding density of certain species. Through these various factors, much of the natural variation in the breeding densities of raptors can probably be explained, but in addition many populations of modern times are below their potential level, as a result of pesticide use or other human interference. It is also possible that some migrant populations are limited while in winter quarters, so that they cannot occupy their breeding habitat fully.

CHAPTER 4

Winter density

Raptors have been little studied in winter but they seem to be no less influenced by food availability at that season than when breeding. In some conditions, the adults stay around their nesting territories and occupy much the same hunting ranges all year. In other conditions, one partner may stay for the winter and the other may leave, or both may leave and then either disperse locally or on a long migration. These different strategies seem to represent progressive adjustments to decreasing availability of prey in winter.

At high latitudes, year-to-year stability in wintering density seems much less common than year-to-year stability in breeding density. This is perhaps because summer food supplies vary less from year to year than do winter ones – in some areas influenced by snow cover – but also because the birds themselves have greater freedom of movement once they are no longer tied to a nest, so can more readily shift around in relation to changes in prey. After breeding, birds soon spread into other areas, and may even become common in habitats from which they were earlier absent. But in the same way that their breeding density may have been limited by shortage of nest-sites below what food availability would permit, in some landscapes their wintering density may be limited by a shortage of perching and roosting sites. Thus the distribution of Kestrels in Ohio depended largely on old buildings or

Illustration: Prairie Falcon and prey

other sheltered roosts, while that of Fish Eagles along lake shores in Uganda was clearly influenced by the presence of trees as perches (Mills 1975, Eltringham 1975). It is not always necessary for a bird to have a good roosting site within its hunting range, however, provided that one is available within easy commuting distance (Mebs 1964).

North temperate regions
A study of Prairie Falcons in Colorado gave results that were typical of those of several winter studies (Enderson 1964). In this open cultivated landscape, the falcons hunted mainly from the tops of power poles. The birds occurred more abundantly in one winter than in the other, and in both winters individuals varied greatly in the time they stayed. In the first winter, the twelve residents spent an average of 58 days in the area, with no more than seven individuals present at once, while in the second winter, the four residents spent an average of only 24 days in the area, with never more than one individual present. All these birds established regular hunting ranges, but many others passed through the area, mainly in November and February, and were seen only once. The presence of Prairie Falcons in the locality corresponded with the presence of their main prey species, the Horned Lark *Eremophila alpestris*, and differences in falcon densities year to year corresponded with differences in lark densities.

European and American Kestrels that were studied in more than one winter showed the same tendencies, again linked with food, namely: (1) big differences in numbers from one winter to another; (2) considerable variation in numbers within the course of a single winter; and (3) individuals staying for greatly varying periods (Enderson 1960, Mills 1975, Cavé 1968, Village 1979). Similarly, Golden Eagles were studied over nine years in an area in south Sweden, and in winters when carcasses of domestic animals were provided as food, more birds stayed for a long period than in winters when no carcasses were provided (Tjernberg 1977).

Other winter studies dealt with the whole raptor population of an area. One of the most detailed was in 96 km² of farmland, with scattered woods and marshes, in southern Michigan (Craighead & Craighead 1956). The population was studied in two winters of differing food supply. Of six species that occurred there, the Red-tailed Hawk, Red-shouldered Hawk, Rough-legged Buzzard, Kestrel and Hen Harrier fed chiefly on *Microtus* voles, while the Cooper's Hawk fed chiefly on birds. In contrast to the studies mentioned above, the whole raptor population of this area became fairly stabilised after the autumn migration. Individuals occupied hunting ranges which varied in extent according to species and food supply. In general, birds held smaller ranges at higher density in the first winter, when voles were plentiful, than in the second winter when voles were scarce (Table 13). Some 96 hawks of all six species stayed through the first winter, but only 27 hawks of four species stayed through the second; juveniles were relatively fewer in the second winter, Hen Harriers were present for a brief period in autumn, and Rough-legged Buzzards absent altogether.

Although numbers were related to food supplies, it was not clear how this came about. Territorial behaviour was not very obvious, but the birds continually responded to one another, often in subtle ways. One bird flying between perches was enough to set in train similar flights by its neighbours, so that all the birds were in continual spatial adjustment to one another. Nor was it clear whether the failure of two species to remain through the second winter resulted from such interactions, or whether it resulted directly from prey densities that were too low to maintain these species. Interactions between species certainly occurred, and in the year when Hen Harriers were present, they consistently hunted those parts of the *Buteo* ranges that were not in use at the time.

In the European Kestrel, spacing and aggressive behaviour in winter were likewise related to food (Cavé 1968). In one area where voles were numerous, birds concentrated and showed no obvious territorial behaviour; but in another area where voles were less numerous, the birds defended individual hunting ranges. This was established by observation and by testing the reaction of resident birds to stuffed specimens, a method previously used by Cade (1955) to confirm winter territoriality in the American Kestrel. It seems, therefore, that territorial or other social behaviour is important in adjusting the wintering densities of some species in relation to food supplies, as it is in influencing their breeding densities.

Tropical regions

Associations between food levels and raptor numbers were apparent in different parts of West Africa. From counts through the year, Thiollay (1978) noted a significant relationship between the numbers of insectivorous raptors (11 species) in particular areas and the numbers of the grasshoppers that formed the main prey ($r = 0.80-0.86$, $P<0.01$). Fluctuations in raptor and grasshopper numbers showed different seasonal patterns in different climatic zones, and it was striking how quickly the migrants accumulated in an area when prey were increasing, and left again when prey declined (Figure 12).

Another study was made in the dry savannah of Tsavo National Park in Kenya (Smeenk 1974). In this area, 12 raptor species were resident through the year, nine others were migrants from elsewhere in Africa, and eleven were migrants from the Palearctic. The resident species were of similar abundance in both years, but most of the migrants reached much the greatest numbers in the year of heaviest rainfall, when all kinds of prey were more plentiful. The totals reached about 106 migrants per 100 km of transect in the wet year compared with about 51 in the dry, the difference being especially marked among Bateleurs, Wahlberg's Eagles and Steppe Eagles. Hence, the main findings of these tropical studies were similar to those of the temperate zone, with annual differences in raptor numbers in particular localities associated with annual differences in food.

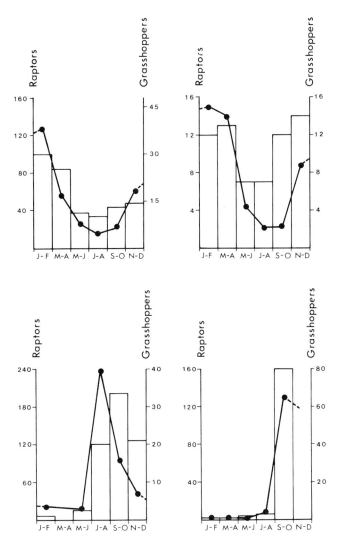

Figure 12. Seasonal changes in the numbers of raptors in relation to changes in their grasshopper (Orthoptera) prey in four regions of West Africa. From top, left to right: the south Guinea zone, north Guinea zone, south Sahel zone and north Sahel zone. The scales are indices of numbers. Re-drawn from Thiollay 1978.

One implication of all such studies is that many raptors are flexible in their migrations, and that at least some individuals vary the distances they move, according to food supplies, so that populations are distributed somewhat differently from one year to the next. This is

especially obvious in some boreal zone populations, such as those of
Goshawks, which reach the temperate zone in large numbers chiefly in
years when prey are scarce further north (Chapter 11). However, all the
migrant species so far studied had greatly fluctuating food supplies;
those that have more stable (or varied) ones might show more stable
winter distributions, with individuals showing greater fidelity to
particular localities. Records of various buzzards and falcons returning
to the same places in successive years will be given in Chapter 10.

In conclusion, all the evidence linking raptor densities with food
supply in winter is again circumstantial, based on correlations between
bird numbers and prey numbers. It is, perhaps, as good as we are likely
to get, but it remains open to the possibility that some unknown factor
might have varied in parallel with food and influenced bird density,
rather than food influencing bird density direct. A shortage of hunting
and roosting perches is the other natural factor which has been
suspected of limiting winter densities, but pesticides and other
unnatural factors are also involved in some populations (Chapter
13–15).

SUMMARY

Studies outside the breeding season in temperate and tropical
localities showed that raptors were more numerous in years when prey
were plentiful than in years when prey were scarce. Numbers varied
more from month to month, and from year to year, than was usual in
the breeding season, perhaps partly because food availability varied
more, but also because the birds had greater freedom of movement then
to exploit temporary abundances. Populations that utilise sporadic
food supplies may be distributed differently from year to year, so that
local numbers fluctuate and at least some individuals may visit
widely-separated localities in different years. In open landscapes, a
shortage of suitable hunting and roosting perches may limit the density
of some populations to lower levels than the available food would
allow. In view of the paucity of relevant studies, these conclusions
should be regarded as tentative.

Sparrowhawk and prey

(Upper left) Ruppell's Vultures at a carcass. This species nests in large colonies on cliffs and forages up to ¹0 km from the colony, specialising on migrant game. (Upper right) The Egyptian Vulture nests solitarily, ¹avenges at carcasses, and is able to break open the eggs of Ostriches and other birds with the aid of stones. ¹ower) The White-headed Vulture breeds solitarily at very low density; it probably takes living prey as well as ¹rrion. Photos: H. Kruuk.

8 Vultures at a carcass. (Upper) Lappet-faced Vultures in threat display; (centre upper) White-headed Vulture and Lappet-faced Vultures put to flight by hyaenas; (centre lower) White-backed Vultures, with two Hooded and one Egyptian Vulture in the foreground, and an Egyptian and a White-headed Vulture behind. In much of Africa up to six vulture species can be found at the same carcass, mostly specialising on different parts of the carcass; (bottom) mainly White-backed Vultures on a Rhino carcass, Ngorogoro Crater, Tanzania. Photos: H. Kruu

CHAPTER 5

Problems concerning nest-sites

Raptors are among the few groups of birds whose numbers and nest success are in some regions clearly limited by the availability of nesting places. To pick an obvious example, most cliff-nesters are restricted geographically to breeding in areas with cliffs. Within such areas, their breeding density may be limited by the number of cliffs with suitable ledges, and their breeding success by the accessibility of these ledges to predators. Other raptors may be limited in open landscapes by shortage of trees, and even in woodlands, nest-sites may be fewer than they at first appear. In a large area of mature forest in Finland, less than one in a thousand trees were judged by a biologist to be suitable for nests of White-tailed Eagles, while in younger forests, suitable open-crowned trees were even scarcer or non-existent. More concrete evidence that lack of good nest sites may often restrict breeding density or success came from the experiments described in Chapters 3 and 16, in which pair numbers increased or success improved following the provision of artificial sites.

Illustration: European Kestrels at nest

Whatever the situation, a good nest-site should offer a sheltered place for the eggs that is reasonably secure from predators. Most cliff-dwelling raptors choose a fairly flat substrate for the nest, if possible under an overhang, or in a fissure or cavity. Protection from predators depends on the sheer precipitousness and inaccessability of the cliff face. Limestone cliffs are especially good for raptors because of the many crevices and ledges they have, but some other kinds of rock may be totally devoid of possible nesting places. Most tree-nesting raptors use good, firm crotches to build their nests in; and large eagles usually choose a tree that is separated from others by its position or height, and that has an open crown, for easy access. Some large raptors place their nests on the very tops of trees, for example the Osprey, Tawny Eagle, Lappet-faced Vulture, Secretary Bird and the Snake Eagles. These nests are open to the sky but give their owners easy access from above. For any tree-nesting species, good security is provided by a tall, smooth-barked tree, devoid of branches for much of its height, and with a thick crown to conceal the nest from below. On the African savannahs, many raptors seem to prefer thorn trees, and presumably gain protection from the spines. Most ground-nesting raptors hide their nests amid tall, thick cover, and some gain extra protection by nesting over water (Marsh Harrier) or on an islet surrounded by water or bog (Peregrines in northeast Europe). Nest-sites do not all reach the ideal standard, however, and in any situation they vary widely in quality as sites for successful breeding.

Protection by situation seems to be a response to carnivorous mammals (or reptiles). Protection against other predatory birds is difficult, but some small raptors put their nests in enclosed situations (small accipiters in thick cover, kestrels in holes) that give some security against larger raptors, as well as mammals. The large raptors, on the other hand, on account of their size, are better equipped physically to defend their nests against marauders. Over much of Europe, there are few climbing mammals, apart from small martens (*Martes*), but in North America the Raccoon *Procyon lotor* seems to be an especially effective nocturnal predator on the contents of raptor tree nests, whereas in Africa, various baboons, monkeys and cats are important.

CHOICE OF NEST-SITE

The choice of nest-sites is influenced by several factors, besides their local availability: (1) whether the species concerned is capable of building nests for itself, as opposed to relying on existing nests or other substrates; (2) the range of situations it can use, whether cliffs, trees or ground; (3) the extent of local competition from other species for sites; (4) the numbers and kinds of predators present, and (5) local traditions of preference. The amount of building possible seems fixed in each species, but the other factors vary in importance from region to region, as described below.

1. Nest-building

The New World vultures (cathartids) do no building, but merely scrape out a hollow for the eggs. The falcons are capable of little more than this. Some early naturalists thought that falcons built stick nests, but no recent biologist has recorded any building activity, apart from rearranging existing material or pecking off surrounding vegetation and placing it in the nest-cup, though the Merlin can form a substantial pad in this way (Newton et al 1978). The falconets and pygmy falcons (*Microhierax* and *Polihierax*) show similarly limited building behaviour. In effect, therefore, all these species need an existing substrate or old nest on which to lay their eggs.

The remaining raptors can all build for themselves, the larger species using their larger nests over longer periods of years than the small ones. Small kites and small accipiters make fairly slight nests, used for only one or two seasons, whereas some large eagles make huge structures, used by a succession of birds for up to a century or more (Chapter 2). Cliff-nesting Griffons and tree-nesting Snake Eagles are exceptions to the trend in that they build very small nests for their size, while some Snake Eagles make a new nest in a different place each year (Brown 1976b). Ferruginous Hawks in North America build exceptionally large nests for the size of bird. In any one area, the nests of different species can usually be distinguished by size, structure or situation, but occasionally the same nest is refurbished and used by different species in different years. Interchanges are especially frequent among various *Buteo* species, between Goshawks and *Buteo*, between *Milvus* kites and *Buteo*, and even between Ravens *Corvus corax* and *Buteo*. An important point, however, is that the larger the raptor and the larger the cliff-ledge or tree-crotch needed for the nest, then the more restricted the number of potential sites available to it. In other words, larger raptors have generally fewer places to place their nests than do small ones.

2. Situations used

Although some raptors put their nests in trees, on cliffs or on the ground, others use only one or two of these alternatives (Tables 14 and 15). Depending on species, the tree-users nest in crotches or in cavities, the cliff-users on ledges or in cavities, and the ground nesters in open situations or in cover. In general, the species that do not build nests use cavities much more than those that do. Cathartid vultures use hollows in trees, potholes in cliffs or tunnels in thorn thickets; and the smaller species can scramble down the hollows in vertical tree-stumps to depths of several metres (Bent 1938).

As a group, the falcons use a wide variety of nesting places, but they include wide-choice species like the European Kestrel and narrow-choice species like the Hobby, which uses only the old tree nests of other birds. The African Grey Kestrel *F. ardosiaceus* uses mainly the domed nests of the Hammerkop Stork *Scopus umbretta*, whereas Dickinson's Kestrel *F. dickinsoni* favours the tops of borassus palms. The five falconets *Microhierax* are all restricted to small tree holes, such as those made by barbets or woodpeckers. The Pygmy Falcon *Polihierax semitorquatus* uses the old nests of a few weaver bird species. In southern Africa, it breeds only in the nest masses of the Sociable Weaver *Philetarius socius* and is not found outside the range of this species, despite large areas of otherwise suitable habitat (McLean 1970). In East Africa it uses the nests of other weaver species which are present in southern Africa but not used.

I do not know why some species are more restricted than others in the range of sites they use for nesting, but there seem to be two extremes of strategy, each with its own merits. A species that is restricted to a specialised site finds less competition from other raptors, but may be confined geographically by the location of such sites. A species using a wide variety of sites is much less confined geographically, but may more often come into conflict with other species, which may wish to use the same site. It is chiefly the smaller raptors (some *Falco, Polihierax, Microhierax*) that are more specialised, and the larger raptors (*Aquila, Haliaeetus, Buteo*) that use a wide range of situations. Thus the range of sites accepted by each species probably depends on its own particular circumstances, and is developed as a compromise between the conflicting needs of a wide choice and limited competition. Other factors are also involved, as discussed later.

3. Local competition

Large species not only oust small ones in conflicts over nesting places, but they also gain from an earlier start in breeding (Chapter 6). In some areas, the stronger species, occupying most of the sites in an area, can reduce the breeding density of the weaker ones (Chapter 3), but in other areas, the stronger merely restrict the weaker to inferior sites. This is illustrated by the findings of an 18-year study on the Colville River, Alaska (White & Cade 1971). Of 80 cliffs in this area, 89% were used for nesting at some time or other, 25% by Ravens, 26% by Gyr Falcons, 50% by Peregrines and 86% by Rough-legged Buzzards. The Ravens and Gyr Falcons began early in the year, and used only the best sites that were sheltered by overhangs. The later-nesting Peregrines occupied a wider range of generally more exposed sites, and the Rough-legged Buzzards occupied an even wider range, overlapping with all the others, and also using very exposed ledges, steep talus slopes or boulder tops on hillsides. Both falcons laid on ledges or in old stick nests built in previous years by other species. Moreover, White (1969) three times found the two falcons together on the same cliff, but each time the larger, earlier nesting Gyrs were on vertical faces, which a person needed a rope to reach, whereas the Peregrines were at the brink of a slope or in other places, which a person could easily walk to. Such site differences seem to result largely from species interactions, the larger, earlier-nesting species getting the best sites, thereby leaving fewer for the other species and leading many individuals to use less good sites.

4. Influence of predation

Several observations indicate the importance of human and other predation on the kind of nest-site that raptors use. Over most of its range, the Peregrine nests only on high cliffs or on islets surrounded by water or bog. But on tundra and other areas where people are few, it will accept lesser crags, earth slopes or even low mounds or boulders on flat ground (Hickey & Anderson 1969, Kumari 1974). 'Generally speaking, the minimum height of cliff acceptable to birds varies inversely with the degree of wilderness' (Hickey 1942). Some easily accessible sites in the Canadian arctic were abandoned following an increase in human population, and the same evidently happened in Britain before 1860, and also in the United States following settlement by Europeans (Fyfe 1969, Ratcliffe 1969, Hickey 1942). In recent decades, Peregrines in Britain have occasionally nested on level ground or on slopes, but the habit has not persisted, presumably because such attempts usually fail (Ashford 1928–29).

The same response seems to occur to other mammal predators, besides man. Use of accessible nest-sites by Bald Eagles in the Aleutians occurs only on islands lacking Arctic Foxes (Sherrod et al 1977). Likewise, ground nesting by Buzzards in Uist, and by Kestrels in Orkney, is linked with the absence on these Scottish islands of wild mammal predators (Newton 1976a). Moreover, the Kestrel habit is fairly recent, as the first known ground nest was found about 1945. The habit then spread rapidly and, once a site was established, it remained in continuous occupation so that by 1955, 19 such sites were known, all in long heather, in cracks in banks or rabbit holes (Balfour 1955). This enabled Kestrels to occupy areas not otherwise suitable, and thereby to increase their numbers. Only a few records of ground nesting are known from mainland areas. Off eastern North America and in the Red Sea, Ospreys nest on the ground on islands devoid of mammal predators, but they disappeared from a treeless island off California, following its colonisation by Coyotes *Canis latrans* (Abbott 1911, Bent 1938, Kenyan 1947). Over most of their range Ospreys nest on trees.

There are thus indications that: (a) some geographical differences in the types of sites acceptable to a species are linked with differences in local predation and disturbance pressures (Peregrine, Kestrel, Buzzard, Osprey, Bald Eagle); (b) minimal requirements have changed with time as disturbance and predation have changed (Peregrine, Osprey); (c) individuals occasionally attempt to nest in less safe sites but seldom succeed (Peregrine); but (d) if they do succeed, the habit may spread, and lead to local expansion in distribution and density (Kestrel). The minimum level of security of nest-sites that any local population accepts may therefore be in continual process of adjustment to predation pressures. In this indirect sense, predators are no doubt responsible for restricting raptor breeding density in areas where safe sites are scarce, for if it were not for their predators, the raptors might accept less safe sites, and thereby breed in greater numbers. The same applies to some other birds limited to safe nest-sites.

5. Local traditions in type of site acceptable

Some nest-site preferences are apparently not connected with predation or competition, and are best explained in terms of local traditions. In some cases, the failure to use a certain type of site outside a given area restricts the breeding range. Localised tree-nesting by Peregrines provides the best example. Over most of the range, Peregrines nest only on crags or islets, as mentioned, but in the Mississippi valley, on islands off western Canada and in Australia they use (or formerly used) cavities in big trees, while in parts of Eurasia, coastal Virginia, Alaska, on islands off western Canada and in Australia, they use (or formerly used) old tree-nests of other birds (papers in Hickey 1969; Jones 1946, White & Roseneau 1970, Campbell et al 1977). Tree nesting enables Peregrines to breed in areas devoid of cliffs, and presumably arose independently in each of these widely-separated areas. In middle-Europe, cliff and tree nesting populations are contiguous (Mebs 1969), and in the Baltic region, cliff, tree and ground nests are interspersed in the same geographical areas (Thomasson 1947, Kumari 1974). It seems inconceivable that birds using different kinds of site do not interbreed, and almost certainly the habit of tree-nesting is not genetically controlled. There is no obvious explanation why the habit is localised when, by nesting in trees, Peregrines could spread over huge areas otherwise not available to them. They are not known to have bred in large parts of North America, Asia and Africa in which tree nests and prey are available (in some such areas non-breeders occur). Competition for tree nests can scarcely be involved, because Peregrines nest fairly early, can hold their own against most other birds, and some of the commonest sites (heron nests) are widespread.

Most other large falcons nest in trees, at least in part of their range, but in old stick nests and not in cavities, and it is not known to what extent traditions are involved. The Australian Black Falcon *F. subniger* nests entirely in trees; the Lanner and Lagger *F. jugger* use trees where cliffs are unavailable; the Saker uses trees mainly in the west and cliffs mainly in the east of its range; and the Gyr uses trees regularly in much of its Asiatic range, but extremely rarely in Europe and North America, except perhaps around the Anderson River in Canada (Dementiev & Gladkov 1954, Bent 1938, Kuyt 1962, R. Fife). On the other hand, the Prairie Falcon is effectively restricted to cliffs, with only one record of a tree-nest involving a repeat after the failure of a cliff-nest (Porter & White 1973). But this species too could probably spread over huge areas if tree-nesting caught on. The restriction of the Pygmy Falcon in southern Africa to areas with a particular kind of weaver-bird nest was mentioned earlier.

Some non-falcon species also show local traditions in their use of nest-sites which cannot be explained by availability. The Red Kite is restricted to trees over most of its range, but also uses cliffs on the Cape Verde Islands (Bannerman 1968); the Tawny Eagle is restricted to trees in Africa but also nests widely on the ground or on cliffs in Eurasia; while the Osprey uses trees and other sites over most of its range, but is restricted to cliffs and rock pinnacles in the Mediterranean region (Terrasse 1977). All these species could probably extend their range if all individuals accepted such a wide range of sites. Such local traditions may depend to some extent on young 'imprinting' to particular site-types, and preferring these in adult life, but research is needed (Chapter 17).

Nesting on man-made structures
It is the open-country species which nest in a wide range of natural sites that have taken most strikingly to buildings and other man-made structures (Table 16). Among small falcons, only kestrels have taken up this opportunity, and several species of kestrels over their whole range use anything from isolated ruins to churches and public buildings in city centres. The German name 'Turmfalke' (Tower Falcon) for *Falco tinnunculus* is testimony to this habit in central Europe. Nesting on buildings is one of the factors that have enabled this kestrel to penetrate towns so successfully. No less than 61 pairs were found nesting within the city of Munich (one pair/5 km), of which 42 were on buildings (Kurth 1970). In London 142 pairs were found within about 30 km of St Paul's Cathedral, a density of one pair/23 km^2 (Montier 1968).

Among the larger falcons, the Peregrine has used old buildings to a considerable extent, either isolated or in small towns, especially in Europe (Hickey 1969); but it has nested in only a few modern cities, despite good nest sites and abundant pigeons as prey. The cities involved show a curiously scattered distribution, implying that the habit arose independently each time; in Eurasia: Salisbury, Heidelberg and Moscow; in North America: Montreal, Philadelphia and New York; in Africa: Nairobi; and in Australia: Perth and Sydney (Hickey & Anderson 1969, S. Davies). Most of these cases persisted for no longer than a few years, a notable exception being the Montreal birds which nested for 18 years, until the DDT era (Hall 1955). Perhaps cities are acceptable only to a small fraction of the Peregrine population, mass human presence deterring the rest. Among the other large falcons, the Lanner has nested on the Egyptian pyramids, and on tall buildings in Salisbury, Durban, Pretoria and other African cities (Kemp 1972, Sinclair & Walters 1976), and both it and the Lagger have used monuments and tall buildings in India (Vaurie 1965); the Orange-breasted Falcon has used churches and ruins in Central America (Brown & Amadon 1968); the Gyr Falcon has occasionally used old Raven nests on disused gold dredgers and other structures in Alaska (White & Roseneau 1970); but the Prairie Falcon has only once been known to attempt to breed on a building (Nelson 1974).

Among other raptors, the use of man-made structures in undisturbed areas is fairly frequent in several species of *Buteo* and *Aquila*, and also in the Egyptian Vulture, Black Kite and Martial Eagle; and occasional instances are known for other species (Table 16). Turkey Vultures and Black Vultures have sometimes nested in disused barns and sheds in the United States, and numbers of Black Vultures also nest in the niches of buildings in Lima, Peru (Brown & Amadon 1968). Ospreys have long used artificial structures, including tree-substitutes such as telegraph poles, and in recent years special platforms. During aerial surveys along the Atlantic coast of North America, platforms for nesting Ospreys were noted in every State from South Carolina to New York, many in use. Around parts of Chesapeake Bay, it seems to have become a status symbol to have a pair of Ospreys on a platform near your home. In 1973, only 31% of the 1,500 pairs around the Bay were found on trees, the rest on duck blinds (29%), channel markers (22%), and other man-made structures (18%), including platforms (Henny et al 1974).

The use of man-made structures enabled all such species to increase their density and extend their range in places where natural sites were scarce or non-existent. In some instances the acceptance of such structures was immediate, but in others it occurred only after many years or only in particular areas. The opportunities seem not to be fully exploited, however, and no doubt an intolerance on the part of the human population in the western world has been partly responsible for deterring at least the larger species.

OTHER ASPECTS OF NESTS AND NEST-BUILDING

Why do eagles and some other raptors build such huge nests? The stick piles, accumulated over many seasons, represent an enormous investment of time and energy on the part of the birds. So bulky that they cannot be concealed, it is possible that such structures have come to serve an important signal function, advertising the nesting territory to others of their species almost as effectively as the owners themselves could do. Such nests, placed in the open crown of a free-standing tree, are visible from several kilometres; and those on the very tops of trees are conspicuous from the air.

The smaller accipiters and kites can make their small nests in a few weeks or less. But large raptors seldom build a completely new nest in a year in which they breed, and spend months over the job. In some such species, pairs have only one nest to which they return for each breeding attempt (eg Crowned Eagle, Brown 1966). In most species, however, each pair has several nests in the territory, and uses different nests in different years (eg *Aquila, Hieraaetus, Buteo*). The birds normally refurbish more than one nest each year, before eventually laying in one. If the attempt fails, they may switch to another nest for a repeat attempt or, if they do not lay again, they often continue through the season to add material to existing nests. Ospreys and some other species are noted for building so-called 'frustration nests' after a failure, completely new nests which take the rest of the season to construct, but which may be used in future years.

Maintaining more than one nest is of obvious advantage. The pair have the opportunity to shift at the last minute if they are disturbed, or to move to a second nest if their first breeding attempt fails early. In another season, a nest may have collapsed or have been taken over by another species, such as (in North America) a Great Horned Owl. In most of the large raptors, if a pair find themselves without a nest at the start of the season, they then do not breed that year, but spend most of the season building one. Some species that remain on their territories all year may nest-build in any month.

The importance of always having a substantial nest available can thus account for much previously unexplained behaviour: the possession by most pairs of several nests, the annual repair of more than one nest – even in years of non-breeding, the continuance of nest-building outside the normal pre-egg period, and the building of frustration nests. For individuals to gain the benefits of their efforts, they have to return to the same territory in successive years, and this seems to be so in those large species that have been studied (Chapter 10).

The use of alternative nests is said to be also a means of avoiding parasites, many of which remain in the nest, ready to attach themselves to the birds in the following year. By shifting between nests, it is claimed, the birds reduce their chance of becoming heavily infested. This may be a subsidiary reason for alternating between nests but to my mind it is less important than the need to maintain several nests in case of loss. Moreover, it does not fit with the finding that Buzzards and Kites are more likely to change nests following a failure than after a success (see Chapter 8), for it is after a success – when chicks have grown in the nest – that the parasite load is likely to be greatest.

Another frequent habit which has attracted attention is the bringing of fresh green leafy sprays to completed nests. In some species, the nest-cup is lined with greenery, beginning before laying (eg Martial Eagle, Black Sparrowhawk), while in others, the nest-rim or the whole nest surface is decked with greenery (eg Buzzard, Harpy Eagle). Whatever the pattern, green sprays often continue to be added for the whole season, whether eggs are laid or not. They are brought even in the nestling period, and placed around the chicks, usually with no attempt to incorporate them into the nest structure. At this stage, they tend to be provided by an adult visiting the nest without prey, as at a changeover, and a great deal is brought by the female in that phase of the nestling cycle between brooding and hunting, when she remains near the nest, with little to do, except wait for the male to arrive with food. One suggestion is that the greenery is a form of nest-sanitation, covering rotten meat and excreta, but this does not explain why it should be done with green, rather than with other material, or why the greenery should be brought long before the eggs are laid and the nest is soiled. Another suggestion is that the greenery serves to maintain humidity, but this is hard to accept as an explanation for behaviour which is as prevalent in damp oceanic islands and rain forests as it is in deserts.

An alternative explanation, that has come to my mind, is that the greenery is another form of advertisement, over shorter range, serving to denote an occupied territory, as opposed to an unoccupied one. Unlike any other material, green sprays change colour quickly, so that the presence of green on a nest immediately signifies recent activity. The advantage to a prospecting bird, when the occupants are not around, is that it does not waste time attempting to establish a territory around a place that is already occupied; and the advantage to the owners is that, by bringing greenery, they might reduce the need to expel intruders by more vigorous means. As the green fades to yellow, the birds keep bringing fresh material and placing it on top, and the fact that they do this in the pre-egg stage, and continue to do so, even if eggs are not subsequently laid, would fit this explanation. It is curious that greenery should continue to be added at later stages, when the nest contents alone should indicate occupancy, but a primary purpose in advertising does not exclude the possibility of subsidiary functions, such as sanitation at the nestling stage.

The new material also helps to fill in the nest cup and change the shape of the nest to a broad platform on which the young can stand and exercise their wings. In addition, by rotting and filling gaps between twigs, the greenery serves to consolidate the structure and lengthen its life. This will be obvious to anyone who has tried to pull to pieces a large raptor nest compared with a mere stick pile.

There seem to be two main groups of nest-building raptors that do not use greenery, the small accipiters, kites and harriers that build concealed one-year-only nests, and the various large species which put their nests on the very tops of trees. These latter species put yellow or other pale material in the nest-cup. Contrasting with the wide dark rim, these pale centres make the nests conspicuous from the air, and new nests are much easier to see than old ones devoid of lining. So in these species, advertising may be a reason for using pale rather than green material. Of course, whether green or yellow, the lining is largely covered when the parent is present. The absence of greenery-bringing in falcons and cathartids can reasonably be attributed to the lack in these birds of any nest-building behaviour. Perhaps, in the end, however, you prefer to believe with Gromme (1935) that the birds use greenery to lend 'an appearance of artistry and colour', or with Hill (1946) that the whole procedure is a 'symbol of the ancient law of instinct that is part of the poetry of nature'.

ANTI-PREDATOR BEHAVIOUR

Security against predators depends not only on a safe site, but also on defence by the owners. The predator, it may be supposed, is less inclined to visit a nest, or tackle a difficult climb, if it must continually fend off an attacking raptor. For obvious reasons, most information on nest defence concerns the reaction of birds to human intruders. When a man approaches a nest, the off-duty bird may sound the alarm, enabling the mate to leave the nest unobtrusively, and the two together may then protest loudly until he leaves. Another initial reaction of the incubating bird is often to squat low, with the head flat against the nest rim. This makes the bird hard to see from the ground and, when coupled with tight sitting, it is easy to assume that a nest is empty. If a bird is surprised on the nest when feeding young, it may freeze on seeing a man, remain motionless until his eyes are averted, then slip swiftly away. This reaction seems especially common in small accipiters, and is another adaptation to avoid drawing attention to the nest, it being less easy to notice a still bird than a moving one.

Once the bird is off the nest, behaviour varies greatly between species and to some extent between individuals, and between stages in the breeding cycle. From one extreme to the other, the common variants include: (a) flying silently from the vicinity of the nest; (b) circling or flying round, calling; (c) diving towards the intruder, calling vigorously, but veering off at some distance from him; and (d) actually striking the intruder with the feet. The main function of the calling is probably to distract the intruder's attention from the nest to the female, but if the male is absent, the female's calling sometimes serves to draw him in to help in the defence. A third function of the calling is to communicate with the young, which squat flat on hearing the alarm of the adults. The purpose of the attacking is presumably to drive away the intruder. Various falcons, Hen Harriers and Goshawks have a reputation for particularly vigorous nest defence, involving diving to within a short distance of the observer, and even striking him. In addition, distraction displays have been recorded from the European Snake Eagle, and also from the Bateleur. The Snake Eagle, on the rare occasions that it performs, is said to drop from the nest and shuffle along the ground as if hurt, in the manner common to certain plovers and game birds (Boudouint *et al* 1953). The Bateleur sometimes perches near the nest and bobs up and down, flapping its limp partly open wings, uttering loud short barks; it also dives at the intruder with rapidly flapping wings, thereby creating a loud and alarming noise (Moreau 1945, Steyn 1973).

For any one species, defence behaviour varies in different parts of the world, apparently correlated with past persecution. In North America, Goshawks almost invariably attack a man who is at the nest, sometimes striking him with the flat of their feet, or raking him with a hind claw as they dash past, tearing skin and clothing. In Europe, with rare exceptions, this is practically unknown. For more than 150 years, Goshawks have been much more heavily persecuted in Europe than in North America, and any birds which came close were easily shot, so that the habit could have been eliminated. In North America and Greenland, Peregrines nearly always come in close, and occasionally strike the observer, whereas in Britain, they fly round calling but generally stay out of gunshot. Within North America, differences are apparent across the continent, as Red-tailed Hawks and other species are much more aggressive in the west than in the east, linked with greater persecution in the east. Also, over much of the continent, Bald Eagles merely circle 70 metres or more above the observer, though occasional birds approach more closely (Grier 1969, Bent 1938), but on Amchitka in the Aleutian Islands, where they have not been exposed to men with firearms, most Bald Eagles stoop to within a metre and occasionally strike a man, beginning their attack when he is as much as 200 metres from the nest (Sherrod et al 1977). They also stoop at anglers carrying fish.

How much these various behaviour differences result from natural selection and how much from learning is not known; but it would be surprising if selection had not acted on the heavily persecuted European populations. On the other hand, Golden Eagles seem to react similarly in all parts of their range, by flying right away. If they always behaved this way, there is little scope for selection by man to change behaviour. Nest defence in some species also varies with nutrition or pesticide level, as discussed in Chapter 8.

Concealment is another form of anti-predator behaviour, and is more evident among the smaller raptors that are least equipped to defend themselves. Small accipiters place their nests in thick woodland cover, and small kites in thorn trees. Most harriers nest in thick ground cover, and incubation is entirely by the tight-sitting, cryptically-coloured female. The Dark Chanting Goshawk *Melierax metabates* brings twigs holding social spiders to its tree-nest. These creatures soon cover the whole structure with silk so that it looks more like a spider colony than a nest (A. C. Kemp). Gabar Goshawks *M. gabar* and Lizard Buzzards *Kaupifalco monogrammicus* also do this in some areas, and the former add pieces of camouflaging bark as well (Brown & Amadon 1968).

The small Swallow-tailed Kite *Elanoides forficatus* of central America is one of the few known raptors in which the chicks do not eject their faeces clear of the nest (Snyder 1975). Instead the faeces accumulate on the nest rim, rather than on the ground, and are continually covered by fresh nest material brought by the adults. This has been regarded as an adaptation to prevent the nest being found by Racoons, which are especially common in this habitat. The nest itself is invisible from below, so droppings on the ground would provide the only visual clue to its presence. No raptor is known to remove droppings like song-birds do, as part of their anti-predator behaviour. But small raptors, in contrast to many large ones, take great care to carry away uneaten prey, perhaps so that it does not attract predators by accumulating on the nest or on the ground below. Some species cache it at a distance and bring it back later (Chapter 9). European Sparrow-hawks cease to carry away uneaten meat when their young can fly.

Certain other aspects of parental behaviour also tend to reduce predation. In the absence of human intrusion, the nests of most species are seldom left unguarded until the young are large. In species in which the sexes take turns to incubate, one partner leaves only when the other arrives to take over. In species in which the female does the incubating, the male may cover the eggs or stand beside them, while the female eats the food he has brought. The eggs of most species are cryptic to some extent, while Honey Buzzards and Bald Eagles have been seen to cover their eggs and small chicks with nest material before leaving (Holstein 1944, Broley 1947).

The young themselves show several types of anti-predator behaviour. On the approach of an observer, they first flatten themselves on the nest, but as he approaches closer, they begin to call. This serves to bring in the parents, if they are not already present. When the observer reaches the nest, the young retreat to the back, and face the observer, with half-spread wings and open bill, striking with their feet at his outstretched hand. This kind of behaviour seems to be fairly general, at least among species of *Accipiter*, *Buteo* and *Falco*. Some Old and New World vultures when provoked, regurgitate their last meal, a steaming, stinking mass, which is repulsive to a man, and may also repel a predator, or at least provide an alternative meal. It has a parallel in the Fulmar *Fulmarus* and other seabirds, in which the regurgitation of an oily liquid derived from food has proved effective in deterring both human and natural predators (Warham 1977). When they are large enough, the young of many raptors move from their nest to separate perches nearby, scrambling back whenever the parent arrives with food. The advantage of the young separating in this way is that a predator is less likely to find them all. Once they reach a certain age, the young leave the nest prematurely if approached by a human or other predator, as do the young of many other birds. In ground-nesting harriers, the natural movement of young from the nest occurs at an early age, and even quite small young have their separate hideouts nearby (Balfour 1957).

These are some of the various forms of behaviour shown by adults and young which, as well as a safe nest-site, serve to reduce the losses to predators. This account has been pieced together from more or less casual observations on several species, and perhaps no one species shows all the features mentioned. The response to human intruders is especially easy to record and this is clearly a field that would repay more study, including comparisons between populations with dispa-rate predation histories. In general, raptors seem to respond to natural predators much more aggressively than to a man, but published observations are few. Black Eagles fly away from a man but attack a baboon on the nest-ledge, and Sparrowhawks keep at a distance from a man but strike at an owl. This may be something to do with the size of the predator, with the extent of the threat it poses, and with the history of human persecution. Raptors also react differently to different natural predators. Peregrines in Alaska begin to attack Golden Eagles while they are still more than 2 km from the nest, but they allow Ravens to approach within about 100 metres before attacking them, and Rough-legged Buzzards to within 50 metres (C. White).

SUMMARY

Nest-sites are an important resource, which often limit the distribution, numbers and breeding success of raptors. The range of nesting places that a species uses is influenced mainly by whether it can build a nest for itself, and the kinds of situation it can use (probably largely inherent in each species); but within this range, choice is influenced by local availability, by on-the-spot competition with other species, by predation pressures, and by local traditions. The range of sites used by a species should not therefore be regarded as static, but as malleable within limits, according to changes in predation pressures and the development of new traditions, both of which have led to changes in raptor distribution. Quality of nest-site can be judged mainly by the security the site provides against mammalian predators and inclement weather. Large nests of eagles may serve an important signal function in advertising the territory to others of their kind, while the habit of bringing greenery to nests may further serve to denote an occupied, as opposed to an unoccupied, nest or territory. Nest defence behaviour against humans varies between regions, apparently sometimes connected with the history of human persecution.

Some of the ideas expressed in this chapter cannot be easily tested, and as such will be open to criticism. But if they help to stimulate further thought, and pinpoint some previously unappreciated problems, they will have served their purpose.

Peregrine

CHAPTER 6

Breeding seasons

As is usual among birds, most raptors breed during only part of the year, when food is most readily available. They therefore breed in different months in different regions of the world, according to climate, but in any one area, particular species usually lay their eggs at about the same date each year. Most species raise only one brood in a year but, if the favourable season is long, the smaller species with short breeding cycles can sometimes raise two broods, and other species have time to raise young from a replacement clutch, if their first fails at an early stage (Chapter 7). Large species have longer breeding cycles, which often extend into the period when food is less plentiful, and the extreme occurs in some condors and tropical eagles, in which successful breeding takes more than one year (Chapter 7).

Brown (1976) recognised two main times of strain in the breeding cycles of raptors, namely the pre-laying/laying period and the early nestling period. In the first of these, the female needs extra food to form eggs and to lay down body reserves (see later). In the second, the male has to obtain more food than at any other stage – enough for himself, the female and the growing young. Earlier he has only himself and the female to feed, while later the female helps with the hunting. To a large

Illustration: European Sparrowhawks copulating

extent, breeding failures occur at these times, for many females on nesting territories do not lay or, if they do, they soon desert their eggs, and the birds that finish incubation most often fail soon after hatch (Chapter 8). Another important time is when the young become independent, for their inexperience would act against them if it was at a time when food was scarce. Which of these periods is most critical probably varies between species, depending on the number of young they raise, and how closely their food needs at different stages match the seasonal changes in prey availability.

Seasonal food changes depend mainly on the activities of prey species – on their migrations, hibernations and breeding – and differ according to the animals involved. In temperate areas, small bird numbers usually increase suddenly in spring, with the arrival of migrants, and then further through breeding to reach a peak in mid-summer, before declining again. Small mammals show a somewhat different pattern, as they increase steadily from a low in spring to a peak in late summer or autumn, and decline thereafter. Insects that are important to raptors (mainly orthoptera and coleoptera) also tend to increase towards late summer, but on the whole they show more variable patterns than mammals. In contrast, reptiles and amphibia are most apparent in spring, soon after they emerge, and become steadily less available through the summer, as they die off or spend more of the daytime in hiding (the young of some such animals are abundant in late summer, but they are mostly too small to interest raptors). As a result of these various trends, raptor species face different patterns of food change through the year, depending on their diets. Affecting all species, however, are the seasonal rise and fall in day-length (= hunting period) and the seasonal growth of vegetation, which makes prey less accessible.

BREEDING IN TEMPERATE REGIONS

In any one region, the larger species tend to lay before the smaller, which helps them to complete their long breeding cycles before food becomes scarce again. The big vultures begin earliest in the year, followed by the eagles, then the medium-sized buzzards and others, and finally the small accipiters and small falcons (Figure 13). This overall trend in timing is modified by food supplies, as is evident by comparing species of similar size that have different diets. Of the three small raptors in England, the Kestrel lays mainly from mid-April to mid-May, the Sparrowhawk in May and the Hobby in June. In this way, the Kestrel has young in the nest at a time when young rodents become plentiful from late May, the Sparrowhawk when young songbirds become plentiful from early June, and the Hobby when young swallows and dragonflies become plentiful from July. These three species reflect a general trend in temperate regions for the rodent-eaters to begin earlier than the bird-eaters, which in turn start earlier than the

Hen Harrier, or Marsh Hawk, at nest with eggs and at nest with chicks. Photos: D. Muir (upper) and F. V. ackburn (lower).

10 (Upper) Hen Harrier, or Marsh Hawk, in vigorous defence of nest, typical of this species. (Lower) Montagu
Harrier about to retrieve a stray chick. Like some other harriers, this species occasionally breeds in loose
colonies, with some males mated to more than one female. Photos: C. E. Palmar (upper) and M. D. England
(lower).

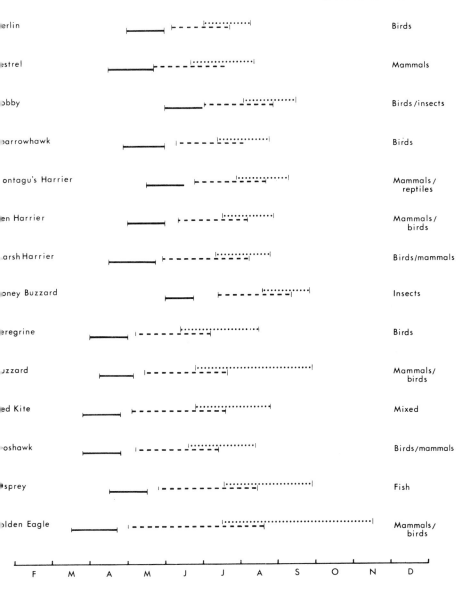

Figure 13. Breeding seasons of raptors in Britain. Species are arranged approximately in order of body-size, and for each species, the lower continuous line shows the period when clutches are started, the middle broken line shows the period when young are in nests, and the upper dotted line shows the period when flying young are fed by parents. The main trend is for large species to begin earlier and have longer breeding seasons than smaller ones, but among species of similar size, timing is also related to diet, shown on the right.

insect-eaters; species with mixed diets are intermediate as appropriate (Figure 13). In northern Europe, the Osprey lays late for a bird of its size, corresponding with the late break-up of ice and the movement of fish into shallows. In southern Europe, the relatively late laying of the Snake Eagle is linked with the appearance of snakes above ground.

As one moves northwards in the temperate zone, the different species tend to begin breeding in the same order, but all start later and closer together, corresponding with the later and more sudden spring. For any one species, the breeding season is shorter in the north than further south. This is achieved by (1) a greater synchrony in the starting dates of individuals and (2) a shortening of their breeding cycles, especially of the pre-laying and post-fledging periods. Take the Peregrine for example. In the southern temperate zone, the first-egg dates in a population span six weeks from late March, but in the southern boreal zone they span five weeks from late April, and in the arctic 3–4 weeks from late May. The southern birds may be paired on their nesting territories for 12 weeks or more before laying, but the most northern ones for less than two weeks. The southern pairs feed their young for more than eight weeks after fledging, but the northern ones for less than five (Bent 1938, Cade 1960, D. N. Weir). In warm parts of Australia, where Peregrines remain on their nesting territories all year, the young stay for as long as 11–15 weeks after fledging (Jones & Bren 1978).

The correspondence between breeding and food supplies is especially evident in Eleonora's and Sooty Falcons, which raise their young on migrant songbirds. The falcons nest on offshore islets, around the Mediterranean and Red Seas respectively, and lay in late July or August, so that their whole breeding coincides with the songbird migration from Eurasia into Africa (Vaughan 1961, Clapham 1964). Their breeding seasons are later than those of any other Palearctic raptors. Moreover, in some desolate parts of the Sahara Desert lacking resident prey-birds, the Lanner Falcon breeds in early spring, with the northward passage of songbirds, and the Sooty Falcon in autumn with the southward passage (Booth 1961, Moreau 1966). Taking these facts together, at least the smaller raptors time their short breeding cycles to coincide with obvious periods of prey abundance. For the larger species, with long cycles, it is more a matter of avoiding the worst months than taking advantage of the best. The condors whose breeding lasts more than a year do not even have this luxury, though they can perhaps ensure that the most demanding stages of breeding do not fall in the usual periods of shortage.

Timing of laying

Whether breeding is timed to coincide with the best months or to avoid the worst, there is still the question of what stimulates laying at the appropriate date. It is helpful here to distinguish between ultimate and proximate factors. Thus a bird may be said to breed in spring because its food is plentiful then, or because its gonads are stimulated to develop under the lengthening days.* Both factors may be involved, but the food supply is the 'ultimate' reason for breeding at a particular time, while daylength is a 'proximate' factor, which brings the bird into condition at the appropriate date (Baker 1938). Through natural selection, birds are supposed to have come to respond to whatever factors provide a reliable indication that breeding will shortly be practicable (Lack 1954). Some factors, such as daylength, may provide the long-term stimulus, while others, such as food availability at the time, may bring about minor modifications in laying date. In this case, food acts as an ultimate and as a proximate factor in the timing of breeding.

Daylength
Daylength gives the most reliable indication of date, because it is the only obvious environmental factor that changes in a predictable manner from year to year. Under experimental conditions, more than 60 species of birds have now been brought into breeding condition prematurely by exposing them to artificially long days in winter, and none has failed to respond (Lofts & Murton 1968). However, only males have been brought into full condition in this way. Females also react, but they need other stimuli, such as a mate and a nest, to induce egg laying.
At least three raptor species have been used in experiments of this type. American Kestrels laid earlier than usual on exposure to long photoperiods, and also came into breeding condition more than once in a year (Willoughby & Cade 1964, D. Bird). In addition, females from two different latitudes, that were exposed to the same natural day-lengths in captivity, laid eggs about a month apart, the northern birds later and under longer days than the southern ones (Porter & Wiemeyer 1972). Such differences between populations are adaptive, and ensure that individuals from each are brought into condition at a date appropriate to the latitude at which they breed. Which helps to explain why migrant raptors on winter quarters have a retarded gonad development at a time when other members of their species resident in the area are already nesting.
The Peregrine and Gyr Falcon are the other raptors that have been subject to experiments involving light and, with appropriate treatment, birds that would normally breed late under long arctic days have bred much earlier in captivity at more southern latitudes, for example near New York (Nelson & Campbell 1973, 1974, Cade 1976, Fyfe 1976, Platt 1977). They laid under the same daylengths in captivity as they would in the wild, 14–15 hours for Gyr Falcons and 15–16 hours for arctic Peregrines. Such Peregrines normally migrate to the southern hemisphere for the winter, so they experience a complicated daylength cycle through the year, with long days in summer and winter. They presumably have a long insensitive period (the so-called refractory period),

* The role of daylength in the timing of bird breeding seasons was first discovered in the nineteen-twenties, by William Rowan who as a younger man did a remarkable pioneer study of Merlins in his native Yorkshire (Chapter 3).

during which they do not reach breeding condition on the long days of the southern summer. This lack of response to long days is important not only in trans-equatorial migrants, but also, at times, in species which reside all year in the temperate zone. Many such species experience the longest days and maximum food supplies long after the usual laying date. Yet if they laid at such a time, they would be too late to have much chance of raising the resulting young. Hence, a response to the annual light cycle by high-latitude birds is important not only in promoting gonad growth at an appropriate time, but also in suppressing it at an inappropriate time. Current theories envisage some free-running rhythm, inherent in the bird, which is brought into phase by the annual light cycle. This internal rhythm helps the bird to respond promptly, or not respond, to external factors, as the need may be.

Food supply
Food supply during winter and spring also influences the laying date

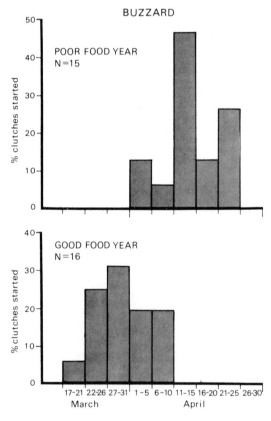

Figure 14. *Laying seasons in relation to food supplies. Above: Buzzard in different years in the same area. Re-drawn from Mebs 1964.*

in many raptors, or even whether laying occurs at all. The field evidence is circumstantial: (1) in species whose spring food supply differs greatly between years, laying dates are earlier in good food years than in poor ones; and (2) in other species, which occupy different habitats in the same region, laying dates are earlier in the habitats that are richest in food. The Buzzard and Kestrel show annual differences, and the Sparrowhawk shows habitat differences (Figure 14), while other examples are given in Chapter 8.

In some other species, the dependence of egg-laying on food supplies only became apparent under exceptional shortages. Although they had built their nests on time, Cooper's Hawks in Arizona laid about a fortnight later in 1971 than in two previous years, associated with an unusual dearth of prey (Snyder & Wiley 1976). Suddenly, a conspicuous wave of migrant songbirds moved into the area and, within a few days, most hawks began laying. The songbirds passed on quickly, however, and several of the hawks did not complete their clutches. At one hawk nest under intensive observation, the feeding of the female

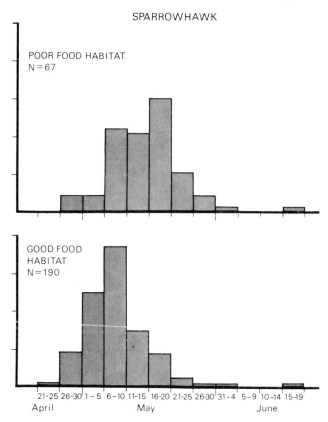

(*Figure 14 continued*) *Above: Sparrowhawk in different habitats in the same region. Re-drawn from Newton 1976.*

by the male increased greatly during the time the migrants were present.

The influence of food on laying date may be modified by other factors, such as weather. Thus the mean laying dates of Dutch Kestrels in different years were correlated not only with vole densities over the previous winter and spring, but also to some extent with temperature and rainfall in spring (Cavé 1968). On any given food supply, warm and dry weather advanced laying, while cold or wet weather delayed it. Cavé assumed that cold increased the energy needs of the birds, so that less of the females' daily food-intake could go towards egg production, and he observed that prolonged rain reduced their hunting efficiency and feeding rate. Similarly, during an 11-year study of Goshawks in Finland, nest-building began early when February and March were warm, and egg-laying began early and clutches were large when a warm April coincided with good prey numbers (Huhtala & Sulkava 1976). Among Buzzards in central Europe, late laying was sometimes associated with late-lying snow, which rendered rodents harder than usual to catch (Mebs 1964, Rockenbauch 1975). Hence in all these species laying was late in poor springs, probably through weather increasing the total food needs of the birds or reducing the accessibility of prey.

Importance of body condition

Extra food is needed before breeding, not merely to enable the female to produce eggs, but also to allow her to accumulate large body reserves for use in the incubation and nestling periods. This belief derives from recent studies of Sparrowhawks, Kestrels and vultures during the breeding season, and also of Tawny Owls *Strix aluco* and some non-raptorial birds. In all these species, egg production was preceded by a great increase in the weight of the female, due to an accumulation of body fat and protein. Females weighed more then than at any other time of year; they lost some weight over the laying period but were still heavy at the start of incubation. Among Sparrowhawks, only those females which put on weight subsequently laid, whereas those that did not put on weight did not lay (though some had built nests) (Figure 15). Males did not increase in weight near the start of breeding, whether they bred or not. Tawny Owls showed similar weight differences between laying and non-laying females, and between laying females and males (Hirons 1976). In both these species, the female was responsible for incubation, while the male did the hunting. In the vultures, however, in which male and female shared in incubation, both sexes increased in weight before breeding (Houston 1976). These are the only raptors I know of whose weights have been followed through the breeding cycle, but in several others, high female weights were noted near egg-laying or early in incubation (W. Tarboton, Village 1979). So in influencing whether eggs were laid, the ability of the female raptor to acquire enough body reserve with which to begin incubation may have been at least as critical as an ability to produce the

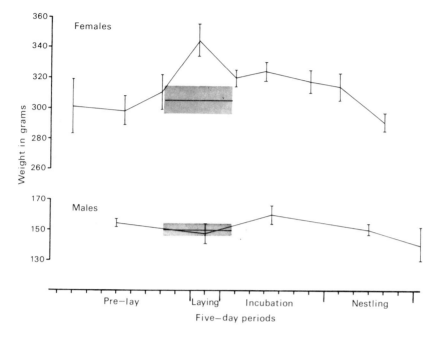

Figure 15. Weights of Sparrowhawks in relation to stage of breeding. Vertical lines show 95% confidence limits on the mean for each period. Broad stippled bands show weights of non-breeders in the laying season, mean and 95% confidence limits. In females, the main weight changes were the increase towards laying (P<0·01), the decline after laying (P<0·01), and the decline in the second half of the nestling period (P<0·01). This last period was when the food demands of the brood were greatest, and when the female participated in hunting. Over the usual laying period, breeding females were significantly heavier than non-breeders (P<0·001). In males, the seasonal changes were less marked, but weight increased slightly towards mid-incubation (P<0·01) and fell thereafter (P<0·05). Over the usual laying period, no difference was apparent between breeders and non-breeders. From I. Newton & M. Marquiss, unpublished.

eggs themselves.

In the Tawny Owl, the relationship between gonad and body condition was examined further, using carcasses of birds that had been killed on roads (Hirons 1976). In females, some development of the ovaries was apparent in autumn, but it increased greatly towards spring. At this season, ovarian condition in different birds was correlated with their overall body weight and with their pectoral muscle weight, which was taken as an index of protein condition. Thus in the weeks before laying, Tawny Owls of both sexes that were in the best physical condition had the best developed gonads, and those in poorest condition had not completed gonad development by the normal laying date.

In the Kestrel, too, ovarian development began in autumn and continued through winter and spring. Cavé (1968) showed experimentally that this process could be accelerated in captive females by feeding them extra food. He fed birds in winter on beef heart, giving one group 200 grams per day and another 40 grams, which was reduced to 16 grams. At intervals, he killed one bird from each group, and each time the well-fed bird was heavier and had larger oocytes. The poorly fed birds did not fall below the weights of wild birds, but they showed little or no sign of reproductive development. All this serves to emphasise the relationship between food intake and body condition on the one hand, and between body condition and laying date on the other.

To judge from field evidence, the female's food intake influences the number of eggs as well as the date of laying (Chapter 8). The proximate mechanism has not been studied in raptors, but in the weaver bird *Quelea* the number of eggs laid by individual females depended on the rate at which their body reserves fell. The largest clutches (4 eggs) were laid by females whose reserves fell slowly (presumably because these females were well fed during laying), while the smaller clutches (3 or 2 eggs) were laid by females whose reserves fell quickly. Once a female's protein condition had fallen to a certain level, any further oocytes were resorbed and laying stopped (Jones & Ward 1976). If the same holds in raptors, the level of body condition at which egg production ceases is normally such that some reserve is left for incubation.

In the Tawny Owl, the reserve that remained after laying helped to buffer the female against any temporary inability of the male to provide food during incubation; it enabled her to stay on the nest instead of leaving to hunt for herself and allowing the eggs to chill. During the nestling period, the female also diverted food to the chicks which, in the absence of body reserves, she would have eaten herself. Evidently, the initial body condition of the female was crucial to the timing and to the success of breeding, while food provided by the male helped to slow the deterioration in the female's condition. By the use of automatic cameras, Hirons (1976) studied the attentiveness of four females, only two of which hatched their eggs. These two received more food from their males, incubated properly and spent only short periods off the nest, whereas the other two received less food from their males, and were less attentive, so that their eggs chilled.

Among Rüppell's Vultures in East Africa, the body reserves of the adults were again important to successful breeding. Houston (1976) measured the combined food-needs of adults and growing chicks and, from the size of the crops on birds returning to the colony, he estimated the amount of food obtained (Figure 16). Comparison of the two figures implied that, for a period during rearing, insufficient food was obtained to satisfy the needs of both parent and chick. Even so, the chicks survived well and had evidently received enough food, whereas samples of adults shot at intervals through the season showed a progressive decline in fat content (Figure 16). The inference was that

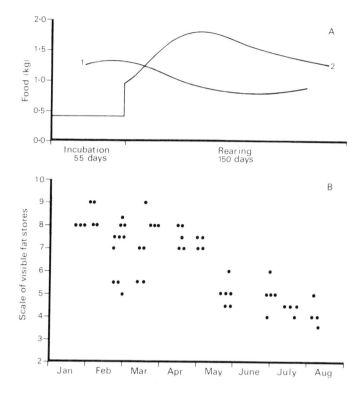

Figure 16. (A) Estimates of (1) food obtained and (2) food required through the breeding cycle in Rüppell's Vultures in East Africa. (B) Decline in fat content of adult Rüppell's Vultures through the breeding season. Each point represents one bird. Re-drawn from Houston 1976.

without this fat the adults could not have raised young. Adult White-backed Vultures also lost fat during the breeding season, while non-breeding immatures did not.

From these various studies, the following implications can be drawn: (1) increased food intake before breeding leads to an improvement in body and gonad condition; (2) this in turn influences laying date and, in some species, the number of eggs; (3) ability to maintain some body reserve influences subsequent success; (4) the extra food is needed more for body improvement than for egg production (but this may vary with the clutch-size); and (5) if food is insufficient, body and gonad development may be incomplete, and laying does not occur. Of course these points apply chiefly to the female, but in most raptors the male is important in supplying food to help the female build up and maintain condition. They are suggestions from limited studies, however, and should not be accepted without more work. The dependence of sex organ development on body condition can be regarded as an adaptation ensuring that birds of both sexes breed only when fit enough to do so.

From findings on other birds, the food required for the initial growth of the oviduct and ovary to the pre-ovulatory stages is very small, while that required for testes growth is negligible. As body weight increases before laying, the fat is stored at various sites around the body, and the protein is stored in the sarcoplasm of the muscles, especially the breast muscles (Kendel et al 1973). It is possible that specific amino-acids are stored selectively, with priority for those that occur at low concentration in food, and so can only be accumulated slowly over a long period. Probably, protein is the most important material needed for egg production, and fat for the continuance of incubation. Contrary to early opinion, however, incubation itself is less demanding than is normal life, because the bird remains inactive in what is usually a well insulated nest.

Other factors influencing laying date

Whether the overall season is early or late, birds in the same population often vary considerably in their laying dates. This variation is not fully understood, but may be due partly to hereditary factors. It may also be due to individual differences in the time birds take to accumulate the reserves necessary for breeding. Compared to the early layers, the late ones may have started in poorer condition, or they may have been less efficient at hunting, or in less good areas, all of which may in turn be partly related to age. The tendency for individuals in their first breeding year to lay late is widespread among birds (Lack 1966), and was confirmed in at least two raptor species. First-year Kestrels in the Netherlands laid, on average, about one week later than older ones in the same area, while first-year Sparrowhawks in Scotland laid about ten days later (Cavé 1968, Newton 1976).

BREEDING IN TROPICAL REGIONS

In the temperate zone, the biological year is governed by temperature. When the weather warms at the end of winter, vegetation of all kinds grows, insects become plentiful, and birds and other animals begin to breed. The whole vast area of the temperate zone is affected similarly, and nearly all animals conform to the same timetable, and produce young in the second quarter of the year. In the tropics, the situation differs. For one thing, the biological year is based on rainfall, and over large areas the climate is marked by the alternation of distinct wet and dry seasons, varying in duration from one region to another. The seasonal changes in cover are enormous, with a rapid growth of vegetation during the rains, and a rapid grazing, drying and disintegration soon after. Most animals breed when the vegetation is green, but some breed at other times, so that the seasonal changes in raptor prey populations are less pronounced than in the temperates. Changes in daylength and temperature are also less pronounced, so that in general, food conditions are more constant through the year than at higher latitudes. In addition, however, as the rainbelts sweep slowly across a continent, areas widely separated are watered months apart. This promotes regional variations in animal breeding seasons of an extent not met in the temperate zone. Rainfall also varies so much from one year to the next that the concept of a 'normal' year is of little relevance. Almost certainly, however, breeding by raptors in the tropics is conditioned as elsewhere, by food supply, but much less information is available.

Some of the most complete data on breeding seasons refer to Zambia, in the southern African tropics, where rain falls in summer, October–April (Benson *et al* 1971). Breeding by raptors is spread more evenly over the year than in the temperate zone, with bigger differences between species and, in any one species, more spread in laying dates (Table 17). Again, however, the large species begin first, laying mainly in April–July (the early dry season), while the small ones lay mainly in August–October (the late dry season and early wet). The young of both groups become independent before the last half of the wet season, when prey populations reach their peak, but when cover is also thickest. Partial exceptions to the general trend in laying dates include the Bateleur, which lays earliest of all, in January–April, and the Black-shouldered Kite, which lays in two main periods, early and late in the dry season respectively.

The tendency for larger raptors to begin before the smaller is also apparent among six species in West Africa, north of the equator. They breed in quite different months from their counterparts in Zambia, but they show a similar relationship with the rainfall cycle, as all six species have young on the wing before grass reaches full height (Thiollay 1975). This is shown for Wahlberg's Eagle in Figure 17. In several regions, some of the large raptors have laying seasons extending over six months or more, so that pairs with eggs or nestlings can be found throughout the year (Table 17; Smeenk 1974). With such long breeding cycles, the precise date that pairs lay probably does not matter so much because they could not anyway accomplish everything within the most propitious season. Fish Eagles are especially variable in the time they breed. On Lake Naivasha in Kenya they lay in any month (but with seasonal peaks), whereas on Lake Victoria and elsewhere they have shorter laying seasons, which can differ markedly in places less than 50 km apart (Brown 1976b). This is often linked with a dependence on different fish species, and possibly with the months in which these fish are commonest in surface waters. Not all breeding by African raptors can be explained so simply however, and Brown (1970) has listed some puzzling anomalies.

Near the equator in East Africa, the rains come in two wet seasons and prey species breed in one or both. Among the raptors, the small species with short cycles breed twice each year, beginning towards the end of each dry spell. In *Melierax* goshawks, this is not merely a case of some pairs breeding at one season and other pairs at the other season,

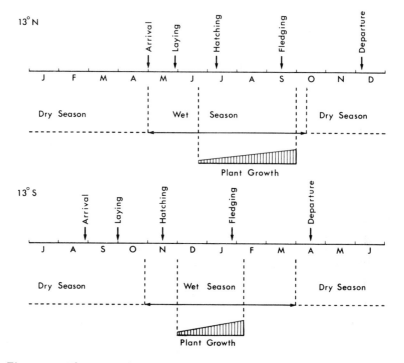

Figure 17. The annual cycle of Wahlberg's Eagle at 13°N and 13°S in Africa. Re-drawn from details in Thiollay 1978 and Steyn 1973.

because twice-yearly breeding was noted in the same marked birds (Smeenk & Smeenk-Enserink 1975). Parts of West Africa also have two wet seasons but in these areas, so far as I know, no raptors are known regularly to breed twice a year. As an overall impression of the dry tropics, perhaps nowhere else is the influence of cover on prey availability more apparent; and it is the availability of prey rather than numbers which seems to be the more critical to raptor breeding.

In equatorial evergreen forest, rain falls in every month, but with seasonal peaks. Practically nothing is known of the raptors in such habitats, but as most other birds there have regular breeding seasons, at least the short-cycle raptors probably do so too. This is borne out in studies of Sharp-shinned Hawks and Red-tailed Hawks in Puerto Rico (H. Snyder). However, in the seasonless environment of some equatorial islands, raptors, like other birds, may be found breeding at any time. In the Galapagos Islands, daylength is almost constant through the year and rainfall is unpredictable. Different pairs of Galapagos Hawks started successive breeding cycles 5–26 months apart (shorter gaps followed failure), and at least some laying occurred in all months (de Vries 1975). One pair laid six clutches in eighteen months, two of which gave rise to fledged young (Figure 18). The variable intervals between successive cycles in different birds probably depended on local food sources, and on individual differences in the time taken to reach condition.

In arid areas of west and central Australia, breeding by the whole bird fauna is irregular and can occur at any time, following the rare and unpredictable rainfall, on which food supplies depend. There may be considerable immigration into areas where rain has fallen. Letter-winged Kites (*Elanus scriptus*) increased greatly in one area in response to a plague of rats, itself the indirect result of heavy rain. Breeding was highly successful, and probably many pairs raised two broods in quick succession (Cameron 1974).

It is unclear which proximate factors stimulate laying in tropical areas. The closer one moves to the equator, the less marked are the seasonal changes in daylength, so that the use of daylength must be minimal or non-existant. Rainfall, or the resulting changes in vegetation and food supplies, have been suggested as timing mechanisms in some other birds. Food availability, acting through its effect on body condition, is perhaps the most likely proximate breeding factor in raptors, but more study is needed. At least this would account for the marked differences in timing between neighbouring regions and between pairs in the same region, and is consistent with the proven influence of food in the temperate zone.

SUMMARY

Small raptor species with short cycles can usually accomplish more of their breeding within the period of maximum food supplies than can large species with long cycles. The general trend in any one area is for larger species to begin before small ones, but this is modified by diet and the time that prey become sufficiently available. In captivity, at least three northern raptors have been brought into breeding condition earlier than usual by exposing them to artificially lengthened days. In

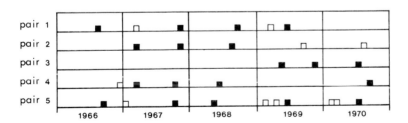

Figure 18. *Laying dates of five pairs of Galapagos Hawks during 1966–70, showing that laying occurred at any time of year, with no marked synchrony between pairs. Open boxes – failed attempts; full boxes – successful attempts. Re-drawn from de Vries 1975.*

the wild, food also influences gonad development and laying date. Breeding females need extra food not only to form eggs, but also to lay down reserves of body fat and protein; females which do not accumulate such reserves do not lay. The ability of the female to maintain good body condition probably influences the number of eggs, and whether the breeding attempt succeeds or fails. In tropical regions, with distinct wet and dry seasons, large raptors again begin before small ones, and most species have young on the wing before the height of the wet season and before cover is thickest. In some areas with two wet seasons, small species with short breeding cycles may breed twice each year. In equatorial areas with unpredictable rainfall, raptors breed in any month (Galapagos), and in very arid areas only after rain has promoted an increase in prey (parts of Australia). In all such areas, increased food availability, facilitating improved body condition and egg production, is the most likely proximate factor controlling the timing of breeding, but further study is needed.

Eleonora's Falcons

CHAPTER 7

Breeding strategies

In this chapter, I shall discuss those features of raptors that limit their breeding rates and influence their life styles. Such attributes include the egg size and the clutch size, the duration of the incubation, nestling and post-fledging periods, the age of first breeding and the frequency of breeding attempts. These features are inter-related with one another, and with environmental factors, such as food (Lack 1968). According to Darwinian theory, the particular combination of attributes shown by any given species is the one which, in the long run, enables individuals of that species to leave most surviving offspring.

In any one population, some features of breeding vary much less than others, the limits of variation being set partly by hereditary factors in each case. Egg size, for example, tends to vary by less than one tenth of the average, as does the duration of the incubation and nestling periods. Eggs below a certain size are usually inviable, while embryos or nestlings whose development slows below a certain rate (through chilling or food-shortage) usually die. On the other hand, clutch-size may vary five-fold or more, according to food-supply and other local conditions. I shall not be concerned in this chapter with the variations

Illustration: California Condor at nest

within species that can be attributed to prevailing conditions, but rather with the differences between species that are apparently inherent. Discussion is centred on the 80 species which breed in the Palearctic and Nearctic regions; tropical species are mentioned only incidentally because they differ in some respects from northern raptors, and are generally less well known.

GENERAL TRENDS

Raptors range in weight from less than 100 grams to more than 14,000 grams. This is an enormous spread compared to that of most other bird orders. Moreover, within the raptors the three main trends in reproduction are related to body size (Newton 1977). The larger the species, (1) the later the age at which it begins breeding, (2) the longer it takes for each successful breeding attempt, and (3) the fewer the young it produces during each cycle. At one extreme, small falcons begin breeding in their first year. They lay five to six eggs at two-day intervals between each egg, and have incubation, nestling, and post-fledging periods lasting about 28, 26, and 14 days respectively, bringing the total breeding period from the first egg to about 80 days. The maximum increase in population possible each year is 3·5–4·0 times the breeding population, assuming the parents survive. On rare occasions when more than one brood is reared each year, the potential rate of increase is even greater. At the other extreme, the large condors do not begin breeding until they are more than five years old. They lay only one egg at a time, and have incubation, nestling, and post-fledging periods lasting about 55, 220, and 210 days. This brings the total breeding period from egg-laying to about 485 days, so that annual breeding is impossible. Ignoring the non-breeding immatures, the maximum possible increase in a population of condors is 50% in two years, or 25% in one year. These relationships between body size and breeding rate hold within raptors as a whole, and within particular genera, as discussed below for different parameters.

Egg size
Raptors lay slightly larger eggs than is usual for birds of their body weight (Rahn *et al* 1975). The trend within the group is the same as for other birds (Lack 1968); namely, small species lay smaller eggs than large ones, but lay larger eggs relative to their body size (Figure 19). In small falcons each egg weighs about 10–12% of the female's body weight; in large accipiters, falcons, buteos and harriers, each egg is about 4–8% of the body weight; and in the largest vultures and eagles, less than 3%. Normally the eggs within a clutch vary little in weight, but in some large eagles the second (last) egg is the smaller and is sometimes up to 10% less in length and breadth than the first (Meyburg 1974).

Goshawks at nest. In favourable conditions some individuals breed in their first year. (Upper) Female in first-
ar plumage; (lower) female in adult plumage. Photos: A. Gilpin (upper) and C. J. R. Ttiwelb (lower).

12 Goshawks are heavily persecuted in northern Europe because they eat game birds. The hawks are usua caught in cage-traps baited with live pigeons (upper), and in pheasant-rearing areas large numbers of hav can often be killed in a short time (lower). In Finland, about 6,000 are known to be killed each year. Photos: R Kenward (upper) and S. Keränen (lower).

Clutch size

The general trend, already mentioned, is for the larger species to produce the smaller clutches. In temperate regions, small falcons, small accipiters, and harriers usually lay 4–6 eggs in a clutch; small kites 3–5; large falcons and large accipiters 3–4; large kites, buteos, caracaras and ospreys 2–3; eagles and small vultures 1–2; and large vultures and condors only one. This inverse relationship between clutch size and body weight is highly significant statistically.*

Clutch size also shows some relationship with other factors, notably diet. This is especially evident among those genera in which species of similar size differ in feeding habits (Table 18). Small falcons that eat rodents tend to lay larger clutches than bird-eating falcons in the same area; and these in turn have larger clutches than the insect eaters. Similarly among the small kites (genera *Elanus, Elanoides, Ictinia*), the rodent eaters lay larger clutches than the insect eaters. Buteos that specialise most strictly on rodents, such as Ferruginous Hawks (*Buteo regalis*) and Rough-legged Hawks, produce larger clutches than other *Buteo* species at the same latitude, while the Common Black Hawk (*Buteogallus anthracinus*), which eats mainly lower vertebrates, seldom lays more than one egg per clutch. Thus in all these species, large clutches are associated with rodent-eating. The buteos that eat most rodents also show the biggest annual variations in clutch size, linked with the annual fluctuations in their prey (Chapter 8). In the small kestrels that eat rodents the annual clutch-size variations are less pronounced (Cavé 1968).

Taking account of both egg size and egg number, the total clutch in large vultures and eagles represents only a few per cent of female body weight, but in small falcons and accipiters it represents more than 60% of female body weight. The possibility of producing a clutch using body reserves is therefore greatest in the largest species, whereas the small ones have no option but to produce their eggs chiefly on food eaten at the time. This does not mean that breeding would be easy for either large or small species, however, for in both groups it also entails the accumulation of large body reserves, as described in Chapter 6.

Laying interval

Small species usually lay their eggs at intervals of two days, medium-sized species at 2–3 days, and large ones at 3–5 (Table 18). This is only a general trend, however, and variations occur among birds in the same population and even between the successive eggs in a clutch (see Balfour 1957 for Hen Harrier). Some other birds of a similar size to raptors produce their eggs at one-day intervals, so raptors are among the slower layers.

---

* On a regression analysis, r = 0·61, N = 68, P<0·001. By grouping female weights into several categories, each with about the same number of species, the association with clutch size was again highly significant ($\chi^2 = 78 \cdot 40$, d.f. 48, P<0·005). Kendall's Tau gave a value of about −0·5, which likewise indicates a strong relationship.

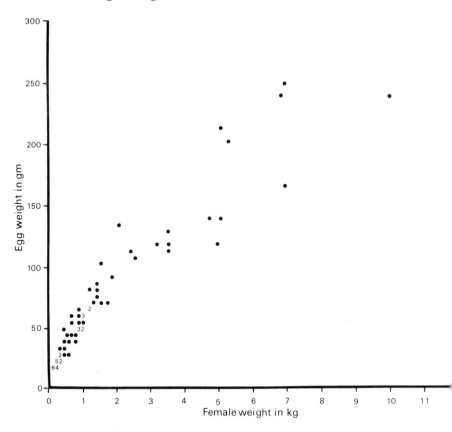

Figure 19. Egg weight (above) and relative egg weight (facing) shown in relation to female body weight in raptors of the Nearctic and Palaearctic regions. Relative egg weight is egg weight as a percentage of female weight. Larger species lay smaller eggs relative to their body weight than do smaller species. From regression analysis, egg weight = 27·997 + 0·026 (female weight), r = 0·95, P <0·001, N = 72. Relative egg weight = 7·819 −0·001 (female weight), r = 0·71, P <0·001, N = 72. Re-drawn from Newton 1977.

Asynchronous hatching and sibling competition

Most raptors begin incubation before they complete their clutches. As a result, the eggs hatch over two or more days and the nestling that emerges last has the biggest disadvantage compared with its siblings. There is relatively greater synchrony in hatching within clutches of smaller species than of larger ones. In small falcons and accipiters, five eggs laid in ten days may hatch within three, whereas in large eagles, the laying and the hatching of a two-egg clutch each requires a 3–5 day interval. The young in all species are thus graded in competitive ability, but in large species the steps between nest-mates are greater. Whatever the species, however, the youngest chick often dies because

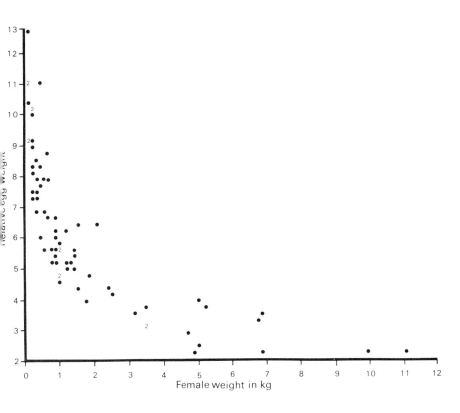

it is unable to compete for food with its larger nest-mates (Mebs 1964, Moss 1976).

The advantage usually claimed for an asynchronous hatch probably applies to raptors, namely that, if food is short, the latest chicks to hatch quickly starve, reducing the brood to the number that the parents can feed (Lack 1954). If all young hatched on the same day, and if food were scarce, it would take longer for the weakest to die, with a correspondingly greater waste of food. Asynchronous hatching can thus be regarded as an adaptation to an unpredictable food supply, enabling all young to survive in times of plenty, but ensuring rapid reduction of the brood to an appropriate level in times of scarcity.

In some large raptors, the oldest nestling repeatedly and mercilessly attacks its smaller nest-mate in what has become known as the Cain-Abel conflict. The smaller chick may eventually die from starvation, injury and chilling, or may fall from the nest in an attempt to escape from its attacker. Such aggression occurs chiefly in the first half of the nestling period, at times when the young are not being brooded. The female takes no direct action to prevent the attacks, and may watch them 'unconcerned' (Steyn 1973). Such behaviour occurs among eagles of the genera *Aquila, Haliaeetus, Hieraaetus* and *Stephanoaetus,* and also in the Lammergeier *Gypaetus barbatus.* It is so prevalent in some populations of the Crowned Eagle *Stephanoaetus coronatus,* Black Eagle *A. verreauxi* and Lesser Spotted Eagle *A. pomarina* that, despite the regular hatching of two eggs, no pair has been found to raise two young (Brown 1966, Gargett 1970, 1971, Meyburg 1974). There thus seems no point in these birds laying two eggs, but the second can at least act as a reserve if the first fails to hatch, as noted in occasional nests (Brown *et al* 1977).*

In the Lesser Spotted and Black Eagles, Cain-Abel conflicts to the death seem to occur irrespective of food availability, and even when there is a continuous surplus of fresh prey on the nest. The young become less aggressive with increasing age, however, and as shown experimentally, if the smaller chick can survive the first few weeks, it has a good chance of fledging. Meyburg (1974) removed the second chick from several Lesser Spotted Eagle nests, fostered each one under another species until it had become feathered and able to feed itself, and then replaced it in the original nest; in all instances both nest-mates were reared. Gargett (1970) achieved similar results in the Black Eagle but only when the nest ledge was large enough for the two young to perch far apart.

In other large eagles, aggression between nest-mates is usual but does not always result in a death. The outcome is influenced by (1) the hatching interval, which determines the initial size difference between the chicks; (2) the feeding rate, which influences both the tendency of the large chick to attack (the chick subsides after each meal) and the ability of the small chick to withstand the attacks; and (3) the amount of time the chicks are left unbrooded, which is when the attacks mostly occur. This last factor is in turn influenced partly by food supply, and whether or not the female helps the male in hunting. Hence, in such species food supply may be a major controlling factor in influencing both the amount of aggression and the outcome. This fits the fact that in Golden Eagles and others, two-chick broods are raised more often in ample food conditions than in poor ones (Chapter 8).

Among the medium-sized species of *Buteo* and *Accipiter,* attacks by older chicks on smaller siblings occur only at times of great hunger, whereas in small raptors such attacks do not occur, even when the young are starving (Balfour 1957, Newton 1976). There thus seems to be a range of variation among the raptors from (1) large species in which aggression among nest-mates is the rule and the smallest chick

---

* Brown *et al* (1977) compared the average production in various eagle species which lay 1–3 (usually two) eggs with that in other species which lay only one, and found that there was no appreciable difference between the two groups. They then concluded that the habit of laying two eggs followed by the elimination of one young by fighting, was mere 'biological waste', and that such two-egg species would be just as productive if they laid only one egg. In my view, this conclusion was unjustified because the comparison between one-egg and two-egg birds should have been made within species and not between them.

'always' dies; to (2) large species in which aggression is usual, and the smallest chick sometimes dies or sometimes survives, depending partly on nutrition; to (3) medium-sized species in which aggression occurs only at times of obvious food-shortage; to (4) small species in which serious aggression among chicks in competition for food does not occur. This trend holds at least within the Accipitridae, the family to which most raptors belong, but I can find no record of serious sibling aggression among the falcons (Falconidae). In conclusion, while asynchronous hatching can be regarded as an adaptation to unpredictability in food supply, the additional tendency in large raptors for the older chick to attack the younger can be regarded as a further adaptation to hasten the outcome and enhance the survival chances of the older chick.

Incubation and nestling periods

Incubation and nestling periods tend to be rather long in raptors, and include some of the longest known among birds (Table 19, Figure 20). Incubation periods last 4–7 weeks according to species, and nestling periods 4–31 weeks. The overall trend is for large species to have longer incubation and nestling periods than smaller ones, but incubation periods also differ between genera to some extent, regardless of body size. Small falcons have shorter incubation periods (28 days) than do accipiters of the same weight (32–34 days); and *Haliaeetus* eagles have shorter incubation periods (35–38 days) than do *Aquila* eagles of similar and lower weights (42–45 days). The Snake Eagle *Circaetus gallicus* incubates for a slightly longer time (47 days) than any of these other eagles, yet is smaller than all except the Lesser Spotted Eagle. Nonetheless, within at least the genera *Falco* and *Accipiter*, the relationship between incubation period and body weight parallels the trend within raptors as a whole (Figure 20).

Incubation and nestling periods tend to be correlated with one another ($r = 0.84$), as in other birds, but for a given lengthening of incubation, the increase in the nestling period is much greater. Nestling periods are often difficult to determine, because for some days before the first flight, the young spend part of their time on nearby perches outside the nest. Also, in species which have been studied in detail, males flew a few days before females, with nestling periods of male and female as follows: Sharp-shinned Hawk, 24 and 27 days; European Sparrowhawk, 26 and 30 days; Cooper's Hawk, 30 and 32–34 days; Hen Harrier, 33 and 38 days; and Crowned Eagle 107 and 115 days (Platt 1976a, Newton 1978, Meng 1951, Balfour 1957, Brown 1966). Most of these species show marked size dimorphism.

Post-fledging periods

The young continue to be fed by their parents for a further period after leaving the nest until they become self-sufficient. Post-fledging periods have been determined for relatively few raptors, partly because of the difficulty of finding the young after they have wandered from the

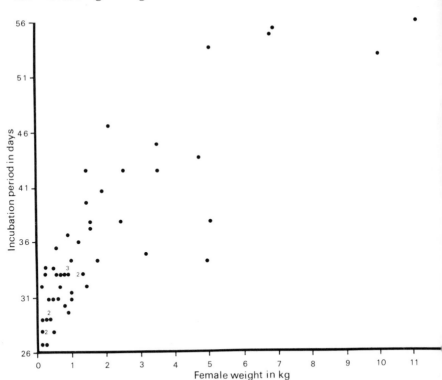

Figure 20. Incubation (above) and nestling periods (facing) shown in relation to female body weight in raptors of the Nearctic and Palaearctic regions. In larger species these periods are longer, especially the nestling period. From regression analysis, incubation period = 30·795 + 0·003 (female weight), r = 0·85, P<0·001, N = 58; nestling period = 29·931 +0·014 (female weight), r = 0·92, P<0·001, N = 58. Examination of particular genera gave the following results: for Falco, incubation period = 27·62 + 0·004 (female weight), r = 0·84, P<0·001; nestling period = 26·94 + 0·013 (female weight), r = 0·88, P<0·001. For Accipiter, incubation period = 32·42 + 0·004 (female weight) r = 0·67, P<0·05; nestling period = 24·46 + 0·014 (female weight) r = 0· 98, P<0·002. For all species, nestling period = −76·31 + 3·63 (incubation period), r = 0·84, P<0·001, N = 57. Re-drawn from Newton 1977.

nest vicinity. Moreover, such periods vary much more among the individuals of a population than do incubation and nestling periods (30–70 days in one population of Red-tailed Hawks – Johnson 1973). They also tend to be shorter in migrant than in resident low-latitude populations of the same species (Chapter 6).

In general, however, the larger species remain dependent longer, and typical post-fledging periods range from 2–3 weeks in small falcons and accipiters, 5–10 weeks in buteos and large kites, and up to several months in large eagles and vultures, sometimes to the start of the next

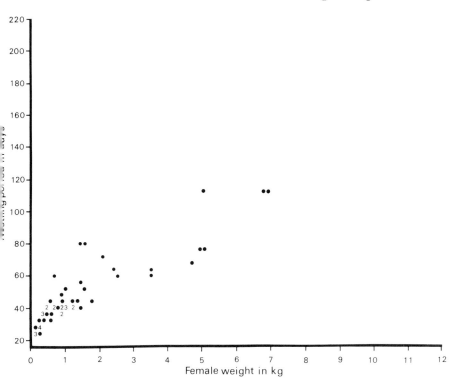

breeding attempt (Table 18). So far as is known, most raptors do not feed their young once they have left the breeding area, and this may be due to the difficulty of maintaining contact, especially in solitary hunters. In some vultures the young leave the nest area with their parents, but these birds do not depend on living prey. Some Old World vultures move away from the breeding colony within two to three weeks after fledging, which led Brown & Amadon (1968) to speculate that 'the period of dependence after the first flight may be short'. However, recent studies have shown that Lappet-faced, White-headed (*Trigonoceps occipitalis*) and Cape Vultures (*Gyps coprotheres*) continue to feed their young for several months after fledging (Anthony 1976; Pennycuick 1976; P. Mundy). Likewise, New World Black Vultures may feed their young away from the nest area for six months or more (Jackson 1975); and the post-fledging period in the California Condor is thought to last at least seven months (Koford 1953).

Growth patterns

Unlike the young of other birds which grow in nests, young raptors hatch with their eyes open and with a complete covering of down. And rather than begging with open mouth for the adult to place food inside, young raptors take food directly from the parent's bill from the start. This may be associated with the particular bill-shape of raptors, which would render the other method of feeding difficult.

The growth patterns are similar to those of other nidicolous birds, in that weight increases slowly at first (and may even drop for a day or two), but then increases rapidly and linearly before levelling off and occasionally dropping slightly around the time of fledging.* The growth of tarsus (mainly bone) and wing (mainly feather) follow a similar pattern, but bone growth seems almost completed halfway through the nestling period, whereas feather growth occurs mainly in the latter half. If food is scarce, the rate of weight-gain slows preferentially to feather growth, so that there is minimum delay in the date of fledging (Houston 1976, Moss 1976). Small species grow much more rapidly than large ones and, whereas small species sometimes reach adult weight at the time of fledging, large species level off well below adult weight, and only increase further after leaving the nest (Figure 21).

In species with marked size dimorphism, the disparity in male and female growth is apparent from an early stage (Chapter 1). Even before feathers grow, the sexes can be distinguished by the size of their legs and feet. The females gain weight for a longer period than the males, but the males become feathered and behaviourally advanced sooner than the females. Thus the greater size of females gives them no marked advantage in the nest, as it is nullified by the more rapid development of the males (Beebe 1960, Newton 1978). By the time the females could become dominant, the males have begun to spend part of their time outside the nest and continue to gain their share of the food through greater agility. Differential development may thus help to prevent an excessive mortality of the smaller males in species with marked size dimorphism.

Breeding frequency

Nearly all raptors outside the tropics raise only one brood each year, because the complete breeding cycle occupies practically the whole suitable period (Chapter 6).

However in some species, a proportion of pairs occasionally raises two broods in a year. This has been noted in rodent-eating species, such as the American Kestrel, White-tailed Kite and Black-shouldered Kite (Howell 1932; Pickwell 1930; W. Tarboton) and also in Harris' Hawk *Parabuteo unicinctus* in parts of the Sonoran Desert where some prey species breed in both spring and autumn. Some Harris' Hawks lay second clutches before the young of the first have fledged (Mader 1975a). Some tropical species also occasionally raise more than one brood in a year (Chapter 6). This of course is possible only if their individual breeding cycles take less than six months, which eliminates

* see Ricklefs (1968) for general review; Cavé (1968) and Roest (1957) for *Falco*; Liversidge (1962), Moss (1976) and Newton (1978) for *Accipiter*; Scharf and Balfour (1971) for *Circus*; Mader (1975a) for *Parabuteo*; Fitch et al (1946) for *Buteo*; Brown & Amadon (1968) and Sumner (1929) for eagles and others; Houston (1976) and Pennycuick (1976) for vultures; Stinson (1977a) for *Pandion*.

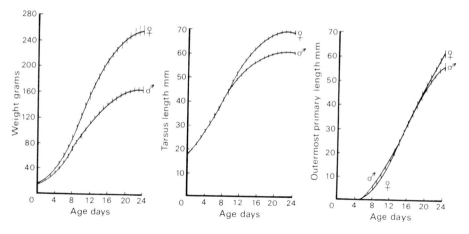

Figure 21. Growth curves for Sparrowhawk nestlings for weight, tarsus length and feather length (means and 95 % confidence limits). Females gain in weight faster than males, and are heavier at fledging. However, young females are still lighter at fledging than adult females, whereas young males are slightly heavier than adult males. Data from more than 50 broods. From Moss 1976.

most species larger than Harris' Hawk.

At the other extreme, successful breeding takes more than one year in some large raptors. The only temperate zone species in this category is the California Condor, but the list also includes the Andean Condor and one or two tropical eagles (Table 19). In these species the incubation and nestling periods are exceptionally long, but are then followed by an extremely protracted post-fledging period. Successful pairs can breed only in alternate years, but pairs that lose their egg or chick at an early stage can lay again the next year. The Crowned Eagle is especially interesting, because in East Africa it has been found to breed every second year, with a post-fledging period of 11 months or more (Brown 1966), while in southern Africa some pairs have been found to breed every year, with a much shorter post-fledging period (Fannin & Webb 1975). In conclusion, breeding frequencies among raptors are apparently determined by (1) the duration of the individual breeding cycles, which is longer in larger species, and (2) environmental seasonality and the duration of the favourable season (Chapter 6).

Raptors which normally produce only one brood in a year sometimes raise young from a replacement clutch, if their first is lost at an early stage. Much depends on latitude, and a given species may lay one or more replacement clutches during the longer season in the south of its range, but not in the north. In temperate regions, replacement laying is usual among small accipiters and in falcons up to the size of Peregrines. It is less frequent among buzzards and kites, and rare among eagles. Some species re-lay only when food is unusually abundant (Chapter 8). The interval between the loss of one clutch and the start of the next in small and medium-sized species is usually around two weeks, but is said to be only 8–9 days in Hen Harriers (Balfour 1957). It has been determined precisely in captive birds: in American Kestrels the re-nesting interval was 11–16 days among 22 pairs, about the same period as between pairing and egg-laying (Porter & Wiemeyer 1972); and in Peregrines and Gyr Falcons it was invariably 14 days (Cade & Temple 1977, Platt 1977). In larger species, the re-nesting interval was longer, with periods of 19 and 29 days for White-tailed Eagles (Fentzloff 1975), and 33 and 41 days for California Condors (Koford 1953).

If the eggs were removed one at a time as laid, some species continued laying well beyond the number needed to make up a clutch. Up to 23 eggs were obtained from a single European Sparrowhawk in a season and up to 18 eggs from a Sharp-shinned Hawk (Walpole-Bond 1938, Bent 1938). In the American Kestrel, four females responded to the experimental removal of eggs as they were laid by producing 26, 18, 13 and three eggs. The same procedure, but leaving one egg in the nest at all times, yielded 23, 22, 23, and eight eggs. The number laid depended on the starting date and in both experiments the bird that laid fewest eggs was the last to begin (Porter 1975). Larger species do not continue laying in this way, and if the first egg is removed from a nest in the wild, the birds either finish the clutch, start again in a new nest, or stop altogether for that year.

Deferred maturity

Most raptors have one or more immature or subadult plumages before acquiring the definitive adult dress. Small species have only one such plumage, but large species have several, replaced every one or more years. As most raptors are seldom found breeding in subadult plumage, it may be inferred that larger species begin breeding at a greater age than small ones, a fact borne out by ringing studies. Small accipiters, small falcons, and harriers usually start to breed at one or two years; buteos, kites and large falcons at two or three years, and Ospreys at three or more. To judge from plumage, large eagles start at four to nine years, depending on the species (Table 18). In captivity, Gyr Falcons in definitive plumage first produced eggs at four years old, Bald Eagles at six, White-tailed Eagles at eight, Griffon Vultures at six, Lappet-faced Vultures at nine, an Andean Condor at eight, and a California Condor at twelve years (Bent 1938, Hulce 1886, Lint 1960, Mendelssohn & Marder 1970, Platt 1977, Richdale 1952).

In the wild, some individual raptors begin breeding before their plumage is fully adult, and others long afterwards, depending on recruitment into breeding populations. Birds breed at a younger age than usual when conditions are especially good, either in favourable areas or years, or when depleted populations leave territories vacant. In the Sparrowhawk in southern Scotland, breeding by birds in their first year was more common in valleys (15% of 131 females), where food was abundant, than in neighbouring hills (5% of 73 females), where food was scarce (Newton 1976). Similarly, five of thirteen breeding female Goshawks were yearlings in a year of good food supplies, compared with none out of 34 in three years of poorer food supplies (McGowan 1975). Comparable figures for Kestrel females were 55% out of 38 in a good vole year and 18% out of 22 in a poor one, and for males 35% out of 37 and 4% out of 25 respectively (Village 1979). Ten pairs of Imperial Eagles in southern Spain bred in adult plumage each year from 1954 to 1959; but when human persecution had reduced the population to two to four pairs at the end of the 19th Century, several birds in immature plumage mated with adults and produced eggs (Valverde 1960). Likewise, in Scotland, the rare instances of Golden Eagles holding nesting territories while in subadult plumage were chiefly in areas of persecution, though most such birds failed to lay (Sandeman 1957). In several other species, birds in immature plumage replaced adults shot on nesting territories in spring, and attempted to breed in the same year (Chapter 3). Species recorded breeding in subadult plumage are listed in Table 20.

These observations confirm that some birds have held nesting territories, mated, and sometimes bred successfully before they would normally do so when conditions were unusually favourable. Where such breeding birds in subadult plumage were sexed, all or most were females (Tables 20 and 21, Chapter 3). Possibly this was because of a general shortage of older females in the population, or because it was easier for females to reach breeding condition at so early an age. Sparrowhawks in Scotland were exceptional in that males outnumbered females among yearling breeders (Newton 1976).

Deferred maturity and the associated plumages have presumably evolved in species in which individuals cannot usually expect to breed before they are several years old. In extreme form, deferred maturity is associated with low annual mortality so that, at any one time, opportunities for breeding are likely to be few, and young birds have little chance in competition with older ones. In populations in which the breeding pairs occupy exclusive nesting territories, and in which all territories are normally occupied, annual recruitment may be limited to the replacement of individuals that recently died. Hence, likely reasons for deferred maturity are the inability to acquire a territory at an earlier age, or insufficient skill in foraging to be able to achieve the necessary body condition and to raise young. This would explain why breeding in subadult plumage has been recorded chiefly when adults have been removed from territories by human action, or when feeding conditions were especially good. On the other hand, if all territories were occupied, individuals could be sentenced to a non-breeding existence for longer than usual, which would account for the presence of some adult-plumaged birds without nesting territories (Chapter 3).

Another basis for deferred maturity may be that individual birds ultimately produce more offspring by waiting till later in life than by starting to breed sooner. If by breeding at an early age birds seriously jeopardised their own chances of surviving to continue breeding in future years, then natural selection would ensure that they waited, even when vacancies were available. This aspect is likely to be especially important in long-lived species, and is discussed again later.

Geographical trends

In certain aspects of breeding, geographical trends occur, as is evident from comparisons between closely related species and between different populations of the same species. Some such trends have parallels in other birds and are of obvious adaptive significance (Lack 1954, Cody 1971). Consider first the duration of breeding. In any one species the breeding season becomes shorter at higher latitudes, where the favourable season is shorter. This shortening of the breeding season is achieved by (1) a reduction in the spread of starting dates of individuals, and (2) a shortening of the pre-laying and post-fledging periods of individuals (Chapter 6). Compared with their closely related tropical equivalents, temperate zone species also have shorter incubation and nestling periods (Table 22).

Another widespread trend in species which lay more than one or two eggs is for clutches to become larger at higher latitudes. Temperate zone species have larger clutches than similar tropical ones (Moreau 1944), and within the temperate zone geographical gradients in mean clutch size have been noted in the Red-tailed Hawk, Red-shouldered Hawk, Kestrel and other species in North America (Henny 1972, Henny & Wight 1970), and in the Common Buzzard, Goshawk and other species in Europe (Picozzi & Weir 1974, Huhtala & Sulkava 1976). In the three American species, mean clutch size increased from south to north, and in the two buteos it also increased from east to west across the United States. The usual explanation of the first trend is that the longer day in the north gives the parents more time to collect food for young, but other factors are also involved (Cody 1966). Clutches tend to fluctuate more at higher latitudes, with many birds laying several eggs in good food years and none at all in poor years; this might partly account for higher mean values in the north, because non-layers are usually excluded from the calculations. For unknown reasons, the Peregrine reverses the usual trend and lays smaller clutches in boreal and arctic regions than in temperate ones (Hickey 1942 for North America; various papers in Hickey 1969 for Europe). How many of these geographical variations are inherent and how many dependent on immediate environmental influence is not known.

OVERALL BREEDING STRATEGY

The various features discussed in this chapter do not vary independently, but occur in particular species as groups of associated characters. As implied at the outset, there is a continuum among raptors between the large clutches, short breeding cycles and early maturity of small species and the single-egg clutches, prolonged breeding cycles and deferred maturity of large species. In addition the small species tend to be short-lived and the large species long-lived.

Why do larger species have longer and less productive breeding cycles? For all species, any breeding attempt is costly in terms of the energy expended and the risks involved, so that the larger, longer-lived species gain less by investing heavily in any one attempt if this will jeopardise individual survival and future breeding attempts (Williams 1966, Goodman 1974). Thus, for species with long life expectancies, natural selection may favour a low reproductive effort in any one season in the interests of better chances to breed in several future seasons. As explained earlier, it may also favour deferment of maturity, giving time for useful experience to be gained. This is the strategy which, in the long run, could enable individuals of large species to leave most offspring. As another factor, the metabolic rate in birds slows with increase in body-size (Lasiewski & Dawson 1967); so the lower breeding rates of the larger raptors could be due partly to their slower metabolism, a given amount of food taking longer to digest and convert to egg or body tissue. This could account for at least the relatively smaller eggs, longer laying intervals, and longer growth periods of the larger species. It is also possible that as body size increases, so does the difficulty of obtaining prey, but on this aspect little information is available.

There is yet another aspect to the problem. Large size confers not only a longer potential lifespan, but also a greater immunity from predation and an increased ability to survive temporary food shortages. Hence, other things being equal, the larger the bird, the more consistently is its population likely to remain close to the level that the environment will support. Under these conditions, reproduction will generally be difficult, and vacancies in the breeding population will be few at any given time. For large species, then, selection pressure will favour the production of well-nurtured young released from prolonged parental care with the greatest chance of competing successfully with others of their species ('K-selection' in the sense of MacArthur & Wilson 1967).

Small species are not only shorter-lived, but also more vulnerable to predation, temporary food shortages, and other extremes. For much of the time, their populations are likely to be well below the level that the environment will support. Reproduction is likely to be easy, with many openings for recruitment into the breeding populations each year. These conditions should favour the evolution of large clutches, high breeding rates, and early maturity ('r-selection' in the sense of MacArthur & Wilson 1967). In conclusion, I suggest that body size and the associated longevity largely create the conditions that lead to the evolution of particular breeding strategies in raptors.

The same reasoning could account for tropical raptors breeding more slowly than their temperate zone counterparts. Most tropical environments supposedly offer more stable food supplies than northern ones, again leading populations to be more consistently near saturation level, so that recruitment is more difficult and K-selection stronger (Dob-zhansky 1950, Pianka 1970). It might also account for the trend in the temperate zone towards larger clutches at higher latitudes and in other strongly seasonal environments (Cody 1960).

The relationship between body size and longevity is part of a general trend among birds and mammals which is largely unexplained. Since it is linked with metabolic rates, this has suggested to some people a mechanical analogy, that the faster the animal 'ticks over', the sooner it 'wears out'.

BREEDING RATES AND POPULATION

The particular strategies shown by different species greatly affect their population dynamics, the growth potential of their populations, and their ability to withstand human and natural predation (Cody 1971, Cole 1954, Richlefs 1973). In large species, population turnover is generally slow, with more overlap between generations and a more stable age structure, all of which tend to dampen fluctuations in numbers. There also tends to be a relatively large non-breeding population, consisting mainly or entirely of immatures. Less than half the total population may breed in any one year, producing only a small number of young. In small species, by contrast, population turnover is rapid, there is less overlap between generations, a less stable age structure, and a high production of young, all of which facilitate fluctuations in numbers. Most individuals that survive a winter will have the opportunity to breed in spring, so that the non-breeding sector remains small. And because of their fast breeding rates, small species can recover from a population low more quickly than larger species.

COMPARISONS WITH OTHER BIRDS

Most trends discussed in this chapter are typical of other orders of birds, but the Falconiformes show greater variation in body weight and breeding rate than any other order. In their population dynamics the small raptors resemble song-birds, and the large raptors resemble certain seabirds. This last analogy extends especially to the small clutches, long breeding cycles, and deferred maturity. In both groups, single-egg clutches are frequent and, when two eggs are laid, often only one young is raised. Long post-fledging periods, in which the young are fed near the nest, occur in some tropical seabirds, notably boobies (*Sula* spp.) and frigatebirds (*Fregata* spp.) (Nelson 1976, Stonehouse & Stonehouse 1961). Moreover, the only other birds whose complete breeding cycles are known to last more than one year are seabirds, namely some albatrosses (*Diomedea* spp.), the King Penguin *Aptenodytes patagonica* and the Great Frigatebird *Fregata minor* (Carrick et al 1960, Richdale 1952, Tickell 1960, Stonehouse 1960, Nelson 1976).

Long-deferred maturity is common, with periods up to ten years in the larger albatrosses and up to six in the frigatebirds and penguins. As in several raptors, females start breeding at an earlier average age than males (Cody 1971).

SUMMARY

Three main trends in the breeding of raptors are related to body size. The larger the species: (1) the later the age at which breeding begins, (2) the longer each successful attempt takes, and (3) the fewer the young produced with each attempt. There is a continuum among raptors, from small short-lived species which have relatively large eggs, large clutches, short breeding cycles, and early maturity, to large long-lived species, which have relatively small eggs, single-egg clutches, protracted breeding cycles, and deferred maturity.

These trends may reflect (1) the greater advantage that large, long-lived species gain from reducing the risks and energy expended in any one breeding attempt in the interests of improved survival for future breeding attempts; and (2) a generally greater difficulty that large species have in obtaining food or in digesting and processing it (metabolic rate slows with increase in body size). Body size as such, and the associated longevity, I regard as the main characteristics that have favoured the evolution of different breeding strategies among raptors. Breeding strategies influence population dynamics, and larger species inevitably have greater inherent stability in numbers and age structure, and a greater non-breeding population, than do small ones. The correlation between body size and longevity is unexplained, as is the case in other animals.

Nearly all raptors outside the tropics raise only one brood each year, although some occasionally raise two. In certain large condors and tropical eagles, each breeding cycle lasts more than one year, and successful pairs breed no more than once in two years. Individuals breed at an earlier age than usual for their species when populations are depleted or when conditions are otherwise exceptionally favourable. This applies more to females than to males. Tropical species have smaller clutches and slightly longer incubation and nestling periods than their closely-related temperate zone equivalents. In species with marked sexual dimorphism, males fly from the nest at an earlier age than females.

CHAPTER 8

Breeding rates

For an understanding of the population turnover of a species, it is important to know not only the potential rate of breeding, but also the rate that is actually achieved, after predation and other factors have taken their toll. This chapter discusses the breeding rates of raptors as observed in the wild, and deals especially with the role of food. In recent decades, pesticide contamination has lowered the production of some populations, but discussion of this aspect is left till later. This is a rather long chapter, reflecting the wealth of information available.

To produce young, a bird has to pass successfully through a series of stages. It must first settle in a particular area, establish a nesting territory and acquire a mate. It must then proceed through nest-building, on to egg-laying, and thence to the hatching and rearing of young. In this sequential process, birds can fail at any stage. Non-breeders include some birds that do not hold nesting territories, and others that do. Among the latter, certain pairs may occupy a territory

Illustration: Buzzard at nest

128

European Sparrowhawk females at nest in the English Midlands, with eggs (upper) and chicks (lower). This small accipiter usually builds a new nest each year and raises 3-6 young at a time. Photos: R. J. C. Blewitt.

14 (Upper) European Sparrowhawk male at regular plucking post on which he prepares food for the fem.
and young. Photo: R. J. C. Blewitt. (Lower) Cooper's Hawks at a nest in Arizona, the smaller male on the right. A
known as the 'Chicken Hawk', this bird declined in numbers in the eastern United States for most of this centu
due initially to excessive human persecution and latterly to use of DDT. After the ban on DDT the species h
begun to recover. Photo: H. A. Snyder.

for only a few days or a few weeks, or may even build a nest, but get no further. Other territory holders may lay and then desert their eggs, while yet others may hatch but not rear their young. Comparing the success of populations, the differences are mainly in the proportions of pairs that reach these various stages. And it is usually in the early stages that failures are most frequent.

Any thorough breeding study should include at least some assessment of the proportions of territorial pairs which lay eggs, and those which subsequently hatch and rear their young, as well as of the number of young in broods. It will also benefit from information on the amount of re-nesting after failures, on clutch-sizes, on egg and chick loss from successful broods, on the growth of the young, and on the causes of failure.

The difficulties of measuring these various parameters have not always been appreciated, and in raptors one is at once up against problems. To begin with, only by careful observation early in the season can non-breeding territorial pairs be found and counted; and then it is often hard to be sure that some of them have not tried to breed in a site unknown to the observer and failed at an early stage. Without care, the observer may find an unrepresentative sample of nests, for instance those that are most conspicuous or accessible, or he may be able to search only at the end of the season, and may find chiefly the successful nests, which survive longest. Second, a large portion of the nest failures in 'developed' regions are often due directly or indirectly to man. Not only is the recorded breeding success then lower than it might otherwise have been, but it is sometimes hard to distinguish natural failures from human-induced ones. The latter are not always deliberate, moreover, as some raptors are discouraged by lumbering and other activities near their nest, and may then desert, or stay away for so long that a predator takes the eggs or young. Indeed the very act of collecting the data may cause some failures. A third problem is that many raptors nest at low densities in remote or inaccessible sites, and this often makes for small samples. This has been overcome to some extent in recent years by use of aircraft, but only certain species can be studied in this way, and only certain kinds of information gained.*

Full assessment of the causes of breeding failure is also difficult, as some kinds are easier to record than others, and almost inevitably some remain unexplained. In judging failures, the possibility of ultimate and proximate causes should always be borne in mind. Thus a bird exposed to food shortage may desert its eggs, which are then eaten by a predator, leaving the broken shells behind. The pre-disposing cause of failure is food-shortage, but the observer may not know this, and record only the immediate cause, desertion or predation, according to whether he happens to visit the nest before or after the predator. These points

---

* Observers have also expressed their results in different ways, which sometimes makes comparisons between studies difficult. Particularly useless for most purposes are those in which success is expressed as young reared from the total eggs laid by a population, without mention of the performance of individual pairs.

should be appreciated when collecting and analysing data, and also when reading work already published. Some accounts of the breeding of raptors are not all they seem.

The most representative results on breeding success have been obtained by those observers who studied a whole population from before egg-laying, and who made special efforts to find all territorial pairs and all nests. The details from a range of studies are summarised in Table 23, in which success is expressed as the mean number of young produced per successful nest, per clutch started, and per territorial pair. The last is the best figure from which to assess production, because it takes account of all kinds of failure, including non-breeding by territorial pairs.

GENERAL TRENDS

As described in the previous chapter, the main trend is for larger raptors to show lower breeding rates than small ones. Most of the eagle species studied produced less than one young per pair per year; the greater production typical of some Bald Eagle populations was due to certain pairs raising 2–3 nestlings at a time (unusual among other eagles), and the low production typical of Crowned Eagles in East Africa was due to the individual pairs producing young no more than once every two years. Among other raptors, production was generally greater than among eagles: various buteonine hawks produced 0·5–1·8 young per pair per year, Ospreys 0·7–1·4, harriers 1·0–2·3, large accipiters and large falcons 1·6–2·7 and small accipiters and small falcons 1·7–3·2. The low production of some pre-1950 Peregrine populations was due to human predation on the eggs or young; while the high production of American Kestrels (3·6 young/pair/year) was due to a high nest success, linked with hole nesting. In quoting these figures, I have omitted the unusually low values from some recent populations whose breeding rates were reduced by pesticide contamination; these are discussed separately in Chapter 14.

INFLUENCE OF FOOD ON BREEDING RATES

In the absence of human factors, much of the variation in the breeding rate of a species can be related to variation in food supply. The main evidence is circumstantial: (1) annual fluctuations in breeding rates are associated with annual fluctuations in prey numbers, while stable breeding rates are associated with stable prey numbers; (2) area differences in breeding rates are associated with area differences in prey numbers; (3) sudden and long-term alterations in breeding rates follow sudden and long-term alterations in prey. These three aspects are discussed in detail below.

1. Annual differences

Annual variations are most marked among species which have restricted diets, based on cyclic prey. Such prey include various rodents (approximately 4-year cycles), hares (approximately 10-year cycles), and game birds (4-year cycles in some regions, 10-year cycles in others). These prey influence several raptors but, as expected, have their biggest impact on the species that most depend on them.

To begin with the rodent feeders: the Rough-legged Buzzard was studied over a period of ten years near Dovre, in Norway (Hagen 1969). In years when voles were extremely scarce, the few Buzzards in the area did not take up nesting territories and nor did they form pairs (Tables 24 & 25). In years of slightly higher vole densities, birds showed territorial behaviour, pair formation and nest-building, but did not lay eggs. At slightly higher (but still below average) vole densities, some birds laid eggs (2–4 per clutch), but soon deserted them and produced no young. At average vole densities, the birds laid; some then deserted their eggs, while others hatched but did not rear their young. At vole densities slightly above average, most birds hatched their eggs, but the subsequent mean production was less than one young per pair; slightly better success was noted under an increasing food-supply than under a decreasing one. At higher vole densities, birds laid clutches of 3–6 eggs and reared broods of 2–4 young, while at extremely high vole densities, they laid 4–7 eggs and reared 4–5 young. Thus, the number of pairs reaching successive stages of a breeding cycle, and the number of young produced, were both related to food-supply. In years when eggs were hatched, the mean production varied between 0·1 and 4·0 young per territorial pair and, taking breeding density into account, the total annual production in the study area varied from 0 to 150 young, correlated with prey densities. In good vole years, egg-laying also tended to be earlier than in poor years, fledging weights were greater, and parental nest-defence was better.

In the same area, vole densities had somewhat less effect on the success of Hen Harriers and Kestrels, which were less restricted to voles than were Rough-legs (Table 25). They had even less effect on Merlins, which ate mainly birds; and no effect on Honey Buzzards which ate mainly insects, nor on Ospreys which ate fish. These latter species could be regarded as 'controls', re-affirming the relationship between breeding and vole numbers shown by the others. The extreme dependence of the Rough-legged Buzzard on good rodent numbers was noted in studies elsewhere in Europe and in North America, but the fluctuations in breeding output were usually less than in the Norwegian area (eg White & Cade 1971).

Annual fluctuations in production were also apparent among Common Buzzards in several parts of temperate Europe (Mebs 1964, Schmaus 1938, Wendland 1952–53, Rockenbauch 1975). They were less extreme than in the Rough-legged Buzzards, probably because the Common Buzzards had more alternative prey when voles were scarce. But again egg-laying was earlier, and clutches and broods were larger, at high than at medium vole densities, and higher still than at low vole densities (Table 26 & 27). Differences were more marked in poor habitats than in good ones. Similarly in Alberta, the closely related Red-tailed Hawk produced significantly larger clutches and broods (2·6 and 2·5) in a good *Microtus* year than in five other years (2·1 and 2·0) (McInvaille & Keith 1974).

Among Kestrels in northern Britain, the production of young over a period of 35 years paralleled the fluctuations in vole densities (Figure 22). This was shown from the numbers of broods ringed each year by amateur bird-ringers, but how much these numbers reflected breeding density and how much nest success was not known (Snow 1968). In a 5-year study in Holland, density and nest success both varied between years, in accordance with vole numbers. The main cause of failure was clutch desertion, which was more prevalent in poor vole years than in good ones (Cavé 1968). In each of these Kestrel studies, the mean brood-size did not differ between years, and mortality of young was consistently low.

The dependence of the Hen Harrier on rodent populations was mentioned above for the Norwegian area. Annual numbers and nest success in Wisconsin were also related to vole densities, and on the Scottish Orkney Islands mean clutch-sizes varied between 3·8 and 5·2 per year, but in this area food was not studied (Hamerstrom 1969, Balfour 1957). Among Black-shouldered Kites in Africa, annual fluctuations in breeding rate were even greater than among northern species. In East Africa, in a poor rodent year, birds settled and progressed only as far as nest-building, but in South Africa during rodent plagues, many pairs raised two broods in quick succession and a few raised three (Malherbe 1963, W. Tarboton, Smeenk 1974, A. C. Kemp).

Turning now to the raptors that eat game birds or hares, breeding success again follows the periodicity in prey populations (Table 27). The main raptor involved in boreal forest is the Goshawk, which eats various grouse species (and in North America also Snowshoe Hares), and on the tundra the Gyr Falcon, which eats ptarmigan (Höglund 1964, Sulkava 1964, McGowan 1975, Cade 1960). In Europe, squirrels are another frequent food of Goshawks, but they have different and less pronounced cycles than do game birds. In Finland, clutch-sizes in Goshawks varied according to game bird abundance, but not according to squirrel abundance (Sulkava 1964). In one area, the average brood-size varied between 1·6 and 2·4 over seven years, correlated with Hazel Hen *Tetrastes bonasia* numbers (Wikman 1977). In Alaska, Gyr Falcon breeding seemed to be especially influenced by ptarmigan numbers in late winter and spring when little or nothing else was available; later in the year other prey species moved north into the falcon's range, so ptarmigan were no longer the limiting factor (Swartz *et al* 1974).

In arid parts of the western United States, the breeding output of Golden Eagles varied according to the numbers of Black-tailed Jackrabbits (*Lepus californicus*), which formed more than nine-tenths of the diet (Murphy 1974, Beecham & Kochert 1975). The rabbit population

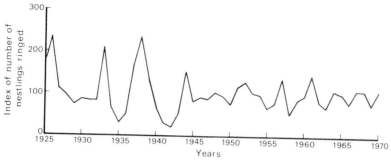

Figure 22. Annual fluctuations in numbers of nestling Kestrels ringed in Britain, expressed as a percentage of an 11-year sliding average. Partly after Snow 1968.

was probably cyclic, and over seven years in Utah the production of 16 eagle pairs increased from 0·6 young per pair in 1967 to 1·1 in 1969, and then declined to 0·3 in 1973. There was more laying and slightly better nestling survival in good rabbit years. In Scotland, too, good years for Golden Eagle breeding often coincided with periods when Red Grouse (*Lagopus l. scoticus*) or Hare (*Lepus timidus*) numbers were high, and in parts of Alaska, a similar relationship probably occurred with ptarmigan (Swartz et al 1974). In the Ferruginous Hawk in western North America, breeding varied widely in response to changes in mammal populations; in one area the annual production varied between 0·6 and 3·0 young per pair, depending on Black-tailed Jackrabbit numbers, and in another area a 47% reduction in success from one year to the next was associated with a 79% decline in Jackrabbit numbers (Woffinden & Murphy 1977, Howard & Wolfe 1976).

Several of the raptors mentioned (together with other species) showed more stable breeding rates in areas where their food supply was more stable (often through being more varied). Compare, for example, the success of Goshawks in northern and temperate Europe, of Buzzards in Germany and southern England, and of Kestrels in northern and southern Britain (Snow 1968). In each case, breeding fluctuated more from year to year in regions where prey was restricted and cyclic than where it was varied and relatively stable. This provides further circumstantial evidence for an over-riding influence of food on the breeding rates of such populations. Moreover, the whole range of variation is paralleled in owls, with the breeding output of several species fluctuating in response to rodent cycles, and that of the Great Horned Owl (*Bubo virginianus*) in the boreal zone in response to hare cycles (Southern 1970, Lockie 1955, Keith 1963). Like the stability of populations discussed in Chapter 3, stability of breeding rate is usually associated with a wide food spectrum, presumably because the chances of total food supply changing violently are lessened the greater the range of species taken.

Whether the annual fluctuations in production are great or small, they are sometimes synchronised over wide areas. In species dependent on cyclic prey, this would be expected because trends in prey numbers are often widespread, and in other species it probably has some common cause, such as the effect of weather on hunting success or body condition. Thus the relatively minor annual variations in the breeding output of Bald Eagles were correlated with one another in four out of six widely separated populations (Sprunt et al 1973). For three numerically stable populations, the annual production varied over eight years in Alaska from 0·7 to 1·2 young per active nest (mean 1·0), over nine years in Wisconsin from 0·7 to 1·1 (mean 1·0), and over twelve years in Florida from 0·6 to 0·9 (mean 0·7). Even species whose breeding rates are normally fairly stable have occasional bad years, associated with obvious food shortage or with adverse weather (see Ogden 1975 for Osprey). Such species then react in the same way as those more often affected, and show more non-laying, or later, smaller clutches, more desertions and more nestling deaths.

2. Area differences

Under this heading I include differences between regions, and between habitats and territories in the same region. Species whose breeding rates are relatively stable from year to year sometimes show area differences which are linked with food. This applied to the Black Eagles of the Matopos Hills in Rhodesia whose diet consisted of 98% hyraxes (Gargett 1977). In areas where hyraxes were abundant, the overall production was 0·6 young per pair per year, while in adjacent areas where hyraxes were scarce, the equivalent figure was 0·3 young. The difference was highly significant statistically (P<0·001), and due chiefly to more of the pairs in the good areas laying.

Among Sparrowhawks in southern Scotland, breeding was better on low ground than on high ground in the same region (Table 28). On low ground, laying dates were earlier, mean clutch and brood sizes were larger, bigger proportions of nests produced young, and nestling survival was greater. In addition, more birds bred as yearlings on the low ground. All these differences were associated with known trends in prey-bird numbers, for on low ground the woods held up to 1,750 songbird pairs/km², but on the high ground no more than 600. Among the various causes of failure, non-laying (having built a nest) and clutch-desertion were significantly less frequent on low ground than on high.

In some studies, young were produced year after year in certain territories, and not at all in others. This was evident for example among Sparrowhawks, Buzzards, Kites, Black Kites, Peregrines, Golden Eagles and Bald Eagles (Newton & Marquiss 1976, Davies & Davis 1972, Fiuczynski & Wendland 1968, Hagar 1969, Watson 1957, Broley 1947). In some such studies, this could be linked with variations in food supplies between territories. Among Buzzards in Devon, those pairs with many rabbits in their territories reared twice as many young after hatching as did pairs with few rabbits (Table 29). However, it was not always clear whether consistently good success was due to the quality of the nesting territory itself and associated feeding areas, or to the high quality of the birds in occupation. Only where the same differences in performance held over many years (kites and eagles), or after changes in the occupants had been witnessed (Sparrowhawk), could one be fairly sure that the territory was the primary factor involved. Perhaps the quality of the pair and the territory were not always independent, for the best birds may have occupied the best territories (see Hagar 1969 for Peregrine), in which case a consistently good production was due to the combined effect of a good pair in a good territory. Other reasons for area differences in breeding success are discussed in Chapter 14 on pesticides.

3. Long-term changes

The best documented case is that of the Common Buzzard in Britain, following the almost complete elimination of its main prey species (the rabbit) by disease. In one Devon valley, rabbits were still plentiful in 1954, but had gone by summer 1955. In 1954, 21 pairs of Buzzards produced 28–33 young, but in 1955, among 14 pairs remaining, only one laid eggs and none raised young (Table 30). By the next year, Buzzards had turned more effectively to other foods and breeding was better than in 1955, but still well below normal. Similar changes happened over much of Britain, and 20 years later, neither rabbit numbers nor Buzzard breeding output had reached the levels known in some areas before disease. On the other hand, in Australia the average clutch of the Little Eagle *Hieraaetus morphnoides*, the Wedge-tailed Eagle and the Goshawk *Accipiter fasciatus* became greater after the introduced rabbit spread into their range (Serventy & Whittell 1951).

Experimental evidence

Evidence for the influence of food on breeding is based not only on associations found in field studies, but also on experiment. When carcasses of pigeons and other birds were left at perching places near to Sparrowhawk nests in a poor habitat, the hawks ate them, and non-laying and clutch desertions were eliminated, clutches were larger, and survival of young was improved (I. Newton & M. Marquiss). In the European Kestrel, the provision of extra food to captive birds lead to marked advancement in oocyte growth; in the wild would presumably have led to earlier laying dates (Cavé 1968; Chapter 6).

Generalisations and species differences

From the studies mentioned, certain generalisations could be drawn (Table 31). In good, as opposed to poor food situations: (a) breeding densities were usually higher, (b) more of the available nesting territories were occupied, (c) more immature-plumaged birds bred, (d) more of the territorial pairs laid eggs, (e) more of the birds that laid eggs succeeded in hatching and rearing their young, (f) mean laying dates were earlier, (g) clutches and broods were larger, (h) nestling growth rates and fledging weights were greater, (i) parental care was better, including defence against predators, and (j) repeat laying after failure was more frequent. Not all species showed all of these trends, but in some species not all of the trends were studied. Also, the extent of annual fluctuations in breeding rate depended largely on the degree of dependence on a particular cyclic prey, with less extreme variation among populations that had alternative prey.

Usually more birds failed near the start of breeding than near the end. This may have been due partly to the increase in hunting time (daylength) with advance of season, partly to an increase in food supply, and partly to the poor-quality birds failing at an early stage, leaving only the good birds for later. At the same time, the most frequent kinds of breeding failure differed between species. In various Buzzard species, non-laying differed widely in extent between areas and between years, as did the clutch and brood sizes; in the Kestrel, non-laying by territorial pairs occurred on only a small scale, and clutch and brood-sizes varied negligibly between years, but desertions varied greatly.* Since all types of failure were more prevalent in poor food conditions, food shortage could often be regarded as the predisposing (ultimate) cause of failure, and the egg addling, egg desertions, predation and nestling mortality as the immediate (proximate) causes.

Food availability probably acts at some stages of the cycle by influencing the nutritional state of the bird itself (Chapter 6). Only when food availability reaches an appropriate level can a bird spend time defending a nesting territory and achieve the condition necessary to produce eggs, and only if food is such that the female can maintain condition can she continue to incubate. Thus many clutch desertions and nestling deaths are apparently due to the inability of the male to keep the female and brood supplied with enough food. In general, it may be harder for yearlings than for older birds to achieve the condition necessary to breed, and this may be one reason why in some species yearlings breed chiefly in good food conditions.

INFLUENCE OF WEATHER

The effects of food abundance may be modified by weather, as when mild springs advance laying in northern regions and cold springs delay it. This can be because weather has affected the availability of prey, or the hunting efficiency and food needs of the birds themselves, as described in Chapter 6. In some species, nestling deaths were greatest in wet weather, partly because of reduced feeding rates and partly because the young became soaked and chilled. This was especially true

---

* Falcon species in general show less variation in clutch-size than other raptors, and even when fed ad libitum in captivity, their clutches are no larger than in the wild (see Cade 1975–77 for Peregrines, Porter & Wiemeyer 1972, for American Kestrels).

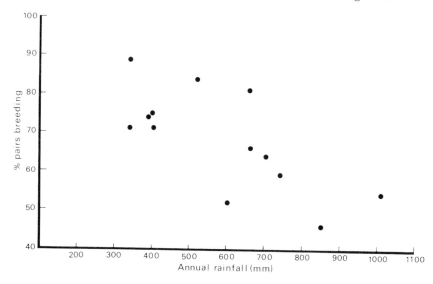

Figure 23. Relationship between proportion of Black Eagle pairs breeding and total annual rainfall, Matopos Hills, Rhodesia, 1964–76. Re-drawn from Gargett 1977.

among large young, when the female was away hunting and not on hand to shelter them in the event of a sudden downpour. Most nestling deaths among Sparrowhawks and Hen Harriers in Scotland occurred on wet days, and annual variations in nestling mortality were associated with annual variations in the number of rain-days in the nestling phase (Balfour 1957, Moss 1976). Among the harriers, broods whose mothers were involved in polygynous relationships and did not receive the undivided support of a male were especially vulnerable. Among Ospreys, a significantly heavier mortality among nestlings in southern Sweden than in central Sweden (48% versus 17%) in one year was associated with unusually wet weather in the south, whereas in two other years, no differences in mortality or weather were noted between regions (Odsjö & Sondell 1976). On the other hand, no differences between wet and dry years were noted in the production of Red-shouldered Hawks in Maryland, but the rainfall was much lower there than in the European areas mentioned (Henny et al 1973). At another extreme, eight of 25 deaths among Golden Eagle nestlings in Idaho were attributed to over-heating. The Golden Eagle is not especially adapted to heat, and these young were in exposed nests that caught the sun at the hottest time of day (Beecham & Kochert 1975).

In other species, correlations between breeding success and weather were noted over several years, with weather apparently acting via the food supply. The best documented case concerns the Rhodesian Black Eagles, in which the proportions of pairs that laid each year varied between 46% and 89%, according to rainfall. Over a 13-year period, seasons with below average rainfall were followed by above average breeding, while seasons with above average rainfall were followed by below average breeding (Figure 23). Also, laying dates were later in years when rainfall was heavy in the three months preceeding laying, the rain having presumably influenced cover and prey availability, but this was not studied. Similarly, the breeding of Peregrines on the Aleutian Islands varied from year to year relative to the spring weather. When seas were rough, the auks on which the Peregrines fed remained far offshore, so that they were less available as prey. In these years, fewer Peregrines nested and clutch desertions were more frequent than usual (White 1973, and verbatim). Summarising, variations in weather acted on breeding mainly through influencing prey availability, hunting efficiency or food needs, but some direct deaths from extremes of weather also occurred.

INFLUENCE OF NATURAL PREDATION

Raptors are not immune to natural predation, though for obvious reasons they suffer less than most other birds. Avian predators on the eggs or chicks include various crows, owls and other raptors; mammal predators include mainly climbing species, such as cats, martens, raccoons, baboons and monkeys. Some predation is associated with food shortage, as discussed above, and some with other factors, particularly: (a) the accessibility of the nest-site, (b) the amount of disturbance from other sources, and (c) the proximity and abundance of particular predators, and the amount of alternative food available to them.

Comparing species, the hole-nesting American Kestrel seems to suffer little predation of nest contents; raptors which use open tree-nests or cliffs are intermediate; while those which nest on the ground suffer most losses, presumably because their nests are vulnerable to the greatest range of predators. The contrast is especially striking from comparisons within species. Among Merlins in northern England, 94% of nests in trees produced young, compared with 63% of nests on the ground (Table 32). Among Ospreys in Connecticut, the success of the whole population was low as a result of pesticide contamination, but nests on raised sites produced significantly more young (10% of eggs laid) than nests on the ground (1% of eggs laid) (Table 32). Among Prairie Falcons in Idaho, nests accessible to mammal predators were less successful than other nests (Table 32), and exposed nests suffered 90% more egg loss than secluded ones. Among Ferruginous Hawks in Washington State, ground nests produced an average of only 0·6 young each, whereas tree nests produced 2·2 (Fitzner et al 1977).

Various *Buteo* and *Milvus* species often nest in the same tree clumps as crows, and egg predation may be accentuated by the observer putting up the raptor and allowing the more brazen crows to take the eggs (e.g. Davies & Davis 1973). Similarly, predation on Peregrine eggs by corvids on Langara Island occurred mainly when the adult Peregrines were pre-occupied in attacking passing eagles. With abundant cover on the cliffs, the crows could approach the nest ledges without being seen (Beebe 1960).

Large owls of the genus *Bubo* are especially important predators, and the Eagle Owl *B. bubo* has been noted as taking the young and adults of almost all the European raptors, up to the size of (and including) female Goshawk (Chapter 12). In North America, the related Great Horned Owl is important not only as a predator, but also from commandeering newly repaired nests even of species as large as the Bald Eagle (Broley 1947). Such owls put a stop to 5% of 619 eagle nesting attempts (18% of all failures) recorded in Florida, and at least twice took over nests after the eagles had laid. The displaced hosts usually built another nest, but did not breed that year. Interactions between Great Horned Owls and Red-tailed Hawks were noted by several observers, and usually the hawks failed when the two species nested close together (Craighead & Craighead 1956, Orians & Kuhlman 1956, Hagar 1957, Houston 1975). The owls either took over the nest before the hawks could lay, or promoted the desertion of eggs, or ate the young. Most predation was on large young no longer covered by the female at night, and young owls were suspected, as well as adults (young owls fledged a month or more before young Red-tails). Luttich *et al* (1971) noted an increase from 5% and 12% to 49% and 36% in the predation on young Red-tails over four years, as a nesting owl population increased from one to nine pairs. McInvaille & Keith (1974) in the same area witnessed no owl predation on Red-tails in the next two years when the owls' favoured prey (Snowshoe Hare) reached its cyclic peak. Similarly, Goshawks took many adult and young Sparrowhawks in one year when the usual Goshawk prey was scarce, but not in three other years (Tinbergen 1946). In another study, Swainsons' Hawks nesting close either to Great Horned Owls or Crows bred significantly less well than did those nesting in tree clumps lacking these predators (Dunkle 1977).

In the studies mentioned above, the effects of predators were especially obvious. In other studies, predators may have accounted for some unexplained losses, but were generally unimportant compared to other factors. Otherwise, the main effects of predators on breeding raptors are hard to assess, namely in restricting species to more or less secure nest-sites, and in some areas thereby limiting their distribution (Chapter 5).

INFLUENCE OF HUMAN PREDATION AND DISTURBANCE

Human predation includes the removal of adults, eggs or young. Its extent has varied from the complete destruction of all nests every year in some study areas (Rowan 1921–22, Owen 1916–22) to nil in others, and, while in general biologists have avoided areas of intense human persecution, it was still a major cause of nest failure in many studies.

For example, it was responsible for one half of all failures in the extensive work on several American species by Craighead & Craighead (1956). Human predation tends to be concentrated in areas of game or stock rearing or in areas of easy access. Thus in Scottish Golden Eagles it was more severe on land managed for grouse and sheep than on land given over to deer; and in American Red-tailed and Red-shouldered Hawks, its extent varied with the distance of the nest from a road (Table 33). In some raptors free of human persecution, an extremely high proportion of clutches was successful, for example 87% of Cooper's Hawks in Pennsylvania, 86% of Buzzards in Germany, 83% of Crowned Eagles in Kenya, 79% of Bonelli's Eagles in Israel, and 79% of Kestrels in Holland (Schriver 1969, Mebs 1964, Brown 1966, Leshem 1976, Cavé 1968).

Leaving aside direct persecution, some raptors have been known to desert their nests following a visit by an observer, or in response to lumbering or other prolonged activity nearby. The influence of such disturbance is especially hard to assess because if the birds desert or reduce their nest attentiveness, the observer can almost always attribute it to something he has done, if only visiting the nest-area. Nonetheless, three main trends can be discerned. First, small species are less sensitive to disturbance than are large ones. Second, in any one species, pairs are most easily caused to desert at the start of a breeding attempt, and least readily when they have young. Third, some desertions attributed to human disturbance may occur among pairs that are insecure for some other reasons, such as food shortage or an unsafe nest-site. Some of the apparent individual variation in response may be due to such factors.

Many raptors prepare more than one nest each year, and may switch from one to another just before laying. Peregrines and some *Buteo* species are especially prone to make such moves. An observer may attribute this to his visiting, but such switches occur naturally, and do not obviously involve reductions in breeding output.

The Ferruginous Hawk is especially liable to desert following a visit to its nest at the egg stage. But at nests that have been climbed to, the frequency of desertions has varied from region to region, and from year to year in the same region, suggesting some pre-disposing or contributory cause. The Ferruginous Hawk preys on mammals that fluctuate greatly in annual abundance, and many pairs would anyway desert in low number years. In a Utah area, seven of 13 nests that were climbed to in one year were subsequently deserted, but none in the second year (Weston 1968). Annual differences were also noted among Red-tailed Hawks in Alberta (Luttich et al 1971). In one year, seven out of 24 nest-trees were climbed during incubation, and each was subsequently deserted; but in a second year only five out of 24 nests in climbed trees were unsuccessful, compared with six out of 19 in unclimbed trees. Thus it was by no means certain how many nests failed from climbing alone, though climbing may have tipped the scales for pairs that had other difficulties.

Gargett (1977) examined the influence of observer visits to Black Eagle nests. Over a 13-year period, some nests were visited at least once each week, others only once a month, and yet others less than once a month, yet the overall production of nests in all three categories was almost identical. Grier (1969) examined the effects of climbing to Bald Eagle nests to ring the young, and found that the subsequent production of 36 nests that were climbed to did not differ significantly from that of 118 nests that were examined only from a distance through binoculars. This disturbance was relatively trivial and short-lived, however, and occurred at a stage when its effects would be expected to be minimal.

Quick checks by biologists are a different proposition to continued human presence. The literature contains examples of nesting territories permanently deserted when areas were opened up for human recreation (Chapter 13). However, in a Minnesota Forest Park, Mathisen (1968) noted no difference in the occupancy or success of Bald Eagle nests in three zones graded according to the amount of human presence. But even in his high disturbance zone, human activity was still low compared to that in some other areas. Nonetheless, some species have accepted a great deal of disturbance, if given time to get used to it. The Peregrines that nested on a building in Montreal bred successfully for years despite the crowds that leaned over and watched them, and the Peregrines on Langara Island bred under conditions of almost constant disturbance from passing eagles, to which they reacted as to a human (Beebe 1960). Some Florida Bald Eagles accepted regular human activity around the nest without deserting; one pair remained and bred after the establishment of a bombing range nearby, though they left for good a few years later (Broley 1947).

What a population will stand in terms of disturbance depends partly on the previous history of human and natural predation, with the birds most tolerant where they can get used to people without being harassed (Chapter 13). Presumably the birds respond to a human as they would to a natural predator, and the value of desertion after a mere visit by a predator, or after an enforced long period off the eggs, is that they avoid the risk of subsequent predation and an investment in a breeding effort which has a poor chance of succeeding. Many natural predators re-visit places where they have found one meal, so that by abandoning the nest the adult wastes no more effort on an attempt which is likely to fail, and also reduces the possibility of perishing itself. Similarly, if the bird was kept off the nest for so long that the embryos were likely to die, it might be more profitable to abandon the breeding attempt and 'take the year off' rather than expend further effort on a probable failure. As in all such responses, there is a balance of risks: the predator may not return, or the eggs may not have addled. The value of taking these risks will vary between species, and between stages of the breeding cycle, and may account for the main trends found. Thus the short-lived small species may have to accept risks in order to leave any offspring at all, whereas long-lived large species can afford to avoid a risk that might preclude a long series of breeding attempts in subsequent years. The widespread tendency to desert more readily from disturbance at the start rather than near the end of a breeding attempt could be explained in terms of investments; the more effort a pair have put into a breeding attempt and the less effort they need to finish it, the greater the incentive to continue. I do not wish to imply that birds make conscious judgements of their chances of success or failure, only that they have evolved to take (or to avoid) certain risks, and to respond appropriately according to circumstance.

INFLUENCE OF PARASITES AND DISEASES

Compared to other factors, parasites and diseases seem to be of minor importance as a primary cause of death among nestling raptors. They are no doubt present in many populations, but probably act mainly as contributory killers in young already weakened by starvation. However, blood-sucking simulid flies *Eusimulium clarum* are thought to have killed seven out of 29 young Red-tailed Hawks that were not short of food (Fitch et al 1946). These flies need standing water to breed in, and the casualties occurred in a year of heavy rain. A 65% mortality among young Prairie Falcons was attributed to heavy infestations of a tick, probably *Ornithodoros aquilae* but at other times many such ticks were found on Prairie Falcons and other raptors, without causing deaths (Webster 1944, Williams 1947). Five broods of Sparrowhawks heavily infested with mites showed no more mortality than many other broods that were only lightly infested (I. Newton & M. Marquiss). Blood-sucking dipterous larvae infected almost all the young in 15 Red-tailed Hawk nests, but caused no deaths (Siedensticker & Reynolds 1972). One brood of Red-footed Falcons died of a contagious eye disease (Horvath 1955), two Grey Hawks *B. nitidus* died of frounce caused by the protozoan *Trichonomonas gallinae* (Stensrude 1965), four out of 129 young Golden Eagles died of trichomoniasis (Beecham & Kochert 1975), one young Black Eagle died from myiasis, an inflammatory reaction to the presence of fly larvae *Passeromyia heterochaeta* (Gargett 1977), and two broods out of 12 broods of Red-tailed Hawks died of an unspecified disease (Siedensticker & Reynolds 1971), as did one brood of Swainson's Hawks (Craighead & Craighhead 1956). These few studies in which disease was noticed should be set against the dozens in which it was not. Disease may be underestimated, however, because dead young are often eaten by their nest-mates or parents, so do not remain for long as evidence.

SOME OTHER CAUSES OF BREEDING FAILURE

In addition to the causes discussed, some complete failures occur through accidents, such as the collapse of tree-nests or the flooding or trampling of ground-nests, and in otherwise successful nests some eggs addle and some chicks die. A selection of particularly detailed results on nest failures is given in Table 34. Even allowing for unexplained failures, the main proximate causes differed widely among species. In European Red-footed Falcons the most important was predation on the eggs and chicks, in British Buzzards it was various forms of human intervention, in British Sparrowhawks it was egg breakage (from pesticides) and human intervention, and in Florida Bald Eagles it was the taking over of fresh nests by Great Horned Owls. Ospreys suffered more than most species from nests blowing down, partly because they often built on exposed dead tree tops. In one study, 7% of 203 nests were destroyed in this way, and in another, nestling losses were reduced from 28% to 7% by the provision of stable platforms (Dunstan 1968, Postupalsky & Stackpole 1974).

Compared to other birds, the amount of egg addling among raptors is often high. In part this addling is linked with food conditions, as mentioned, and perhaps with the nutritional state of the female. In Buzzards, addled eggs formed 5% of all incubated eggs in good food years and 15% in poor ones (Mebs 1964). In Sparrowhawks, addled eggs were more frequent in poor than good food areas, in late than in early clutches, and in repeat clutches than in firsts (Newton 1976, Owen 1926). In Black Kites, addled eggs were more frequent in small clutches than in large ones (Fiuczynski & Wendland 1968). Other estimates of the proportions of incubated eggs that were addled (or infertile) included the following: 4% in Red-tailed Hawk (Luttich *et al* 1971), 7% in Kestrel (Cavé 1968), 9% in Hobby (Fiuczynski 1978), 10% in Goshawk (McGowan 1975), 11% in Prairie Falcon (Ogden & Hornocker 1977), 12% in Red-footed Falcon (Horvath 1955), 10–27% in different populations of Eleonora's Falcon (Vaughan 1961), and about 33% in Hen Harrier (Balfour 1957). Addled eggs were no more frequent in large raptors than in small ones, but their effects were much greater in large species. Thus, one addled egg made little or no difference to the output of a small species laying 4–6 eggs in a clutch, but for some large eagles or vultures it meant total failure for that year. Moreover, some eagle pairs produced inviable eggs year after year, incubating them for up to twice the normal period (eg Broley 1957).

Compared to many other birds, nestling mortality in raptors is also high and again partly associated with food-shortage. The evidence comes from comparisons of different years and different areas, and from comparisons of broods of different sizes. In the Buzzards, broods of two young suffered 29% mortality in good food years and 40% in poor ones (Schmaus 1938), and in another part of Germany the corresponding figures for broods of three were 26% and 42% (Mebs 1964). In the Sparrowhawk, the trends were as with addled eggs, with nestling deaths more frequent in poor than in good food areas, and in late than in early broods (Newton 1976, Moss 1976). In the Kestrel, mortality was greatest in the largest broods, especially late in the season, but larger broods still produced the most young (Table 35).

In many other studies, nestling mortality was slight and independent of brood-size; much of it occurred soon after hatch, and was probably due more to the inability of late hatched young to compete successfully with their older siblings than to food-shortage as such. Estimates of overall nestling mortality include the following: 3% in Hobby (Fiuczynski 1978), 14% in Galapagos Hawk (de Vries 1975), 17% in Prairie Falcon (Ogden & Hornocker 1977), 17% in Kestrel (Cavé 1968), 6% and 20% in Buzzard (Picozzi & Weir 1974, Mebs 1964), 27% and 38% in Red-tailed Hawk (Luttich *et al* 1971, Fitch *et al* 1946), 17% in Broad-winged Hawk (Rusch & Doerr 1972) and 32% in Golden Eagle (Beecham & Kochert 1975). Besides food-shortage, causes included soaking (Sparrowhawks,) overheating (Golden Eagles), or predation and disease, as discussed above.

Little is known about mortality in the period between leaving the nest and becoming independent of parental care. Towards the end of this period, mortality is hard to estimate because one is seldom sure whether young that disappear have died or left the area. As broods are often noisy at this stage, they are especially liable to be found and shot. Hickey (1952) estimated from ring recoveries that 25% of young Hen Harriers in the United States died before 1 August. For most of this period, the young were dependent on their parents. Likewise, considerable mortality at this stage was recorded among three species of accipiters in western North America, but practically none among Red-tailed Hawks (Snyder & Wiley 1976, Johnson 1973).

SEASONAL TRENDS IN BREEDING RATE

In most populations studied, the spread in laying dates was considerable, and individuals laying near the start of the season bred more successfully than those laying later. The late layers produced smaller clutches, with fewer viable eggs, and were also more prone to complete failures. In years when the whole population bred late, average clutch sizes and success were lower; and where repeat clutches were laid, they were smaller and less successful than firsts. These trends were evident in such diverse species as Sparrowhawks, Kestrels, Buzzards, Peregrines and Bald Eagles (Newton 1976, Owen 1926, Cavé 1968, Mebs 1964, Herbert & Herbert 1965, Broley 1947). They thus conformed to a widespread trend in birds, for which no good explanation has been given.

Among Kestrels in Holland, early-hatched young survived better than late ones, not only in the nest, but also after fledging, as was found from ring recoveries. Of 303 recoveries (dead) of young ringed up to 30 June, 53% were reported in the first year after fledging, and 47% in the second year. Of 118 recoveries of young ringed after 30 June, 65% came in the first year and 35% in the second. The change in the ratio is statistically significant, implying that more late-hatched young died in their first year. From dates of recoveries, this was especially marked

(Upper) Common Black Hawk feeding chick at nest in Arizona. This raptor of Central America breeds
side rivers and feeds on crabs, frogs and other aquatic animals. (Lower) Harris' Hawk at nest. Found in deserts
Central America, this species is often polyandrous, with some females mated to two males. Photos: H. A
yder.

16 Two hawks of North-American woodlands. (Upper) Red-shouldered Hawk, a common inhabitant of da*
woodlands; (lower) Broad-winged Hawk, a migrant which winters in South America. Photos: D. Muir.

between fledging and 31 August (Table 35; Cavé 1968).

Since the earlier pairs of these various species were more successful, it may be wondered why all the pairs did not breed early. In the Sparrowhawk and Kestrel, part of the spread in laying dates was due to first-year birds laying later than older ones (Chapter 6). But such age-linked differences were too small to account for the whole seasonal spread and, within each age group, the first birds to begin breeding were still most successful. The decline in clutch-size was especially hard to understand because it occurred while food and daylength were increasing. One possibility already mentioned is that birds differed in their hunting abilities, and the inefficient ones took longer to reach breeding condition than the others and for the same reason produced fewer eggs and more often failed (Chapter 6). This could also explain the average differences in performance between old and young birds, the latter being less experienced.

INFLUENCE OF AGE ON BREEDING RATES

In many species of birds, individuals breeding for the first time produce fewer young than do older, more experienced, ones. This trait has been confirmed in some of those raptor populations in which it has been examined, but not in others. Yearling Sparrowhawk females laid slightly later, smaller clutches than did older ones in the same habitat, but showed no difference in subsequent success (Newton 1976). In American Kestrels, no difference in clutch size or success occurred between yearlings and 3–4 year-olds in captivity, but year-lings less often laid repeat clutches (Porter & Wiemeyer 1970, 1972). Among Hen Harriers in Wisconsin, 11 out of 14 (79%) yearling females produced young, compared with 25 out of 32 (77%) older females, so in this respect the two age-groups were equal (Hamerstrom 1969). In Golden Eagles, some birds in immature plumage held territories but mostly failed to lay, and in Peregrines immature females either failed to lay, or laid smaller clutches than older birds (Watson 1957, Sandeman 1957, Hickey & Anderson 1969). There is one recent record from Scotland of a female Peregrine in yearling plumage raising young, and two from Alaska (I. Hopkins, C. White).

For some raptors, it has been suggested that clutch size and breeding output decline again in old age. The only evidence known to me is for the Peregrine and from egg collections containing series of clutches from what was assumed to be the same female. In those I have seen, the eggs became smaller and paler, as well as fewer, as the bird aged.

SOCIAL INTERACTIONS

One further factor known to influence the breeding of raptors is the interaction between pairs in dense populations. This aspect was studied in the Black Eagles that bred at high density in the Matopos Hills in Rhodesia. The different pairs usually had several nests in their territories, and avoided using closely adjoining sites from which the brooding birds could see one another. When they did use such close nests, one or both pairs often failed, so that their success was significantly lower than that of the remaining birds. In addition, the disturbance caused by a new pair establishing a territory was sufficient to reduce the breeding attempts and the breeding successes of neighbours. Persistent intrusions of a single adult into an established territory had a similar adverse effect on the pair in occupation (Gargett 1971, 1975).

Among Fish Eagles in Uganda, reduced production of young was noted in places of highest breeding density, again associated with greater interaction between pairs (Thiollay & Meyer 1978). In other studies, the interactions involved different species, as when Bald Eagles in process of establishing territories depressed the breeding of neighbouring Ospreys (Chapter 2). Similarly, among young Ferruginous and Swainson's Hawks in Alberta, nest failures in both species increased the closer the nests were. This was attributed to the observed increase in aggression between close nesting pairs, which was often followed by nest desertion (Schmutz 1977).

OTHER BEHAVIOUR RELATED TO BREEDING PERFORMANCE

Some of the variation in parental and defence behaviour that occurs within species is linked with food, and perhaps with the nutritional state of the female, as already discussed. Several attempts have been made to relate the behaviour recorded at one visit to a nest with the subsequent performance of that nest. Thus, the usual reaction of an incubating Bald Eagle to a survey plane was to sit tight on the nest (Whitfield et al. 1974). Of seven nests where the incubating adult stood up and flew from the nest on the first survey, five had failed by the second survey at the chick stage. The difference in success between this group of nests and the rest of those seen on the first survey was statistically significant (p<0·001). Fyfe et al (1976) found that the defence behaviour among Merlins was deficient at nests which subsequently failed, but in this case both defence and nest failure were linked with high pesticide content in the eggs. Moss (1976) recorded whether the female was present or absent during daily visits to Sparrowhawk nests to weigh the young. At nests where the female was most often absent, nestling growth rates were poorer; this was attributed to food shortage near the nest forcing the female to hunt elsewhere. Hence, in all these species, events could often have been predicted from the behaviour of the female, long before they happened.

Some raptors respond to a breeding failure by moving to an alternative nest within their home range. Two studies of Buzzards and one of Red Kites in Britain showed that birds were more likely to return to the same nest if they had bred successfully there the year before than if they had failed (Table 36). In Golden Eagles, frequent alternation between several nest-sites has been associated with regular distur- bances, and shifts away from certain areas with the onset of some new human activity (D. N. Weir, Murphy 1974).

BREEDING RATE AND POPULATION TREND

How does the breeding rate of a raptor influence its future breeding numbers? Many raptor populations that have been studied produced more young than were needed to replace adults lost naturally and the breeding population was limited by the carrying capacity of the environment. This was shown by the presence of 'surplus' birds, which were able quickly to replace birds killed on nesting territories and breed themselves (Chapter 3). Thus most populations could probably withstand a small reduction in breeding rate without declining, but in some the reductions were so great and so widespread that insufficient young were produced to offset the usual mortality of adults, so that the populations declined. This was true in several species influenced by organo-chlorine pesticides. In some such cases no change occurred in adult mortality, and the decline in numbers could be attributed entirely to measured declines in breeding rates (Chapter 14). In other words, the breeding rate influenced subsequent breeding numbers only when it fell below a certain level.

Breeding rate was also important in influencing the rate at which the depleted Red Kite population in Wales recovered between 1950 and 1970 (Davies & Davis 1973). Over this period, broad correlations were found between the numbers of chicks fledging in one year and the number of unpaired birds in the second year (r = 0·54, P<0·02), and between the numbers of unpaired birds in the second year and of breeding adults in the third (r = 0·68, P<0·01). By implication, if more fledglings had been produced, the population would have increased faster. Breeding success may similarly be involved in the natural population changes of those northern raptors that depend on cyclic prey. Good breeding during periods of food abundance probably contributes to rapid population increase, and poor breeding during periods of food scarcity to population decline (Sulkava 1964).

With knowledge of the annual mortality and of the age of first breeding, it is easy to calculate the average number of young that each pair must produce annually to maintain a stable population (Henny 1972). This so-called 'recruitment standard' has been calculated at 1·33–1·38 for Red-tailed Hawks, at 1·15 for Buzzards, at 0·95–1·30 for Ospreys and at 0·91–1·58 for Hobbies (Henny 1972, Mebs 1964, Henny & Wight 1969, Fiuczynski 1978). Such figures are valid only if mortality of full-grown birds remains the same through time, which was the case for those species in which it was checked (Chapter 14). Recruitment standards can then be used to assess the losses in production that a species could sustain without declining, or to predict population trends, when actual breeding rates are known.

FURTHER COMMENTS ON NON-BREEDING

In this last section, it seems worth re-stating the factors that lead to non-breeding, for the non-breeders are the most neglected fraction of any raptor population. This is chiefly because without the ties of nesting they are harder to find and study. They consist of at least four categories of birds, as defined by the reasons for their non-breeding. The first two include birds which live away from nesting territories. Some of these evidently fail to breed because they are unable to obtain enough food (and so reach the necessary body condition) to defend a nesting territory and proceed with laying (e.g. Rough-legged Buzzards in poor food years). Others fail to breed because, although they may achieve the necessary condition, they are unable to acquire a nesting territory, because all suitable territories are already occupied by other birds. The emphasis is on a suitable territory, because territories vary in quality and although a given bird may be able to breed on a particularly good territory, it may be unable to breed on a poor one (Newton & Marquiss 1976).

The other two categories of non-breeders consist of birds which have nesting territories, yet still fail to produce eggs. Some such birds are in pairs, and may or may not build nests, but will proceed no further. Since such pairs are more frequent in poor than in good food conditions, food is again a major factor, and probably acts by preventing the female from achieving condition. In large eagles, the relationship between non-breeding and poor food is less direct, and intermittent breeding seems the rule for many pairs, whatever the food supply. Among ten African species, it was unusual for pairs to breed more often than two years out of three, and there was a significant tendency for pairs which had raised a young in one year not to lay the next year (Brown 1970a). Perhaps, in such large birds, breeding is so prolonged and exhausting a process that they must often take a year off to recover. Particular detail on this point was obtained for Black Eagles in Rhodesia (Gargett 1977). In this species, the probability that nonbreeding in one year was followed by a breeding attempt in the next was 2·1:1; the probability that a failure in one year was followed by a breeding attempt in the next was 4·2:1; and the probability that successful rearing was followed by a breeding attempt in the next year was 1·7:1. These ratios were significantly different from one another, yet were consistent with an average of 68% of pairs breeding each year, or with each pair attempting to breed on average in two years out of three. Other factors associated with non-breeding in this population included the building of new nests (as opposed to the repair of old ones), disturbance by neighbours, change of mate, and the appropriation of the nest-site by another species.

The fourth category of non-breeders are birds that live alone on nesting territories, apparently unable to attract a suitable mate. Such singles are especially prevalent on poor territories (e.g. Peregrine, Hagar 1969), and in populations depleted by human action (e.g. Golden Eagles, Sandeman 1957). Again the emphasis is on a suitable mate because, in species studied, individuals were choosey in respect of mate, and would not necessarily pair with the first bird of appropriate sex to come along (Chapter 17).

In many species all four categories of non-breeders contain a greater proportion of immature-plumaged birds than would be found in the breeding population, so immature-plumaged birds evidently have greater difficulty in breeding than older birds. It is usually assumed that this is due to immaturity of the gonads, but since gonad condition is linked with body condition and since immature-plumaged birds breed mainly in good food conditions, it is hard to tell whether gonad or body condition is the primary factor involved. Competition with older birds may also prevent immature-plumaged birds from breeding in areas where all nesting territories are already occupied (Chapter 7).

SUMMARY

Large raptors generally have lower breeding rates than small ones. Most large eagles produce less than one young per pair per year, most medium-sized raptors produce 1–2 young per year and small ones up to three or more. Within species, in the absence of human intervention, much of the variation in breeding success is associated with variation in food supply. This is shown mainly by the large annual fluctuations in the breeding rates of species dependent on cyclic prey, and by the area differences in breeding rates of species exposed to area differences in prey numbers. Experimental provision of extra food has advanced the oocyte development of captive Kestrels and improved the breeding performance of wild Sparrowhawks.

Food shortage is often a pre-disposing factor in nest desertions and other failures. Natural predation causes some losses irrespective of food, and varies in relation to the situation of the nest and other factors. In some regions human predation is a major cause of breeding failure. Disease and parasites are generally unimportant causes of nestling deaths. In some species, birds that lay early in the season are more successful than birds that lay late, and old birds are more successful than young ones. In a few raptors nesting at high density, aggressive interactions reduce breeding rates, both within and between species. The main causes of non-breeding in birds of breeding age are inability to acquire a suitable territory or mate, or inability (through food shortage) to achieve the necessary body condition. In some eagles the chances of non-breeding are increased following a successful attempt the year before.

CHAPTER 9

Behaviour in the breeding season

More is known about the breeding behaviour of raptors than about any other aspect of their lives. This is because the need to return to a nest ties an otherwise wide-ranging bird to one place where it can readily be located and watched. Almost everything we know about some species was learnt by observation at the nest. In nearly all species, the sexes take different roles, the male being responsible for most of the hunting and the female for most of the incubation and parental care (Chapter 1). It is convenient to divide the breeding cycle into successive stages, from the occupation of territories and pair formation, through the pre-laying, laying, incubation, nestling and post-fledging periods, at the end of which the young become independent. Only a few species have been studied in detail through all stages, but many species have been studied in one or two stages. Most conform to the same broad pattern, permitting considerable generalisation.* The statements made

* To write this chapter I have consulted mainly the following: for vultures and condors, Gailey & Bolwig 1973, Houston 1976, Koford 1953, Suetens & Groenendael 1966, 1972, Whitson & Whitson 1969; for Osprey, Siewert 1941, Green 1976; for kites, Davies & Davis 1973, Meyburg 1967, Snyder 1975; for eagles, Brown 1952–53, 1955, 1955a, 1960,

Illustration: Brown Snake Eagle at nest

in this chapter apply to all species studied in particular respects, except where stated otherwise. I do not intend to describe behaviour in great detail, but rather to discuss aspects relevant to the main theme of the book.

OCCUPATION OF NESTING TERRITORIES AND PAIR FORMATION

In some populations, the breeding pairs remain on their territories all year, and the approach of breeding is marked by increased aerial display and by the pair spending more time together around the nest-sites. In other populations, it is chiefly the males that remain on territory in winter and that spend most time around the nest site; the females winter elsewhere in the same general area, or in a different, more distant area. In yet other populations, both sexes vacate the nesting territory in winter, but within pairs the male is usually first to return in spring. All three patterns are found in a single species, or even within a single population, depending partly on the ability of the home range to support the birds in winter (Chapter 4). Year-round residence seems the rule wherever conditions permit, and the other strategies represent progressive adjustments to reduced food-availability around the nesting place in winter. The presumed advantage of year-round residence on the nesting territory is that it gives security of tenure ; but it may also help to keep the same mates together for successive breeding attempts.

The greater attachment of the male to the nesting place in winter has been noted in Gyr Falcons (Platt 1976), Peregrines (Herbert & Herbert 1965) and European Kestrels (Village 1979); and the earlier return of males in spring has been noted in Merlins (Newton et al 1978), American Kestrels (Smith et al 1972), Red Kites (Brown & Amadon 1968), Swainson's Hawks (Schmutz 1977), Marsh Harriers (Brown 1976a), Montagu's Harriers (Robinson 1950) and Hen Harriers (Hamerstrom 1969). This is not invariable, however, and the female of a pair may occasionally arrive first, especially in northern populations where the whole arrival period is short. Among Merlins and others, it has also been noted that males are less likely than females to change territories between years (Chapter 10). These facts have led to the belief that, in most raptors, the male is responsible for holding the territory between

---

1966, 1976b, Broley 1947, Fentzloff 1975, Gargett 1971, 1975, Meyburg 1971, 1974, Rettig 1978, Steyn 1973, Willgohs 1961; for harriers, Balfour 1957, Hamerstrom 1969, Hosking 1943, Johannesson 1975, Laszlo 1941, Robinson 1950, Watson 1977, Weiss 1923; for accipiters, Holstein 1942, 1950, Kramer 1973, Newton 1973a, 1978, Owen 1916–21, Schnell 1958, Siewert 1933, Snyder & Wiley 1976, Tinbergen 1946; for Honey Buzzard, Holstein 1944; for falcons, Beebe 1960, Cade 1960, Enderson 1964, Enderson et al 1973, Herbert & Herbert 1965, Jenkins 1978, Keicher 1969, Nelson 1970, Newton et al 1978, Platt 1976, 1977, Rowan 1921–22, Schuyl et al 1936, Tinbergen 1940, Willoughby & Cade 1964; for buteos, Fitch et al 1946, Holstein 1956, Mebs 1964, Tubbs 1974, Weir & Picozzi 1975.

seasons, or for re-establishing it each spring. In some Peregrine populations, single males held territory all summer in the absence of a female, whereas single females showed no special attachment to any particular cliff, but moved around and visited other cliffs where pairs were already established (Hagar 1969). In another long-term study, whenever the female of a pair was killed, the male usually continued to hold the cliff and tried to attract passing females by displaying; but whenever the male of a pair was killed and no replacement appeared before the breeding season, the female usually deserted (Herbert & Herbert 1965 gave several incidents). Similarly in Gyr Falcons, lone males held cliffs on their own through the breeding season, but lone females did not (Platt 1977).

In the accipiters, opinions differ as to which sex is back first in spring, but most favour the female. In migrant populations of the European Sparrowhawk, females return earlier than males, and in migrant and resident populations it is usually the females that are seen in display over the territory (Moritz & Vauk 1976, Newton 1973a). In all these species, the important point is that the sexes behave differently, one tending to stay and hold a territory and the other moving around from one to another. In this way birds can be most sure of getting both a territory and a mate. If both sexes behaved in the same way, whether moving or staying, they would be more likely to miss getting a territory or to miss one another until it was too late to breed.

Once on territory in spring, the male soon begins to show two aspects of behaviour, namely, displays and nest-inspection (falcons) or nest-building (others). Depending on species, displays consist of conspicuous perching or calling near the nest-site, effective over short distances; and of aerial manoeuvres over the territory, which are visible over long distances. Aerial displays are often spectacular and eye-catching, and involve undulating and fast diving flights. In some species, they end on or near the nest, thereby emphasising its location. Aerial displays probably serve to advertise over long distance the fact that the territory is occupied, and so assist in spacing by discouraging other birds from settling close. They may also help a lone bird to attract a mate but, if so, this is not the only function because the displays continue long after pair formation has occurred. At this later stage, they are especially likely to be given when an intruder appears near the nesting area or after a fight with one. In at least some species, the presence of the female is not necessary for the male to begin nest inspections or nest-building, nor for him to start bringing prey to the nesting territory. Both activities usually occur in the mornings.

In migrants, it is not necessarily the first male to visit a territory that stays there, nor the first female; but the Hen Harriers studied by Hamerstrom (1969) were probably extreme: 'we actually trapped 29 harriers, 11 males and 18 females, near a small marsh (big enough for only one territory) in our attempts to catch and mark the 'resident pair'. The female which finally brought off young in the marsh in July was first caught on 17 April; her mate, however, was not one of the 11 males previously banded there'. One wonders whether this is typical of harriers or whether, in this case, the trapping caused some of the birds to move on.

In at least some species the males have special displays which are directed to passing females, and which differ from the territorial displays just mentioned. If a female lingers, one of the earliest responses involves the male drawing the female's attention to nest-sites. Male Ospreys display on the nest itself, Peregrines and Gyr Falcons have a special ledge display, American Kestrels and European Sparrowhawks have been seen to lead the female from site to site within the territory, while European Kestrels have a special 'V flight' which they use only to approach potential nest-sites (Willoughby & Cade 1964, Nethersole-Thompson 1943, Tinbergen 1940). Considering the importance of nest-sites in limiting breeding (Chapter 5), the early prominence of such behaviour is perhaps not surprising. If the female stays, she may begin to visit nest-sites on her own, or with the male in attendance, and may also begin to defend the area against other females. During the early stage, however, either partner may be ousted by another of the same sex, as witnessed for example in American Kestrels, Peregrines and Ospreys (Cade 1955, Herbert & Herbert 1965, D. N. Weir). The mated male may still display at, and even feed, other passing females (Hagar 1969). Also, the female may leave after some days of her own accord, especially if feeding conditions are poor (Black-shouldered Kite – J. Mendelsohn). All these conclusions depend on the fact that the sexes can be distinguished, and in raptors this is a good deal easier, especially for the experienced observer, than in many other birds.

In general, pair-formation in raptors is associated with nesting territories, potential breeding places containing nest-sites. For some migrant populations, however, some slight evidence suggests that pairs may occasionally form, or stay together, away from breeding areas: (a) some birds seem to be already paired on their day of arrival on nesting territories, as noted in Honey Buzzards, Swallow-tailed Kites and some northern Peregrines (Holstein 1944, Snyder 1975, C. M. White); (b) some birds are seen in pairs in places where they cannot breed through lack of a nest-site but may attempt to displace neighbouring pairs that have nest-sites, as noted in Fish Eagles, Bald Eagles and Kestrels (Thiollay & Meyer 1978, C. M. White, Village 1979); (c) some migrants are seen as pairs in winter quarters, as noted in Peregrines and Hobbies (A. C. Kemp). As they stand, these scant records are hard to interpret, and more information is needed.

Nesting territories are defended mainly by display, as mentioned above, but also if necessary by more vigorous means. Typically, the defender flies towards an intruder in direct, purposeful flight, and if the intruder does not leave, there may be a fight in which the birds grapple with one another, or even lock talons and crash to the ground. This has been seen, for example, in Kestrels (myself), Peregrines (Herbert & Herbert 1965, Kumari 1974), Golden Eagles (Gordon 1955) and Buzzards (Gilbert 1951). (Such fights should not be confused with the talon-gripping and wheeling seen between the sexes in mated pairs of Fish Eagles and others, Brown 1976). Territorial defence is generally strongest at the start of the season and wanes later, but at any time it also varies with the behaviour of the intruder and the mood of the occupant. In various *Buteo* and *Aquila* species, neighbouring pairs get to know and tolerate one another to some extent, whereas strangers are attacked immediately and are buffeted from territory to territory as they pass through a breeding area (eg Buzzards – Mebs 1964, Black Eagles – Gargett 1975).

That territories or mates are strongly contested is shown by the seriousness of the fights which sometimes occur at the start of the season, and in which an intruder may occasionally succeed in replacing an occupant. Fatal combats at nesting territories have been noted between male Peregrines (Hall 1955), female Prairie Falcons (R. Fyfe), female Ospreys (Cash 1914), male Golden Eagles (Brown 1955), male White-tailed Eagles (Banzhaf 1937, Willgohs 1961), male Black Eagles (Gargett 1971), and female Sparrowhawks (I. Newton & M. Marquiss). In some such instances, the death of one or both combatants was preceded by several days of fighting. In addition, Dixon (1937) several times saw one female Golden Eagle replace another, which then lived for some time on the edge of the home range, continually harried by the newcomer; Petersen (1956) saw a succession of Kestrels try to take over a nesting territory from the resident pair; and Gargett (1975) witnessed persistent intrusions by single adults into Black Eagle territories. There is thus plenty of evidence for serious conflicts around nesting territories; but whether the birds are competing for a particular territory, a particular mate, or merely for the opportunity to breed is usually hard to tell.

In populations in which pairs reside all year on their territories, there would seem to be little scope for choice of territory or mate. In species such as eagles and buzzards, vacancies in breeding populations occur infrequently in the absence of human intervention, and the onus for a newcomer is on getting into a territory quickly after the death of an owner. A successful bird thus inherits a territory and a mate simultaneously. But in populations which take up nesting territories separately each spring, individuals have greater opportunity to select a new territory or a new mate. In fact these conditions would hold in any population in which individuals move around between one year and the next.

Nesting territories vary in quality, that is, in the conditions they offer for raising young (Chapter 8). This depends partly on the actual nest-sites and the security they give against predators and inclement weather, and partly on the local food-supply. Raptors can almost certainly respond to territory quality, and prefer the better ones (Newton 1976a; Chapter 3). In some species the attraction of the partners to a particular nesting place seems at least as important in leading to pair formation as their attraction to one another. In the Peregrine, some males were seen to occupy and display over inferior cliffs, and occasionally attract a female for a few days, but not long enough to breed (Hagar 1969, D. N. Weir). Potential mates also vary in quality and in the chance they offer of successful breeding, and where evidence is available, mating is by no means random within the local population. If the two sexes compete among themselves for nesting territories, this could passively ensure some selective mating, the most competitive birds of each sex holding the preferred territories. But there is also some active selection of a mate, and it is known from captive breeding that individuals of either sex can be more easily induced to pair with certain partners than with others. Some captive Peregrines would not mate with one partner they had had for several years, but did so readily with another (R. Fyfe). In the wild, evidence for mate selection is provided by the fact that in the polymorphic Swainson's Hawk individuals mated with a bird the same colour as themselves significantly more often than was probable had mating been random (P<0·05, Dunkle 1977), while in other species pairing was selective by age. In the European Sparrowhawk, yearlings paired with yearlings, and older birds with older birds, much more often than chance probability, and pairs consisting of two old birds showed the best success (Newton *et al* 1979). In accipiters and harriers, eye-colour changes with age for some years after the definitive plumage is acquired, so the birds have at least one means of telling ages, but whether they actually use this method is not known (Snyder & Snyder 1974, Balfour 1970).

I have gone to some length to suggest that raptors select and compete for nesting territories and for mates because, although many people might accept the idea without question, the outcome has an important influence on the success of individuals. It also helps to explain much of the early season behaviour, yet has so far been largely neglected by students of raptors.

PRE-LAYING PERIOD

In addition to territorial defence, the main activities in the period before laying include nest-site inspection, nest building, various courtship procedures and copulation, and the feeding of the female by the male. Nest building takes place chiefly in the mornings, and it is the male who is mainly responsible. He may build or refurbish the entire structure himself; or he may bring the material while the female works it into place. The female may add the lining to the nest on her own. The extent of female participation seems to vary between pairs, however, and if for some reason the male is reluctant to build, the female may do it all herself. This is usual in polygynous harriers.

The equivalent behaviour in falcons is nest scraping, in which the bird lies in a suitable substrate and makes a hollow with its feet and body. The scrape has special postures or displays associated with it; in large falcons they are performed initially by the male and later by the pair together (Platt 1977). The two partners may make separate scrapes to begin with, and only later does the female come to accept one of them for laying. In some species that build nests, the male often begins by adding material to more than one old structure, so again it could be the female who makes the final choice. The female raptor may brood on the chosen nest for a week or more before the eggs are laid. This is possibly to check the safety of the nest, for if the bird is disturbed at this stage, she will often shift hurriedly to another site, if one is available (Nethersole-Thompson 1943). The same sometimes applies if the birds are disturbed during building.

The nest itself seems to be of special significance in some large eagles and vultures and in Ospreys, perhaps because, being so huge and hard to replace, its importance as a resource is that much greater. Placed in the top of a large tree, the nest forms a convenient platform from which to display and survey the surrounding area. Such birds spend much of their time perched on the nest, even before laying begins, and if the nest is destroyed, they usually do not breed that year but spend much of the season building a new one (Chapter 5). In species which remain paired on their territories all year, occasional nest building may occur in any month.

Most copulations occur during bouts of nest building. Some species copulate on the nest structure itself, others on nearby perches and yet others in both situations. Copulations seem often to be initiated by the female soliciting without prior display by the male, though any display may be too subtle for the observer to detect. Copulations take place several times each day throughout the prelaying and laying period, which in some species lasts for three months or more (see Brown 1966 for Crowned Eagle). In Gyr Falcons in two years, copulations occurred up to 38–39 days before laying. In the Loch Garten Ospreys, they occurred over 16–21 days in different years, and in each year involved more than a hundred attempts, though not all were successful. In Goshawks they occurred 500–600 times in the pre-lay period (Holstein 1942). Copulations thus occur over a longer period and much more often than might be thought necessary for fertilising eggs; so perhaps they are important as pair-bonding behaviour as well.

Courtship procedures among raptors may involve mutual soaring and other aerial displays, which are different from the territorial advertisement mentioned earlier; they may also include chasing, loud calling or the adoption of special perching postures. Some *Buteo* species soar in wide circles, with the mates flying close, one behind and slightly above the other, often with legs dangling, occasionally 'wing touching' and uttering piercing screams (see Fitch *et al* 1946 for Red-tailed Hawk, Weir & Picozzi 1975 for Buzzard). Accipiters fluff out the white undertail coverts during many interactions in the early stages of breeding.

The feeding of the female by the male becomes especially important as the egg-laying season approaches. Usually called 'courtship feeding', the term 'supplementary feeding' is preferable, because it serves to help the female accumulate the body reserves necessary for breeding

(Chapter 6). During this phase the female may eventually quit hunting altogether and remain near the nest, becoming very heavy prior to laying. The rate at which the male can supply food influences both the timing and number of eggs laid, as well as the subsequent success (Chapter 6). The male brings the food to the nest or to its vicinity, and transfers it to the female. Some species pass the prey from beak to beak, especially when it is small, others pass it foot to foot, and whether they do it on a perch or on the wing seems to depend on the flying abilities of the species concerned. Perched transfers are usual in most species, and the regular places where these occur are marked by a litter of prey remains. Aerial passes are usual in harriers, in which the female turns upside down and takes the food in her feet from the feet of the male (Watson 1977). The female raptor sometimes begs at this stage, which causes the male to fly off, presumably in search of more prey.

In some studies the male provided all the female's food from the time of arrival (Ospreys, Black Kites); in others the male started some days later and provided a steadily increasing proportion until the female was wholly dependent by the time the eggs were laid, but with considerable variation between pairs (Green 1976, Meyburg 1967, Tinbergen 1940). Resident Peregrines in western Canada started supplementary feeding about a month before laying (Nelson 1970), but in the other migrant species mentioned the period was less than this.

A few days before the eggs are due, the female raptor develops 'egg-laying lethargy', during which she spends most time dozing while slumped on a perch with feathers fluffed, and moves only with infrequent laboured flight. Catching her own food at this time would be difficult for species that hunt agile prey, and she does not take food from the male on the wing, only from a perch. Her behaviour rapidly improves after laying begins.

INCUBATION PERIOD

Raptors fall into three groups, in the division of incubation between the sexes. First is the group in which the male normally plays no part. It includes species in which the male is so small in relation to the female that he could not cover the eggs effectively (eg European Sparrow-hawk and American Sharp-shinned Hawk); it also includes the ground-nesting harriers in which males lack the cryptic colouring of females and are often polygynous. Second are those species in which the male relieves the female when he brings food to the nest, and may continue to sit for a period while the female eats and rests nearby. These include the majority of species so far studied, from small falcons to large buzzards and eagles. In some such species (eg Kestrel) the male sits only while the female feeds, but in others (eg Merlin, Peregrine, Gyr Falcon, Osprey) and male may sit for stints lasting 2–3 hours, and totalling a third or more of the daylight period (Tinbergen 1940, Newton et al 1978, Enderson et al 1973, Platt 1977, Green 1976,

Garber & Koplin 1972). Third are species in which the sexes share incubation more or less equally, and in which the female leaves the nest area to obtain her own food. These include the large *Gyps* vultures, which have food sources best suited to self-foraging (Houston 1976). Within any one species, individual pairs vary to some extent in incubation routine, but the female seems to control the situation and terminates the male's stints. In the first two groups mentioned above, the female seems usually to incubate at night, though male Ospreys, Little Sparrowhawks *A. minullus*, Merlins and American Kestrels have been found to do so occasionally (Green 1976, Liversidge 1962, Rowan 1921–22, Willoughby & Cade 1964). At changeovers, the returning bird in some species may bring a fresh sprig of nest material.

Raptors are often said to begin incubation with the laying of a particular egg, say the first or the last, or the last but one. Such statements are extrapolations from the spread in hatching dates, and recent studies have shown that, in species which lay several eggs, incubation does not always begin abruptly but may increase gradually with successive eggs. For example, a Prairie Falcon pair incubated for 15, 22, 64, 64, 90 and 90% of the daylight hours for the first six days from the laying of the first egg. The female was on the nest at dusk and dawn on five out of six occasions, suggesting that, even before steady incubation began, the eggs were covered at night (Enderson *et al* 1973). They may not have been heated as much as during incubation proper, however, for during the laying period the birds seemed to sit higher on the eggs.

Behaviour during incubation is characterised by intermittent dozing, and frequent turning of eggs. Turning is achieved by a special action in which the parent rises, peers at the eggs, and sweeps the bill gently between them towards the belly. In this way the eggs are also shifted in position relative to one another. Turning is especially important in the first half of incubation to prevent membranes from sticking together prematurely, while changing the positions of the eggs may help to ensure a more even distribution of warmth. Temperature control also entails the bird adjusting its behaviour according to ambient temperature, and varies from merely shading the eggs from hot sun, as in tropical goshawks *Melierax* (Smeenk & Smeenk-Enserink 1975), to continuous incubation found in arctic Peregrines and Gyr Falcons (Enderson *et al* 1973, Platt 1977). In the latter, eggs were normally left uncovered for 2–4 minutes at changeovers, but at one nest, when the air temperature was −35°C, eggs were exposed for only 20–45 seconds, the new bird coming to lie beside the incubating one, just prior to a switch (Platt 1977).

From time to time, the incubating bird on a stick nest reaches down among the eggs and tugs at the nest bottom. Similar behaviour occurs in other birds, but I do not know its purpose. Incubating raptors also nibble at surrounding vegetation, and snip off pieces that are in their way. Falcons on a cliff ledge reach out and scrape debris towards the nest with their bill, or pick up and deposit objects on the nest-rim.

Watching the birds, one gains the impression that this is done merely to relieve boredom, but it also functions to build up the nest rim, which in turn helps to retain the eggs when the bird leaves in a hurry. A heaped-up nest-rim is a sure sign that incubation has occurred in a falcon scrape. When moving around near eggs, some raptors hold their feet in a special position, with the back toe turned forwards between the front ones, which are bunched together with the claws pointing inwards. They also turn their eggs with the bill open (Nelson 1976, Fentzloff 1976). Both actions reduce the risk of spiking the shells. As hatching approaches, the female takes up a different incubation stance, and appears to apply less pressure to the eggs. She also becomes reluctant to leave the nest, even at the approach of the male with food. The eggs pip about 24–48 hours before they hatch, but even earlier the chick can be heard tapping and cheeping inside. For a few hours after hatching, the chick lies limp and exhausted, too weak even to raise its head.

NESTLING PERIOD

After the hatch, the male in many species seldom visits the nest until the young are large, and plays at most a negligible part in brooding. For example, four Peregrine males did one third of the daytime incubation, but only 1·2% of the brooding and 3·7% of the feedings (Enderson et al 1973). To begin with, however, the male continues to catch all the prey, plucking and decapitating it, and bringing it to the nest vicinity for the female. She in turn takes it to the nest, clamps it under her feet, and tears off pieces for the young to take from her bill. Anyone who has watched a large eagle feeding its tiny chick cannot fail to be impressed with the gentle and delicate way in which this is accomplished. In the first few days, the young are normally satiated after receiving only a few small pieces, and the female eats the rest of the prey herself. As the young grow, they eat other parts of the carcase as well, and increasingly the female tends to eat only those pieces which the young refuse or are unable to swallow. The prey is plucked less and less as the young grow, and towards the end may be merely dumped on the nest for the young to deal with themselves. Small prey tend to be eaten completely, whereas the hard parts of large prey tend to be left (hence studies of diet based only on the remains found at nests are usually biassed in favour of the larger items). Prey that are unfinished at one meal may be cached by the parent away from the nest, on an old nest or in the fork of a tree, and brought back on another occasion. The caching of surplus food has now been recorded in a wide range of species, from small falcons to large eagles (Schnell 1958, Angell 1969, Brown 1966, Platt 1977). It is especially marked in Eleonora's Falcon which feeds from the waves of migrant song birds that pass its breeding haunts, and may have to rely on the cache on occasional days when the flow of migrants is broken (Vaughan 1961).

In some species the male occasionally feeds the young directly (eg Merlin, Peregrine). In others he does not normally do so and, if the female deserts, the young may die surrounded by carcases which the male has brought but not distributed (see Bannerman 1956 for Sparrowhawk). This has given rise to the idea that the males of such species are incapable of feeding young. But in some studies after the female disappeared, males were seen to attempt to feed the young and, given enough time, became proficient (eg Sparrowhawk, Cooper's Hawk, Kestrel and White-tailed Eagle, Tinbergen 1940, Snyder 1974). The time taken by the individual male to learn will be crucial if such young are to survive.

During the nestling period, three overlapping phases of female behaviour can be recognised which broadly correspond to the changing needs of the young. In the first phase the female broods or shades the young almost continuously, except when collecting and distributing food from the male. At this stage the young are small and defenceless, and unable to control their own temperature. In the second phase, the female does much less brooding, but remains near the nest, and continues to take food from the male and feed the young. She may take up a prominent position from which to watch for the return of the male with food, in some species flying out to take the food before he reaches the nest area (Platt 1977).

The female may also obtain some food for herself at this stage, but is dependent on whatever prey happens to be available near the nest. By this time the young are larger, better able to keep warm and to call loudly on the approach of a predator. This calling brings the female swiftly to the nest, ready to defend her brood. In the third stage the female leaves the nest vicinity for part of the time to hunt elsewhere. At this stage the young are feathered, and can feed themselves on prey brought to the nest; they are also better able to defend themselves by use of feet and bill and, towards the end, by fleeing from the nest. They are left on their own for long periods, but may still be shaded during heavy rain or strong sun, by the female standing over them with spread wings. By this time, the roles of the parents are less distinct, and hunting becomes the main activity of both. However, the female may continue to tear up food for the young, even though they can do it themselves, but the male merely dumps it on the nest (Jenkins 1978, Newton 1978). In some studies, the males made less frequent visits to the nest at this stage than they did when the young were small. Their importance as food-providers declined as that of the females rose (harriers, Schipper 1973; European Sparrowhawk, Newton 1978).

The transition between these phases is gradual and, even within one species, occurs at a different stage from nest to nest. The duration of brooding is especially variable (depending partly on ambient temperature, Enderson et al 1973), as is the point at which the female stops distributing food for the young. For example, four Kestrel females stopped feeding their young at ages 15–31 days, and three Sparrowhawk females stopped at 19–32 days (Tinbergen 1940, Newton 1978).

(Upper) Swainson's Hawk at nest. This species breeds on the prairies and deserts of western North erica and winters on the pampas of southern South America; it makes one of the longest migrations known ong raptors. Photo: D. Muir. (Lower) The Galapagos Hawk has not known serious human persecution and is ame on some islands that pictures like these can be taken at close quarters without a hide. Photos: M. P. ris.

18 Common Buzzards, male and female at the nest (upper); an immature (lower) feeding from a dead rab Because they readily take carrion, Buzzards are especially easy to poison; and are limited in Britain gamekeepers. Photos: R. J. C. Blewitt.

At some nests the female started to hunt before the young were half grown, and at others she did no appreciable hunting until after the young had flown.

Much of the variation in hunting and feeding patterns probably depends on the male's success at obtaining food for the female. In the first phase, the female is wholly committed to brooding, and is not free to hunt, even if the male's provision is not enough, but later the female has more flexibility. She hunts mainly, it seems, when the male's efforts are insufficient to feed the brood, and whether she then hunts near the nest or elsewhere depends partly on the availability of prey in the nest vicinity. By flying long distances to hunt, the female may get the extra food, but at the same time she may increase the risk to her young from predators or inclement weather. In Scotland in areas poor in food, whole broods of Sparrowhawks died during sudden downpours when the female was away hunting (known from radio telemetry) (Moss 1976).

Raptors tend not to recover eggs which have been displaced out of the nest cup (though the Merlin and the American Black Vulture have been seen to do so, Rowan 1921–22, Stewart 1974), but some at least recover stray chicks. Richard Fyfe has a film of a Prairie Falcon hobbling out of the nest and recovering in its bill a chick which had rolled out, and England (1976) photographed similar behaviour in a Montagu's Harrier. The habit that some raptors have of bringing green leafy twigs to nests with young is discussed in Chapter 5.

POST-FLEDGING PERIOD

The precise age at which the young make their first flight is often hard to determine without a continuous watch, because for some days beforehand they spend part of their time on nearby perches, and afterwards they may return to the nest from time to time. They also fly prematurely if disturbed by a human or other predator. Some species continue to use the nest as a feeding platform and the adults take food items there. This is true of accipiters, Crowned Eagles and also of Honey Buzzards (Newton 1973a, Brown 1966, Holstein 1944). All these species take large prey items (pieces of wasp comb in the case of Honey Buzzards), and the nest forms a secure platform on which to rest the prey. When food is passed direct to the young, they may take it to the nest themselves. Away from the nest, the young do not remain together, but perch separately around the vicinity. A predator that finds one young is thus less likely to find the rest.

The young soon begin to fly considerable distances to meet the incoming adults with prey and, if the adults always approach from the same direction, the young gradually drift further from the nest. Families of Peregrines have been seen more than one kilometre from the nest cliff, and it is likely that many Osprey broods are lured to feeding areas some distance away. As the young become stronger on the wing, they also begin to range further from the nest of their own accord. This makes it difficult to determine the end of the post-fledging period, because it is hard to be certain exactly when the young leave.

Except when delivering prey, the adults tend to stay away from their young after the young have left the nest, perhaps partly to avoid being continually pestered for food. When hungry, the young of some species become extremely aggressive towards their parents, which may serve to drive the parents out on another hunt. In harriers and accipiters, the female becomes much more important as a food provider, making more frequent visits and with larger prey than the male. In some pairs studied at this stage, the male made no more than a token contribution to the food needs of the flying young. On the other hand, in eight pairs of Black-shouldered Kites, the female stopped feeding the fledged young, leaving the male to rear them (J. Mendelsohn). If desertion by either parent at this stage depends on body condition, as seems possible, variation between and within species might be predicted, depending on food.

The old falconry literature contains accounts of adults 'enticing' their young away from the nest with food, and subsequently teaching them to hunt, but only recently have biologists witnessed behaviour that could be interpreted in this way. When their young were feathered, adult Peregrines were seen to fly from perch to perch around the nest with prey, until one of the young eventually flew over to take the prey. The reluctance of the adults, especially the male, to get close to hungry young at this stage may be partly to avoid being gripped, an accident seen occasionally in Peregrines. The 'teaching' aspect is apparent from the way in which food transfers develop. Initially the adults take food to their young and pass it over in a perching position or leave it on the nest. But once the young can fly well, they begin to catch the prey in the air, or fly up to the approaching parent, and remove it from the parent's beak or feet. Young Kestrels can manage this within two weeks after leaving the nest (Tinbergen 1940). Later still, the adult may surrender the prey only after a fast chase, or may even release bird-prey alive for the young to catch. This latter behaviour has now been seen in the European Sparrowhawk, Cooper's Hawk and Peregrine (I. Newton, McElroy 1974, S. Sherrod). Since the accipiters often do not kill their prey at first grasp, little or no specialisation of behaviour is required for them to release the prey alive, but in Peregrines it entails the suppression of the killing neck-bite, which these birds normally administer immediately after making a catch. Peregrines have been seen to catch and release the prey several times before the young eventually caught it. They have also been seen to fly low over the ground, flushing prey, which were then caught in a stoop by the young, or to dive at and miss birds that they could easily catch, leaving them for the young to take in a following stoop (Beebe 1960). All such individual actions are open to various interpretations, but taken together, it is hard to avoid the conclusion that the adults behave in a way that encourages the young to perform actions associated with hunting.

The whole sequence was studied in Ospreys, whose food passes became more demanding as the young birds developed (Meinertzhagen 1954). The adults began by perching near the young with a fish, and flying away with it when the young approached, apparently encouraging the young to follow. The adults then repeatedly dropped the fish, stooping to catch it again, while the young soared and flew round overhead. Soon the young began to stoop at the fish themselves, and became proficient at taking it from the air or from the water surface; finally they began to submerge and catch their own fish. The whole process took only six days.

As young raptors become stronger on the wing, they begin to develop for themselves behaviour patterns which characterise the hunting of adults. Young Peregrines circle and stoop at one another, and make mock attacks at inanimate objects, followed by trial attacks at prey, and finally they make actual attempts to hunt for themselves (Nelson 1970). Similarly, young Hobbies of a certain age circle and stoop at one another over the nest wood (Schuyl *et al* 1936); young Kestrels begin to hover and attack pine cones and other objects, which are then worked over with the bill (Tinbergen 1940); young Sparrowhawks perform fast-flying chases among the trees (I. Newton); while young vultures spend long periods in 'practice gliding' over the nesting areas (Houston 1976). Even in the nest, young Honey Buzzards begin to show the scraping movements which the adults use when digging out wasp nests, with the result that the nest bottom is often torn out (Holstein 1944).

To begin with, these various behaviour patterns are directed at a variety of objects and seem to occur chiefly at times when the young are not hungry; when the young are hungry they utter begging calls, or sit and watch for the arrival of a parent. Presumably the young learn by trial and error which objects represent food and which do not; and among the food items, which are the most profitable to hunt. Easily caught prey, such as insects, may be taken before more difficult prey, such as birds. Three broods of Hobbies began to take insects for themselves 4–6, 8–9 and 11–12 days respectively after leaving the nest, but all continued to receive some bird-prey from their parents for three weeks after fledging and, by the end of this period, they had still not begun to kill birds for themselves (Schuyl *et al* 1936). Young buteos were likewise seen to kill prey within a few days after leaving the nest but continued to receive supplementary food for several weeks (Angell 1969, Matray 1974). Two Red-shouldered Hawk broods began to hunt at about two weeks after leaving the nest; they took mainly insects and other small items to begin with, and added mammals and other vertebrates later. Their proficiency improved considerably during the first two weeks of hunting, but they were fed by their parents for 8–10 weeks after fledging (Snyder & Wiley 1976).

The learning process in the young does not depend entirely on parental participation. From the falconry practice of 'hacking back' the young of Peregrines and many other species have long been known to achieve independence on their own; the young are placed in an artificial nest and provided with food until they have learnt to fly and hunt for themselves. Perhaps in natural conditions parental teaching serves mainly to hasten the learning process.

From existing information, it appears that in the post-fledging period the young of some species take little or no food for themselves (accipiters, Honey Buzzards), others have begun to take easily-caught but not difficult prey (Hobbies), and others have begun to take the same prey as the adults (buteos, Peregrines). Clearly, more information is needed on the development of hunting skills, and on the contribution of prey they have caught themselves to the total food eaten by young of different ages.

The duration of the post-fledging period was studied in particular detail in the Red-tailed Hawk, using radio-telemetry on 41 young in 16 broods in Montana (Johnson 1973). The individual young stayed in their parents' territories for periods of 30–70 days after leaving the nest, and brood-mates left up to 31 days apart. Some young made no movements from the home territory until they left permanently, while others left and returned of their own accord up to five times. Most went only a few kilometres but one went 27 km before returning, and periods away ranged from less than one day to seven days. Three young migrated at 34, 57 and 70 days after fledging. In Buzzards resident in Scotland, individual young stayed in their parents' territories for 2–7 months, but all had left before the next breeding season (Picozzi & Weir 1976). The striking feature in both these studies was the enormous individual variation in post-fledging period within the population, but it was not noted whether the long-staying individuals were fed for the full period that they remained.

The post-fledging period is often said to be terminated by the adults driving the young away, but the only well authenticated cases occurred where the young were still present when the adults began another nest (White-tailed Kites – Pickwell 1930, Black Eagles – Gargett 1975). In the sedentary Crowned Eagle in East Africa, in which the post-fledging period may last nearly a year, the parents continue to bring food to the nest until the young no longer collects it, a situation which arises only when the young has become self sufficient (Brown 1966). In populations which migrate after breeding, the young presumably have no option but to become independent by this time. For most raptors there is no evidence that the young migrate with their parents and plenty to the contrary, but more information on this point is desirable. So far only in the Old World and the New World vultures are the young known to accompany their parents for a long period after leaving the nest area (Chapter 7).

VULTURES AND CONDORS

Vultures differ from most other raptors in several aspects of breeding, with remarkable convergence between the Old World and the New

World stocks. Their main food – small slippery pieces from the soft tissues of large animals – does not lend itself to being carried in the feet, and both Old and New World species carry food in the crop and feed their young by regurgitation. Moreover, both groups feed mainly liquid food to begin with, followed by an increasing proportion of solid matter. The young take food directly from the bill of the parent. Black Vultures (*Aegypius*) have been seen to regurgitate three kinds of substance for the young: a clear liquid (possibly water), a viscous brown liquid (probably well digested food from the stomach) and more solid material (probably freshly eaten food from the crop) (Suetens & Groenendael 1966).

In the Griffons (*Gyps*), the crop is a simple elastic diverticulum of the oesophagus which, when full, holds up to a fifth of the bird's body weight in food, bulging out between the breast feathers (Houston 1976). The adults can feed small young several times a day, even though they themselves have not eaten for up to 24 hours (Mendelssohn & Marder 1970). They give saliva along with the food. The composition of the saliva is not known, but rearing young by hand is more successful if the young are given a digestive enzyme preparation along with the food, at least during the first two weeks. Rearing is less successful if food is moistened only with water. It is not known if other vultures add saliva as the Griffons do, but the swelling of the jaws, mentioned by McMillan (1968) for parent California Condors, may be caused by the growth of salivary glands.

Birds that eat only the soft parts of animals are liable to suffer from shortage of calcium (the soft tissues of ungulates, for example, contain only 0·01% calcium, Houston 1978). Griffons overcome this problem by bringing to their chicks small splinters of bone obtained from carcases eaten by hyaenas or other large predators. In areas where such bone splinters are not available, the young often suffer from weak and broken bones, and the adults bring all sorts of inappropriate objects to the nest, such as bits of pottery and glass (Ledger & Mundy 1975, Mundy & Ledger 1976). This is a special problem in ranching areas of South Africa where the bone-breaking mammalian carnivores have been eliminated, and where one fifth of young Cape Vultures *Gyps coprotheres* suffer from bone defects.

The possibility that parent Griffons might secrete extra calcium from their salivary glands was eliminated by Houston (1978), who injected captive adult Griffons with radioactive calcium, but was subsequently unable to detect it in the young. Vultures in other genera and other carrion feeding birds are less likely to suffer from calcium shortage, because their diets include the whole bodies of small animals. The Lammergeier has the remarkable habit of feeding its young largely on fresh bones, which are swallowed whole, marrow and hard parts both being digested. Unlike other vultures, this species carries food to the nest in its feet (Suetens & Groenendael 1972).

Vultures and condors also differ from other raptors in that the sexes take similar roles in incubation and parental care, with one parent on the nest while the other forages. There are usually one or two changeovers each day (Koford 1953, Houston 1976, Stewart 1974). It is not clear to me what is the advantage of this system over one in which the sexes take different roles, but it may be linked with long distance hunting and the inability of one bird to bring back enough partly digested food for two others.

SNAKE EAGLES AND OTHERS

Snake eagles (*Circaetus*) also diverge in some respects from the usual pattern of raptor behaviour. These birds swallow each snake as caught and regurgitate it to the young. Because snakes often wriggle after capture, they may be more easily carried in this way and less easily stolen than if held in the feet, for a dangling snake might be especially easily seen and pirated. Typically, the adult arrives at the nest with part of a snake, often still wriggling, protruding from its beak. It either pulls out the snake with its foot, or offers it to mate or offspring, which seizes the tail in its beak. Then one bird leans back, the other pulls, and out comes a length of snake; the process is repeated until the snake lies in the nest. If the snake is still writhing, an eaglet may 'kill it' with its feet, and make twisting pecks up and down the spine, evidently an innate act because the young would not have seen a parent do this (Steyn 1973). The snake is then picked up and swallowed head first. Very large snakes are torn up and fed to small chicks; but it is amazing how large a snake a downy chick will engulf (Brown 1976b).

In fact, snake eagles of all ages can swallow very large prey. A full-grown Brown Snake Eagle, *C. cinereus*, which weighs about two kilograms, can swallow a snake 150 cm long or a Puff Adder as thick as a man's arm. After such a meal the crop does not bulge, because the snake passes directly to the stomach. When Steyn (1973) fed a captive bird on rats or other meat, it developed a bulging crop, just as other raptors would, but when he fed it a snake the crop did not bulge and a taut swelling appeared in the lower abdomen. The excreta of snake eagles is unusually dry and solid; the young do not eject it clear of the nest, as do most other raptors, but onto the nest edge or just beyond. Another feature of snake eagle young is the early development of the dorsal plumage. This is also shown by the Secretary Bird, certain vultures and others that have tree-top nests exposed to the elements. Only after its back feathers are grown is the chick left unshaded from the hot sun for large parts of the day.

The only other raptors known to me which regurgitate food for their young are the African Long-crested Eagle *Lophoaetus occipitalis*, the Bat Hawk *Machaerhamphus* and the Secretary Bird. The first two species must usually obtain their food in less than an hour around dusk or dawn, and the swallowing of several items into the crop (rodents and bats respectively) prevents long interruptions of hunting in the short time available (Jarvis & Crichton 1978, Black 1978). Other raptors and owls, which carry individual prey to their nests, hunt over a much longer period during the day or night. The Secretary Bird feeds its young on large numbers of insects and other small items, and it would be impracticable for such a large bird to journey to the nest separately with each titbit. It carries larger items in the bill (A. C. Kemp). Honey Buzzards feed their young on small wasp grubs, but they carry pieces of comb to the nest, and pick out the grubs one at a time (Holstein 1944). Hence, in all the species which depart from the usual pattern of raptor behaviour, an explanation linked with feeding habits is readily apparent.

BEHAVIOUR OF SUB-ADULTS

It remains to comment on the year-by-year development of breeding behaviour in raptors which do not lay until they are several years old. In some medium-sized and large raptors, the sub-adults sometimes take up vacant territories, form pairs and make nests, without proceeding any further (Golden Eagle, Sandeman 1957; Peregrine, Hagar 1969). The advantages of this include (1) the acquisition of a nesting territory as soon as one becomes available, thus reducing the risk of not getting one if they waited another year; (2) pair formation, a process that may take longer in large raptors than the reunion of individuals previously mated (for Black Eagle see Gargett 1977, for sea-birds see Coulson 1966, Fisher 1975); (3) learning the best available feeding areas without the strain of raising young, and (4) a nest built in preparation for breeding in a later year (Chapter 5).

In the typical breeding cycle of captive Gyr Falcons, certain behaviour appeared in regular sequence through the season, beginning with nest-scraping and then ledge displays by the male, visits by the female to the displaying male on the ledge, food transfers from male to female, mutual ledge-displays, copulation and then laying (Platt 1977). Birds kept together from their first or second year progressed further along this sequence each year, but different pairs progressed at different rates. One successful pair reached the male nest-scraping stage in their second year, mutual ledge-displays in their third year, and egg-laying in their fourth. In their first few years, they reached any given stage at a later date than breeders did, and might still be nest-scraping, for example, when breeders had eggs. All this is reminiscent of behaviour observed among known new pairs in the wild, though in some species occasional individuals reached the laying stage at a younger age than normal in favourable circumstances (Chapter 7).

FEEDING RATES AND FOOD CONSUMPTION DURING THE BREEDING CYCLE

Only in a few species has food consumption been measured through

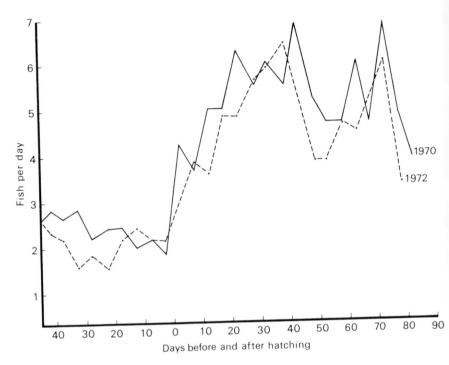

Figure 24. Fish brought per day to an Osprey nest at Loch Garten, Scotland. Continuous line – 1970; broken line – 1972. Re-drawn from Green 1976.

the whole breeding cycle. These include the much-watched Ospreys at Loch Garten in northern Scotland, where the male brought all the food for the female and young from the time he arrived until the young left the nest (Green 1976). This amounted to 2–3 fish per day in the early part of the season, increasing steadily after hatch to about six per day just before the young left the nest; the supply then dropped around the time of fledging and increased again when the young were on the wing (Figure 24). The same pattern held in different years. The size of fish brought through the season did not change, so the numbers gave a good indication of the food consumed at different stages.

A similar picture was obtained for Sparrowhawks which were fed *ad libitum* in captivity and which raised four young (Hurrell 1973). Food consumption approximately doubled after hatch, then increased to a peak in the late nestling stage and fell slightly as the young left the nest. In wild Kestrels, however, the pattern of food consumption varied among the four pairs, though all ate voles throughout (Tinbergen 1940). One female received the biggest increase in food from the male before incubation began, another early in incubation, and the remaining two early in the nestling period (Figure 25). Perhaps these females differed in their initial condition and in their pattern of deposition and usage of body reserves. They all probably accumulated reserves at some stage before hatch, because in the early nestling period they gave almost all the food they received from the male to the chicks. In consequence their own consumption dropped greatly from what it was in late incubation.

Food eaten during the nestling stage has been studied in a much wider range of species, but findings have varied greatly between species and between different pairs of the same species. They fall into two main groups. In the first, the total food consumption (a) increased during the nestling period, as the young grew, and (b) differed between nests according to the number of young present; also (c) the female did little or no hunting until the young were fledged. This pattern was found in studies of Peregrines, Marsh Harriers, Ospreys, Kestrels and some Sparrowhawks (Enderson *et al* 1973, Johanesson 1975, Green 1976, Tinbergen 1940, 1946). In two Kestrel pairs, which nested only 100 metres apart and took the same prey, the peak in food consumption occurred at different dates, but at the same stage of nestling development.

In the second group, the total food consumption (a) showed little or no increase as the young grew; (b) did not differ significantly between broods of different sizes, nor in the same broods, before and after their

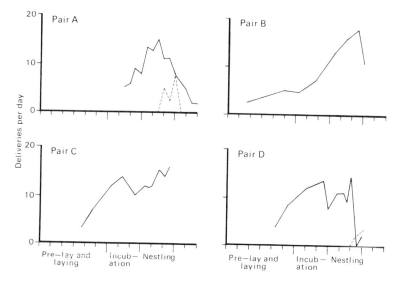

Figure 25. *Voles brought per day to four Kestrel nests in the Netherlands. Continuous line – brought by male, broken line – brought by female. One female received the biggest increase from the male before incubation began, another early in incubation, and the remaining two early in the nestling period. At only two nests did the females hunt before the young flew. Re-drawn from Tinbergen et al 1940.*

depletion by mortality; also (c) the female started to hunt before the young were fledged and sometimes before they were half grown. This pattern was noted in other studies of Sparrowhawks, in Sharp-shinned Hawks and Cooper's Hawks (Newton 1978, Snyder & Wiley 1976). Moreover, Snyder & Snyder (1973) detected no changes in the day-to-day food intake of a Cooper's Hawk brood in which they altered the number of young experimentally every few days. In yet other studies, the results showed some of the above features, but not others.

The first set of findings probably arose under conditions of food abundance, and the second under conditions of scarcity. In the first, the male was apparently able to supply enough food for the female and young, and to alter his hunting effort to match the changing needs of the family; in the second, the male was unable to supply enough food, so the female helped with the hunting, and the total amount of food obtained was set by the foraging success of the parents, rather than by the needs of the young. In some such cases wide variation in total food consumption occurred among broods of the same species – between 5,900 and 18,000 g for broods of 3–4 young of the European Sparrowhawk in the 32 days after hatch (Newton 1978). Tinbergen (1946) noted that the young of this species fledged successfully on widely differing rations, and suggested that, when food was plentiful, some broods were 'overfed'. This may have helped their survival over temporary periods of shortage.

So far, I have written chiefly in terms of food consumption, with little mention of feeding rate, the frequency with which prey items were delivered to the nest. Where parents managed to keep up with the needs of their young, feeding rates were higher when prey-size was smaller. Hobbies, for example, brought insects to the nest at a much faster rate than they brought birds, but they took insects only when these were readily available near the nest (Schuyl et al 1936). This raises a related point that, when hunting near the nest, it may be profitable for the parents to bring both large and small prey, but when hunting at greater distance only large prey, because of the time taken to travel back and forth. Possibly, small prey caught at a distance are eaten by the adults themselves, in which case food studies at the nest might accurately reflect the food of the nestlings but not of the adults.

Comparing species, feeding rates were found to vary between extremes of 57 prey deliveries per day among Red-footed Falcons taking insects to their young, to only two deliveries per week in South American Harpy Eagles taking mammalian prey large enough for several meals. Harpy Eagles showed the lowest feeding rates yet found among raptors, with one prey delivery per week during incubation, and one per 3·5 days after hatch (Rettig 1978). African Crowned Eagles also brought prey (usually antelope) infrequently, with one delivery per 3·5 days during incubation, one per 1·6 days during the nestling period, and one per three days at post-fledging. These were average rates however, and the fledged young sometimes went up to 13 days without food (Brown 1966).

In some species, prey deliveries were depressed during heavy rain (in the Sparrowhawk by one third, Newton 1978). This was partly because hunting was less efficient, and partly because the female had to cover the young instead of hunting herself. In fine weather, no diurnal rhythms in prey deliveries were noted among Sparrowhawks and Kestrels, but in Hobbies feeding on insects, prey deliveries began around mid-morning and increased till evening, corresponding with the activities of the insects themselves (Newton 1978, Tinbergen 1940, Schuyl et al 1936).

Where parents keep up with the needs of their young, how are hunting efforts regulated? The chicks can communicate their hunger to the female by begging, and the female in turn can communicate these needs to the male. The females of some species have loud begging calls which they give from the nest when hungry (Tinbergen 1940 for Kestrel, Haverschmidt 1953 for Marsh Harrier, Rettig 1978 for Harpy Eagle); female accipiters also have a special 'dismissal scream' which they give when the male has delivered food (Schnell 1958, Newton 1978); and female Hen Harriers when hungry have been seen to chase the male away on another hunt (Balfour 1957). For much of the nestling period, the female thus acts as intermediary between the young and the male, and can modify her own behaviour according to circumstance. Once the young no longer need brooding continuously, the female is free to help with the hunting if the male is unable to supply enough food.

SUMMARY

Despite the divergent ancestry of different types of raptor, most species that have been studied behave in a remarkably consistent way, presumably through convergence. Exceptions to general trends can usually be explained plausibly from the feeding habits of the species concerned. In falcons and some other species, males are more attached to nesting territories than females are, more often remaining overwinter or returning first in spring. Aerial displays may be important in the advertisement, defence and spacing of nesting territories. Other early displays by the male to the female are centred on nest sites. Potential nesting territories and mates vary in attractiveness and, in at least some populations, are strongly contested. Serious fighting sometimes occurs between territory occupants and intruders, and occasionally an intruder takes over.

The main activities in the period between pair formation and laying include territory defence, nest-building (scraping in falcons), courtship and copulation, and the feeding of the female by the male. Copulations are frequent and may help to strengthen the pair-bond, while the provision of extra food by the male helps the female to accumulate the reserves necessary for breeding. Such provision continues during the laying, incubation and early nestling periods. During the nestling period, the behaviour of the female changes to meet the needs of the growing young. The stage at which she leaves the nest area to hunt varies between pairs; it probably depends on food supplies and whether the male can feed the family unaided. In the post-fledging period the adults of some species appear to 'teach' their young to catch prey, though the young are capable of learning for themselves. Unlike most raptors, vultures and a few other species regurgitate food for their young.

In species studied for the whole breeding cycle, food consumption doubled after hatch, and then increased to a peak in the late nestling stage. It also varied slightly according to the number of young in the brood. This probably occurred in conditions of abundant prey, where food consumption through the season was set by the changing needs of the family. In other studies, food consumption showed no obvious increase during the nestling period and no differences between broods of different sizes. In such cases, consumption was apparently set by the hunting success of the parents, and was below the ideal intake of the young.

Sparrowhawk

CHAPTER 10

Fidelity to breeding areas

In many bird species, as may be judged from ring recoveries, individuals tend to return to breed for the first time in the general area where they were born; and once they have bred, they often return to the same territory in successive years. As a rule, this holds to a greater extent for one sex than for the other, but is true of both resident and migrant, solitary and colonial species. Raptors conform in all these respects, insofar as can be judged from the available data, and also illustrate some other aspects of homing behaviour, not yet shown in other birds.

Most information on the distances moved by young birds between their natal and subsequent breeding places comes from the recoveries of birds from national ringing schemes, reported by members of the public. The assumption usually has to be made that any bird of appropriate age recovered in the breeding season was in fact nesting at the locality concerned. Great care must be taken to separate the immatures, for ringing has already shown that these birds may summer in areas partly different from the breeding adults of their population. So far several such traits have emerged: (a) the immatures may remain all summer in 'winter quarters', eg some first-year Steppe Buzzards; (b) they may migrate north later and spend less time on the breeding areas

Illustration: White-tailed Eagles over breeding area

173

than adults, eg some Broad-winged Hawks; or (c) they may migrate north but stop short of the breeding areas, eg some second-year Ospreys (Olsson 1958, Österlöf 1951, Matray 1974). Moreover, different populations of a species may behave in different ways, as may the different age-groups among the non-breeders of a single population. In the Osprey, most recoveries of yearlings in summer came from the 'winter quarters', most recoveries of second-year birds were from further north but not as far as the breeding areas, and only in later years were most recoveries from the breeding areas themselves. This pattern held in the Old World and the New (Österlöf 1951, 1977, Henny & van Velzen 1972). In the Black Kites which migrate from Europe to Africa, a more protracted pattern was found, as the majority of summer recoveries came from progressively nearer the birthplace from the first to the fourth and later years (Table 37). In several migrant species, occasional adults were reported in summer on 'winter quarters'; and it was usually assumed that these birds were skipping a breeding attempt, not nesting far to the south of their usual area.

FIDELITY OF YOUNG TO BREEDING IN NATAL AREAS

It is convenient to start with the European Sparrowhawk, because more records are available for this than for any other raptor. The species is non-migratory in Britain and, when the young become independent in August, they disperse outwards in various directions from the nest. They begin breeding in their next summer or the one after, in their first or second year.

The breeding season recoveries of this species were mostly clustered around the birthplace and became sparser with increasing distance (Figure 26). Thus about 76% of recoveries were within 20 km of the birthplace, and almost all were within 50 km. Treating all the recoveries together, their density fell off exponentially in concentric circles out from the birthplace, a drop of about 99% between the 0–5 km zone and the 21–25km zone. This pattern was established by the first autumn and hardly changed thereafter, implying that in this species most birds dispersed within a few weeks of fledging to near where they would later breed. I therefore included all first-year birds in the analysis just described, even though some of them may not have nested till later in life.

This picture is based on the national ring recoveries, but work in a confined study area over several years showed another interesting point, namely an apparent avoidance of breeding in the natal territory. Birds bred all around their natal territories, but not actually in them, even though these territories were often vacant at a suitable time. Out of 79 recovered, only one individual bred in the territory where it was born, but only after previously breeding in another territory (Table 38). Similar tendencies have been noted in other kinds of birds, and they may serve to reduce inbreeding. Among Peregrines and Prairie Falcons, however, known young have been seen to return to their natal cliff in the following spring, only to be driven away by the adults (R. Fyfe). Whether in the absence of the adults they would have stayed and bred is an open question.

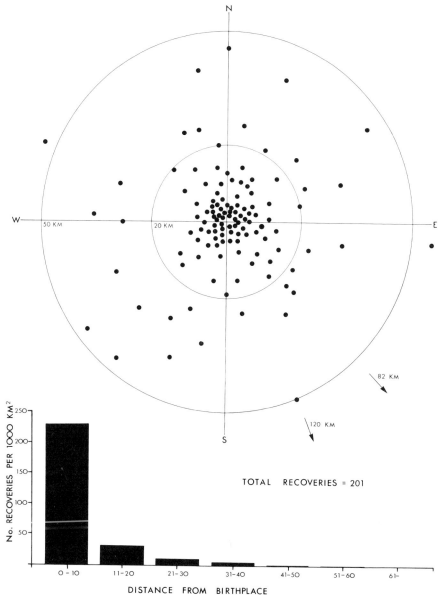

Figure 26. Upper: breeding places of Sparrowhawks in relation to birthplace (in centre), as shown by British ring recoveries. Lower: density of ring recoveries in concentric circles out from the birthplace.

The breeding season recoveries of other raptors also thin out with increasing distance from the birthplace, but at different rates in different species. A circle drawn at 50 km radius of the birthplace would include almost all the breeding season recoveries of British Sparrowhawks, 89% of British Kestrels, 75% of British Merlins, 71% of Swedish Goshawks, 62% of German Buzzards and 43% of Fennoscandian Ospreys of breeding age (Table 39). Using data from the European and Soviet ringing schemes, Galushin (1974) calculated the mean distance between birthplace and presumed breeding place for eleven raptor species (Table 40). This distance was much greater in some of the irruptive species, which depended on sporadic food sources, than in the others. The most extreme was the Rough-legged Buzzard, with a mean dispersal distance of nearly 2,000 km (number of recoveries was not given).

Within species, geographical trends occurred which accorded with regional differences in the stability of food sources. Mean distances between birthplace and breeding place were greater for the Kestrel (277 ± 57 km) and Common Buzzard (295 ± 105 km) in northern and eastern Europe than they were in western and central Europe (146 ± 59 and 60 ± 11 km), and greater still than they were in Britain (Table 39). Among Goshawks in northern Europe, movements were longer in poor food years than in good ones (Haukioja & Haukioja 1970). Presumably birds had to search further, on average, in poor food years before they found areas with sufficient food (Chapter 11).

In two species, dispersal between birthplace and breeding place was examined separately for each sex, and in both species females moved furthest. In the Sparrowhawk this was evident from the British ring recoveries, in which two-thirds of males had moved less than 10 km from their birthplace, and two-thirds of females had moved more than 10 km (Table 41). In the Hobby it was evident from a local study near Berlin, in which 85% of 180 male Hobbies seen breeding had been raised in the area, compared with only 11% of 174 females (P<0·001). As the sex ratio among nestlings ringed in the area was about equal, this difference implied a much greater fidelity to birthplace among males than among females (it was too great to be due to differential mortality) (Fiuczynski 1978). Such sex differences in dispersal may occur because males and females are exposed to different ecological conditions; or they may be inherent, perhaps again serving to reduce inbreeding. Thus despite the many intensive studies that have been made of colour-ringed populations of other birds, hardly any brother-sister or parent-child matings have been recorded, and when they did occur, they were less productive than matings between unrelated individuals (Lack 1954, Greenwood et al 1978).

FIDELITY OF ADULTS TO PREVIOUS NESTING AREAS

Because most raptor species are found nesting in the same places

9 (Upper) Rough-legged Buzzard (male) at nest in Norway. Breeding success varies greatly from year to year according to the supply of voles and lemmings which form the prey. Photo: A. N. H. Peach. (Lower) Immature Steppe Eagles feeding on emerging termites in East Africa. Groups of wintering eagles often feed in this way and move from area to area with the rainbelts which bring out the termites. Photo: W. Leuthold.

20   *Golden Eagles in W. Scotland. Female at nest with chick (upper); bringing fresh greenery to nest with young (lower). Photos: C. E. Palmar.*

year after year, it has often been assumed that individual adults are faithful both to their nesting territories and to their mates in successive years. This view gained credence from the fact that certain individuals, which were recognised by some peculiarity of behaviour, plumage or egg-type, occupied particular territories for long periods. There are many examples for Peregrines and some for African eagles (eg Herbert and Herbert 1965, Brown 1966, 1972). However, studies with ringed birds, in which identification was certain, have revealed wide differences in these respects between populations. Such studies were made by biologists working in restricted localities. They entailed trapping and marking the occupants of certain territories, then checking for the presence of these birds in future years. They tell us about the turnover of birds at particular territories and about shifts between territories within study areas, but of course they miss any birds that might have moved outside study areas between one year and the next. Data from national ringing schemes have so far provided nothing of value on this aspect, chiefly because most birds were recovered only once, usually dead.

Studies with marked birds are so few that they can be described individually. Ten pairs of Greater Kestrels remained on their territories over a 3-year study period in the Transvaal, with only one replacement of a male which died (Kemp 1978). Thus these birds showed not only great fidelity to territory and mate, but also exceptionally high survival. Among Peregrines in south Scotland, two males and 12 females were trapped on the same territory in more than one year, and only one female had changed territories between years. This entailed a movement of 29 km over three intervening territories (R. Mearns). Fidelity to territory was also noted in individual Hobbies near Berlin (Fiuczynski 1978), and in various *Buteo* species. In *B. galapagoensis* on the Galapagos Islands, one pair stayed together on the same territory for five years (de Vries 1975). Four migrant Broad-winged Hawks *B. platypterus*, two of which were mates, returned to the same territories in eastern North America in consecutive years (Matray 1974).

More extensive data for the European Sparrowhawk, revealed two trends (Table 42). First, in habitats rich in food females remained on the same territory from one year to the next, but they often changed territories in habitats poor in food. Second, they more often changed territories after failed breeding attempts than after successful ones. Males seemed to behave in the same way, but too few were re-trapped to be sure. In the poor habitats, it was rare for pairs to remain together between years, and in the few instances recorded, the pair stayed on the same territory. In the area concerned, some territories were closer than one kilometre, but many females moved up to five kilometres between years, and some up to ten kilometres, with one at 27 km (Table 38). Recoveries showed the same exponential fall-off as those of birds breeding for the first time, but over shorter distances. Evidently, fidelity to a previous nesting area was much stronger than fidelity to a natal area.

Among Merlins in Alberta, males more often returned to the same nesting territory than did females. Of 12 adult males re-trapped, nine were on the territory where they had previously bred, two were on territories less than four kilometres away, and the other was 12 km away. But of ten adult females, only two were on the same territory, three were on territories less than 15 km away, three were 15–30 km away, while two were more than 100 km away (Hodson 1975). These were generally greater distances than recorded in Scottish Sparrowhawk females but, unlike the Sparrowhawks, Merlins were migratory in this region.

Results on two other species deal with the return of birds to study areas rather than to specific nesting territories. Among Hen Harriers in Wisconsin, only 28% of birds marked as breeding adults were known to return to a 16,000 ha area in later years, and only one female was found to mate with the same male more than once, but she had also mated with another male in the interim (Hamerstrom 1969). Birds that bred successfully in one year were more likely to return than birds that failed, but this might have been due to differential mortality, rather than to any enhanced tendency to return after a success. Broadly similar findings were made on the same species in Orkney, where one female occupied the same nesting territory for six years, while another used five different territories in the same period (Balfour 1962). Likewise, in northeast Scotland, four out of five marked females returned to a study area in successive years, but only one kept to the same nesting territory (Picozzi 1978).

In the Kestrel, ringed adults were often present in the same locality in successive breeding seasons, but not necessarily in the same nest-boxes (which were close together). Fidelity to the area from year to year was as high as 70% when numbers were on the increase, and as low as 10% when they were on the decline. These trends were in turn related to vole densities, and fidelity was lowest in the poor vole years (Cavé 1968).

The habit of breeding in the same territory year after year is probably advantageous, so long as it is a good territory, because birds are likely to be most successful in places with which they are familiar. In view of the relationship to food conditions noted in European Sparrowhawks and Kestrels, however, even species which on present evidence show great fidelity to nesting territory might be expected to behave differently in more extreme conditions. As a further speculation, year-round residence on territories may be more conducive to constancy in partners from year to year than in the case of seasonal residence, when the partners have to rely on meeting again each spring. Thus change of mate among migrants may sometimes be due to divergence in arrival times, leading the first arrived partner to take a new mate before last year's mate returns. Also, long-lived species may in general show more fidelity to territory and mate than short-lived ones, and non-irruptive species may show more fidelity than irruptive ones, but to check these aspects more studies are needed.

FIDELITY TO PREVIOUS WINTERING AREAS

The fact that some raptor populations vary in distribution from one winter to the next according to prey numbers implies that many migrant individuals are not faithful to particular wintering places (Chapter 4). The same individuals may migrate as juveniles but not as adults, or they may migrate in some years but not in others. Nonetheless, there are a few records of migrant raptors on the same wintering range in successive years. They refer to Steppe Buzzards in South Africa, Rough-legged Buzzards in south Sweden, and Common Buzzards in Europe (Siegfried, in Moreau 1972, Sylvén 1979, Olsson 1958). One white Buzzard was seen in the same place for eight consecutive winters, and another distinctive bird for 12 winters (de Bont 1952, Schuster 1940). Four out of 21 marked American Kestrels returned to the same places in two winters, and four out of 27 Prairie Falcons returned to a study area, but other individuals in these studies might well have moved elsewhere (Mills 1975, Enderson 1964). Clearly, much more information is needed on the fidelity or otherwise of migrants to wintering areas.

SUMMARY

Raptors tend to settle and breed in the neighbourhood where they were born and, in the species studied, ring recoveries (and hence dispersed birds) declined exponentially in successive circles out from the birthplace. Within species, geographical or annual differences in dispersal distances also occurred, probably dependent on the distribution of good food sources and of vacant territories. Individuals in some irruptive populations tended to disperse further than those in other populations. In Sparrowhawks and Hobbies, females dispersed further on average than did males.

Among the adults, considerable variation in behaviour was found, between species and within species, according to conditions. In the Peregrine and some other species, individuals used the same territories in successive years. In the European Sparrowhawk and Kestrel, high fidelity to a previous nesting place was associated with good food conditions, and a tendency to move with poor food conditions. In the Merlin, females more often changed territory than did males. Fidelity to a wintering area in successive years is probably much less marked than fidelity to breeding area. The few records refer to various buzzards and falcons.

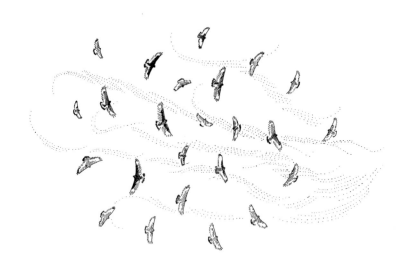

CHAPTER 11

Movements

Movements provide an important means by which birds adjust their local densities to periodic changes in food. For convenience, they can be divided into the following types.

First, there are the dispersal movements discussed in the previous chapter. Even in populations which live throughout the year in the same general area, the young disperse in various directions from their natal to their subsequent breeding locality. In some species the young move further in years of food scarcity than in years of plenty, presumably because the young have greater difficulty at such times in finding an area with sufficient food to support them (Chapter 10).

Second, there are regular local movements, by which birds may spread from high ground to low ground after breeding, or from a restricted to a wider range of habitats, and back again for the next breeding season.

Third, there is migration, which can be described as a massive shift of birds twice each year between regular breeding and wintering ranges. Compared with local movements, migration usually involves a longer journey in a more restricted direction. But both local and migratory movements occur primarily in response to seasonal changes in food. The latter are in turn caused by the alternation of warm and cold seasons in temperate regions, or of wet and dry seasons in the tropics.

Illustration: Buzzards on the move

Fourth, there are invasion migrations (or irruptions), in which the proportion of birds leaving the breeding range, and the distance they travel, varies greatly from year to year (the directions are the same). Noted mainly in northern regions, such movements occur in response to annual, as well as to seasonal, fluctuations in food.

Fifth, there is nomadism, in which birds drift from one area to another, residing for a time wherever food is temporarily plentiful. The areas successively occupied may lie in various directions from one another, and no one area is necessarily used each year. This kind of movement occurs among certain birds (including raptors) that live in some Australian and southern African deserts, where the infrequent and sporadic rainfall leads to unpredictable changes in food supply. In years of widespread drought, many species (again including raptors) move from the dry centre of Australia towards the edges (Glover 1952, Serventy 1953).

These different kinds of movements grade into one another, but in any given population, one kind usually prevails. Movements have been studied by observations (made directly or with radar); by bird counts made at particular places in different seasons; by use of ring recoveries; or in recent years by the use of radio transmitters to follow individual birds on their journeys. Watching hawks on migration has become a favourite pastime for hundreds of birdwatchers, and in many countries the concentration points are now well known.

Migration is the most spectacular of bird movements. Involving seasonal shifts of millions of individuals, it produces twice each year a massive re-distribution of birds over the world's surface. Some regions, such as the Arctic, support certain species only in the breeding season, while others, such as parts of the tropics, provide a temporary home for species which breed elsewhere. In temperate regions, some species leave for the winter, and others come in. For individual species, the summer and winter ranges may be overlapping or separated geo-graphically, and one or the other may be greater in area (Figure 27).

Nearly every raptor species that has been studied performs some kind of migratory movement in at least part of its range. The longest journeys are made by those populations which fly regularly between eastern Siberia and southern Africa (eg Eastern Red-footed Falcons), or between northern North America and southern South America (eg tundra-breeding Peregrines) (Figure 28). Even on the shortest routes, this entails some individuals flying more than 30,000 km on migration each year. Journeys almost as long are made by the Western Red-footed Falcons, Lesser Kestrels and Steppe Buzzards that travel between western Siberia and southern Africa, and by the Swainson's Hawks that travel between Alaska and Argentina. Long sea crossings are made by some species, for example by the Eastern Red-footed Falcons (India to southern Africa) and by the Lesser Sparrowhawk *Accipiter gularis* and Frog Hawk *A. soloensis* (Japan to East Indies). Long desert crossings are made by the many species that travel between Eurasia and tropical Africa.

The diversity of migration is also remarkable. It is not just the north-south affair that many people think of. East-west movements are common in some species, for example in the Prairie Falcons which breed in the mountains of western North America, and each autumn

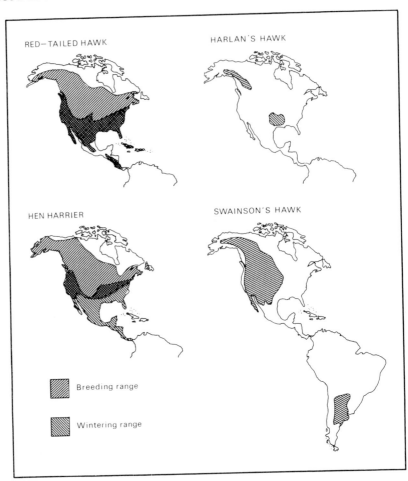

Figure 27. Breeding ranges (shading right to left) and wintering ranges (shading left to right) of certain raptors to indicate variations in migration patterns.
    (a) Red-tailed Hawk – population spread over a larger area in summer than in winter, with no obvious southwards extension in winter.
    (b) Hen Harrier – extensive breeding and wintering ranges, overlapping in the middle.
    (c) Harlan's Hawk – separated and restricted breeding and wintering ranges on the same continent.
    (d) Swainson's Hawk – separated and extensive breeding and wintering ranges on different continents.

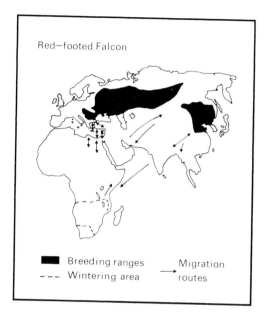

Figure 28. Breeding and wintering ranges, and migration routes of Red-footed Falcons. Partly after Brown 1976.

spread eastwards across the prairies (Enderson 1964). East-west movements also occur in some tropical species, as rainbelts sweep across Africa, changing prey availability as they go (Brown 1970, Moreau 1972). The Bald Eagles of Florida show another unexpected pattern: the young are raised in winter, then move northwards and northwest for up to 2,200 km spending from May to September in the northern United States and Canada (Figure 29). They therefore travel north in spring and south in autumn with the conventional migrants but, unlike them, they have been reared in the south before doing so. Most of the adults also leave Florida for a time, but it is not clear from ringing whether they make the same journey as the juveniles.

In temperate regions, some raptors disperse in various directions from the nest, including northward, immediately after the breeding season, before moving south again on their autumn migration. Such movements have been noted, for example, among Kestrels, Buzzards and Ospreys in Europe (Schifferli 1965, Olsson 1958, Österlöf 1977). They perhaps give the young some knowledge of the surrounding area, which may be of use to them when they return to breed in a later spring.

MIGRATION IN RELATION TO FOOD

Migration might be expected to occur in those populations which survive in greater numbers if they leave their breeding areas for the non-breeding season than if they remain there for the whole year (Lack 1954). The usual reason why breeding areas become unsuitable during part of the year is lack of food. Such food shortages occur for raptors because many avian prey species migrate from the area for part of the year, while other kinds of prey hibernate or become hidden under snow. Hence the purpose of the autumn exodus from northern regions is fairly obvious. The reason why birds leave their wintering areas to return in spring is less obvious, because many such areas seem able to support the birds during the rest of the year. But if no birds migrated north in spring, the northern lands would be almost empty of many species, and a large seasonal surplus of food would go largely unexploited. Under these conditions, it is easy to appreciate that any individuals which moved north to an area of abundance might be better able to raise young than if they stayed in the south and competed with the species resident there. The migratory habit can thus be regarded as a product of natural selection, ensuring in the long term that individuls adopt whatever pattern leads them to survive and breed most effectively.

The role of food supplies in controlling the movements of raptors is evident from: (1) the regional and annual variations in the proportions of birds that leave for the non-breeding season; (2) the locations of wintering areas; and (3) the timing of movements. Other factors are involved to a small extent, as will be discussed.

1. Regional variations in proportions of birds leaving

Like many other widespread bird species in the northern hemisphere, many raptors are completely migratory in the north of their breeding range and completely sedentary in the south, while in the range in between some individuals leave and others stay (partial migration). In general, therefore, the extent to which any population migrates for the winter broadly corresponds to the degree of reduction in prey supplies. Moreover, in any one population, more birds stay in years when prey is plentiful than in years when prey is scarce (Chapter 4); in mild winters than in cold ones; or in good habitats than in poor ones; again implying an influence of food on the numbers that stay or leave (Craighead & Craighead 1956, Mebs 1964, Thiollay 1978).

In partial migrants, it is common to find that a greater proportion of adults than of juveniles stay behind, and more of one sex than of the other. In some falcons, the males stay in greatest numbers, and in accipiters the females (Platt 1976, other references in Table 3). In Goshawks, this sex difference is much more apparent in juveniles than in older birds (Mueller et al 1977). These trends are evident from counts on breeding, migration and wintering areas, and also from ring recoveries.

2. Annual variations in movements

In some temperate zone raptors the distance moved varies from year

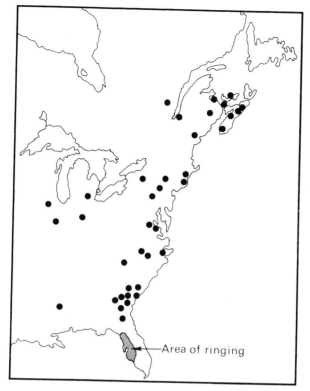

Figure 29. Northward migration of Bald Eagles raised in Florida, as shown by ring recoveries. Shading shows ringing area. From Broley 1947.

to year, with the majority of individuals staying in the north of the wintering range in years when prey are plentiful there, and moving further south in years when prey are scarce. This is most pronounced in raptors that depend on fluctuating (cyclic) prey. Their so-called 'invasion movements', in which every so often they appear in large numbers well south of their usual range, follow periodic crashes in their prey. The effect of food shortage is accentuated because the raptors themselves tend to be numerous at such times, as a result of good breeding and survival in the previous few years, when prey were abundant (Lack 1954, Keith 1963, Sulkava 1964).

The main invasions by the Rough-legged Buzzard occur at about 4-year intervals, whereas those by the Goshawk in North America occur roughly every ten years, corresponding with the 4-year and 10-year prey cycles respectively (Tables 43 and 44). Invasions may occur in only one autumn, or in two successive ones if prey remains sparse. Invasions by Rough-legged Buzzards often occur in the same years as those by Snowy Owls *Nyctea scandiaca* (both species eat rodents), while in parts of North America the invasions by Goshawks show a similar periodicity to those by Horned Owls (both eat hares and grouse). In neither the Old World nor the New are the lows in prey populations synchronised over the whole range, so invasions by any one species tend to come in different years in different regions. Those by the Rough-legged Buzzard and Snowy Owl are also more marked in North America than in Europe, presumably because the birds are more numerous in North America, having a greater area of breeding habitat (tundra) on that continent.

In Goshawks the big invasions often contain relatively more adults and fewer juveniles than the smaller migrations of other years. The many migrants captured in Wisconsin during 1962–63 and 1972–73 included more than 50% adults, whereas of the fewer migrants caught in other years adults comprised less than 20%. This may be partly because fewer adults wintered in the north in poor food years, and because breeding in such years was generally poor. The 1972–73 invasion was also earlier than usual, and was the only year out of 25 in which females outnumbered males, instead of the other way round. Perhaps in this extreme year, the proportion of males in the overall population had been reduced by differential mortality, itself the result of competition with larger females for an extremely limited food source (Mueller *et al* 1977).

To judge from the annual differences in migration, many individual raptors must winter in widely separated areas in different years, as discussed in Chapters 4 and 10. On their breeding grounds, too, some rodent-eating species concentrate to breed wherever food is plentiful at the time, so that their local populations fluctuate greatly from year to year (Galushin 1974). This implies that at least some individuals breed (as well as winter) in widely separated areas in different years. So far, however, ring recoveries lend no support to this idea because they nearly all refer to birds handled only once as adults. In all their migratory behaviour the raptors show parallels with certain seed-eating birds that depend on fluctuating tree-seed crops (Newton 1972a).

3. Location of wintering areas

Like the migratory habit itself, the location of the wintering areas can be explained in terms of natural selection, the birds from each population wintering in whichever region they can reach and survive best in. Suppose that the birds from a certain breeding area have heritable tendencies to fly in particular directions at migration time and back again in spring, but that these directions differ from bird to bird. Some birds will then reach suitable areas and many will survive to breed again, others will reach less suitable areas and fewer will survive, and yet others will reach unsuitable areas and die. Thus those with the most rewarding directional tendencies will perpetuate themselves, and in this way the migratory habits of a population could become fixed. To judge from ring recoveries, most populations migrate in restricted directions to their respective wintering areas, but among the Kestrels that breed in Britain the directions show more spread, with some recoveries to the west in Ireland and others to the south-southeast in Europe; there are several examples of brood-mates migrating in different directions (Snow 1968). Only in populations (like some truly nomadic ones), which on balance are as likely to find food in one direction as in any other, is no directional preference likely to become fixed by natural selection.

(a) Wintering areas of different species. The forty raptor species that breed in the western Palaearctic fall into three groups: (i) those that stay within Europe, including the south side of the Mediterranean, 17 species; (ii) those that winter wholly (or almost wholly) within Africa south of the Sahara, 13 species; and (iii) those that winter partly in southern Europe and partly in Africa, 10 species (Table 45). These divisions reflect feeding habits, and the extent to which particular prey remain available at high latitudes in winter. Species that stay entirely within Europe or that winter partly in Europe and partly in Africa have diets based on warm-blooded prey – on birds and mammals, obtained alive or as carrion. Most of those that winter entirely within Africa have diets based on cold-blooded prey, which are inactive or otherwise unavailable in the northern winter. Such raptors include six insect-eaters (five small falcons and Honey Buzzard), two reptile-eaters (an accipiter and a snake eagle), one fish-eater (Osprey) and three generalists feeding mainly on cold-blooded prey (one harrier, one eagle and one kite). The Egyptian Vulture and Booted Eagle also winter in Africa, despite more varied diets, but as their food has not been studied in detail, it may include more cold-blooded prey than textbooks suggest. From further east in the Palaearctic, where winters are harsher and warm areas more restricted, several other raptor species enter Africa in strength, for example the Peregrine.

This system is closely paralleled in the New World, where again it is the species that depend largely on cold-blooded prey that withdraw most completely from Canada and the United States in autumn, namely the Osprey, Mississippi Kite, Broad-winged Hawk and the Swainson's Hawk (which feeds more on insects in winter than on mammals, its other main prey, Bent 1938). Hence, on both continents, the migrations and wintering areas of raptors are as might be predicted from their feeding habits.

Sometimes, it seems, the winter distribution of one species is influenced by that of similar species. Nielsen & Christensen (1970) have pointed out that the four migrant eagles from Europe can be considered as two species-pairs, and that in each case the larger species winters north of the smaller, with the division in the eastern half of Africa. In the first pair, the Imperial Eagle stays mainly north of Ethiopia, while the smaller Steppe Eagle winters mostly further south. In the second pair, the Greater Spotted Eagle stays mainly north of Ethiopia, while the Lesser Spotted Eagle winters further south. Thus each pair together fits the trend known within species as 'Bergman's

Rule', for the larger-bodied forms to occur at higher latitudes than smaller-bodied forms. As another example of closely related species occupying mainly different areas, four small falcons from the Palaearctic live almost entirely on insects while in southern Africa, namely the Lesser Kestrel, Hobby and the Eastern and Western Red-footed Falcons. The individual breeding ranges of these species total some 17 million km^2, yet in Africa they are all concentrated within about three million km^2. They overlap extensively, but each has 'a geographical area of maximum density which is distinct from the others'. They in turn differ in main area from a fifth small falcon from the Palaearctic, the Common Kestrel, which occurs further north and eats a mixture of insects and rodents (Rudebeck 1963, Cade 1969, Moreau 1972). The other two insectivorous falcons from the Palaearctic, the Eleonora's and Sooty Falcons, winter in equivalent habitat in Madagascar, but mainly on opposite sides of the island. Most of these small falcons have fairly similar feeding habits, so it is perhaps competition for food between the species that leads them to occupy mainly different areas.

(b) Wintering areas of separate populations of the same species. Within each species, the birds from different parts of the breeding range usually migrate to different parts of the wintering range. Populations that breed furthest west tend to winter furthest west, and vice versa. Allowing for the uneven distribution of land masses, this holds in the Old World both in Europe and in Africa, reflecting the more or less parallel migration of populations, revealed by ring recoveries (Moreau 1972, Österlöf 1977). Superimposed is another trend, for the northernmost breeding populations to winter furthest south, passing over the intermediate populations. The Peregrines that breed furthest north in North America migrate furthest south in South America, while those which breed in intervening areas migrate much shorter distances or, nearer the tropics, not at all (Enderson 1965, White 1968). Likewise, in the Old World it is the Peregrines breeding on the tundras of west Siberia that make the longest journey to southern Africa (Vaurie 1965). The same trend is shown on a more modest scale even by the Kestrels breeding in Britain, as those from the south of the country reach northern France, while those from the north reach southern France and Spain (Mead 1973). Similar 'leap-frog' migrations occur in Buzzards and Turkey Vultures and may be widespread in raptors, as in some other birds (Salomonsen 1955, Moreau 1972, Stewart 1977). They presumably arise because populations breeding progressively further north have to migrate ever further south to find unoccupied wintering areas.

(c) Wintering areas of different sectors of a population. In many raptors, as already indicated, the juveniles migrate further than the adults to winter in partly different areas. Among Palaearctic Steppe Eagles wintering in Africa, most adults stay in the northeast, but many juveniles fly further south (Brooke et al 1972, Smeenk 1974, Steyn

1973). Among Sparrowhawks from Denmark, the older birds winter further north than the first-year and second-year birds (Schelde 1960). Sex differences occur among Sparrowhawks in eastern Europe, males moving further than females, with the result that the winter ranges of the sexes overlap but do not exactly coincide (Belopolskij 1971). A similar sex difference was noted in the dispersal movements of Goshawks in northern Europe (means of 70 km for males and 40 km for females, $P<0.005$), but in this species the distances moved varied greatly from year to year, depending on food (Sulkava 1964, Haukioja & Haukioja 1970). Such age and sex differences do not occur in all raptors, however, and juvenile Ospreys are found no further south on African wintering areas than are adults of their population (Österlöf 1977).

(d) *Changes in wintering range.* At least two raptors have changed their wintering habits as conditions have changed. Since the nineteen-fifties Red Kites have overwintered in increasing numbers in several parts of Europe from which they formerly had migrated. Reports have come from places as far north as southern Sweden, and from France across Switzerland and Germany. The birds are usually associated with rubbish dumps, and in some localities include many adults as well as immatures, and in other localities almost entirely immatures (refs. in Juillard 1977). On the other side of the Atlantic, small numbers of Swainson's Hawks have begun to winter in Texas, southern Florida and a few other areas in the southern States, while the rest of the population passes on to traditional wintering areas in southern South America (Browning 1974).

4. Timing of movements

Movements are closely tied to the seasonal changes in prey, and in general the adults seem to spend as long on their breeding areas each year as food permits, thus gaining the maximum time for breeding. In any one breeding area the species that depend on cold-blooded prey are normally resident for a shorter period each year than are those that depend on warm-blooded prey. This is because of a more restricted period of prey availability for the former species rather than the longer journeys they have to make. For any one population, arrival in and departure from the breeding areas often coincides precisely with the major re-appearance and disappearance of food. For example, Prairie Falcons return to their breeding areas in northern Idaho in January/February, when ground squirrels emerge in numbers. After breeding, the falcons leave the area again in June/July, when the squirrels disappear below ground for the hot dry period.

Raptors that eat birds often migrate at the same time as their prey. The precise relationship between the passage of Sparrowhawks over the Courland Spit in the Baltic and the passage of their various finch prey is shown in Figure 30. At this locality, annual variations in the timing and duration of Sparrowhawk passage were correlated with similar variations in the prey passage, and extra Sparrowhawk move-

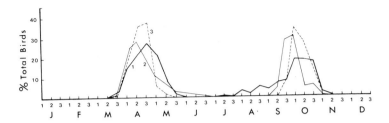

Figure 30. Migration of Sparrowhawks over the Courland Spit in the Baltic in relation to the migrations of their main prey-species. Shown as a percentage of all birds trapped in spring and autumn respectively. (1) Thick line – Sparrowhawk; (2) thin line – Chaffinch Fringilla coelebs; (3) broken line – Brambling F. montifringilla. Re-drawn from Beloposkij 1971.

ments in some years were associated with extra movements by Cross-bills Loxia curvirostra (Belopolskij 1971). This suggests that at least some of the hawk movements were stimulated by prey movements, rather than both being stimulated by the same weather. A correspondence between the migrations of Sharp-shinned Hawks, Gyr Falcons and Peregrines and their avian prey species has been noted in North America.

As with other aspects of migration, age and sex differences occur in the timing of passage. In long-distance migrants, in which the whole population leaves for the winter, it is sometimes the adults that depart first in autumn, travel faster, and reach their wintering grounds first. In Black Kites from Europe, the adults leave 3–4 weeks before the juveniles and in Ospreys the adults leave two weeks earlier; in Peregrines from arctic North America the adults pass various observation points earlier, and linger less than the juveniles; and in Steppe Buzzards from Siberia, the adults are first to arrive in South Africa and first to leave again for breeding areas (Schifferli 1967, Hunt et al 1975, Broekhuysen & Siegfried 1970). On the other hand, in short distance partial migrants it is usually the juveniles – the cohort that leaves in greatest numbers – that are first to depart in autumn. Among Sharp-shinned Hawks and Goshawks in Wisconsin, the peak in juveniles on autumn migration occurs about a month before the peak in adults (Mueller & Berger 1967, Mueller et al 1977). In Sparrowhawks migrating over Heligoland (Germany), the median dates of autumn passage for the different age and sex classes (and the numbers trapped) were as follows: juvenile male, 2 October (203); juvenile female, 11 October (121); adult male, 19 October (29); and adult female, 25 October (20). In spring the reverse occurred, with the adults preceding the immatures: adult female (only one bird caught); adult male, 8 April (6); immature female, 11 April (11); immature male, 26 April (14). Assuming these dates reflected migration, the adults spent longer on their breeding areas than immatures, and females longer than males (Moritz & Vauk 1976).

Discussion

While the ultimate reason for birds leaving their breeding areas for part of the year is food shortage there, this need not always provide the immediate stimulus for departure. Songbirds are thought to react to proximate environmental factors in such a way that they leave their breeding areas before food becomes scarce and while there is still time for them to lay down fat for the journey. This situation presents a parallel with the timing of breeding, discussed in Chapter 6.

The proximate factors stimulating migration in raptors have not been studied, but they may differ between complete, long-distance migrants and partial short-distance ones. In the former, individuals may respond to daylength changes and leave before food becomes scarce; the adults may depart first because they are able to reach the necessary body condition sooner than the recently-fledged juveniles, and may then migrate faster because they know the route. But in those partial migrants, in which the proportion of birds leaving varies with circumstance, individuals presumably respond to food conditions at the time (possibly via changes in their body composition, Chapter 6). Only in this way could the variations from year to year, from one habitat to another, and from one age or sex group to another, be explained. For such partial migrants, food shortage would then act as both an ultimate and a proximate factor in influencing whether any particular individual migrates; it might also influence the date of migration and the length of journey.

This does not mean that birds wait until they are starving before they move on; they might leave if they were having difficulty in getting food or living space in competition with other birds. As winter approaches and competition increases, adults should in general fare better than immatures, thus forcing more of the immatures to move away. Within age groups, one might expect the larger females to be dominant over males (if conflicts occur); but whereas in some species the females stay in greater numbers, in others the males stay. Factors other than food and dominance relationships could be operating here, however. First, whether males or females stay seems to correlate with which sex is most attached to the territory, apparently the male in falcons and the female in accipiters (Chapter 9). Second, in highly dimorphic species, in which the sexes take different foods, whether one or other sex leaves may depend on the availability of their respective prey species. Neither of these suggestions is based on much information, so both are tentative.

In complete migrants and in partial migrants, it is the adults that are first to return in spring; presumably adults in each group are under the same pressure then, to establish a nesting territory as soon as possible. This is less of a problem for young birds because they may not be breeding or, if they are, they may benefit by waiting until the adults are already established rather than arriving early, taking a territory and risking being ousted at a later date (young birds ousted by older ones has been noted in Marsh Harriers, H. E. Axell). Also, whatever the best time for breeding, most young birds may be obliged to return later than adults because of an inability to achieve the necessary body condition in time. For this suggestion, however, I know of no evidence one way or the other.

The role of inheritance in migration needs clarifying. In populations which are completely migratory, it has usually been assumed that all individuals are inherently programmed to migrate, though the precise timing of movement may vary with conditions. In populations which are partly migratory, there are two possibilities: that some individuals are programmed to migrate and others are not; or that the same individuals are programmed to migrate in certain conditions and not in others. The two possibilities are not mutually exclusive, but as discussed above, the second possibility best accounts for the observed variations in raptor migration.

Summarising, raptors like other birds move so as to keep themselves in optimal habitat for as much of the year as possible. Regional, annual and seasonal variations in movement patterns can all be broadly linked with variations in feeding conditions, as can the location of wintering areas. In this last respect, the distributions of other populations are apparently influential (probably through competition for food), as are the relationships and experience of individuals within populations.

ACCOMMODATION IN WINTER QUARTERS

The general southward withdrawal towards the tropics means that raptors are much more concentrated in winter than in summer. In the African savannahs and grasslands, moreover, the Palaearctic migrants during their sojourn greatly outnumber the residents. An important question is how all these birds are accommodated, more densely than in their breeding areas, in lands already richly endowed with raptors of their own. The immigrant species do not distribute themselves evenly over the continent; only the Honey Buzzard enters closed woodlands, whereas nearly all the remaining species are found on savannahs and grasslands where they tend to segregate, either geographically or ecologically, from the majority of African species (Thiollay 1978).

In Africa, the major strategy of many immigrant populations is itinerancy in relation to local flushes of food, as provided by termite emergences, locust swarms, *Quelea* colonies or the insects and other small creatures disturbed by bush fires. Termites are eaten not only by the insectivorous falcons, but also by the large Steppe and Lesser Spotted Eagles, European Black Kites and others (Chapter 2). *Quelea* colonies and their associations of other breeding birds are exploited especially by Steppe Eagles and Lanner Falcons. Uncommitted to a nesting locality, the visitors have much greater freedom of movement than the residents, many of which remain on territories all year, or are actually breeding at the time (Brown 1970, Moreau 1972). This limits the effectiveness of the residents in exploiting transient food supplies,

21   *Black Eagle at nest. This eagle breeds at unusually high density in the Matopos Hills, Zimbabwe Rhodesia, notable for the great abundance of hyraxes which form the eagle's prey. Photo: Peter Steyn.*

22 Black Eagle female at nest (upper); and sibling aggression (lower). This is one of several large rapto
species in which the first-hatched young attacks and usually kills the second in what has become known as the
'Cain-Abel' conflict. Photos: Peter Steyn.

and hence their ability to compete for them with the migrants. The movements of the migrants are far from random, however. Some species seem to follow rainbelts, which drift across the continent and induce changes in prey populations. Such raptors may thus be on the move for much of the time between leaving their breeding areas in one year and returning there the next, on a long circuitous journey, more or less repeated year after year. In a sense they ride the crest of a wave, continually shifting from one temporary abundance of food to another.

For some other species, too, the situation seems fairly clear. The Honey Buzzard in the woodlands probably has no native competitor for its staple diet of Hymenoptera, and south of the equator it arrives at the time of the seasonal flush. The Osprey as a breeder is extremely scarce and local in Africa, and as a migrant is largely coastal, thus avoiding to some extent the resident Fish Eagle which occurs mainly inland. Other species, such as the four Palaearctic harriers, meet practically no opposition in Africa. In the north and west there is no resident harrier, while in the south and east the two resident species are so restricted in distribution that the immigrant Pallid and Montagu's Harriers have the drier grasslands almost to themselves (Brown 1970). On the other hand, the various immigrant harriers overlap extensively with one another in habitat and food, and often share the same roosts. Exploiting superabundant resources may reduce the need for them to segregate.

Cliff-nesters such as Peregrines, apparently avoid competing with the residents, mainly by occupying areas devoid of cliffs, including coastal flats. Non-breeders of local stock may also occupy such areas, but the two populations are at least partly separated. The same applies to the immigrant Steppe Buzzards, which in southern Africa occur in open featureless terrain, thus avoiding the local Jackal Buzzard *B. rufofuscus* which at that season occupies the hills, and the Mountain Buzzard *B. oreophilus* which is confined to forest. Similarly the Palaearctic Snake Eagles occupy the dry grasslands of the Sahel zone, just south of the Sahara, and mainly north of the various related African forms (P. Ward). The Egyptian Vultures from Europe also occur in these dry grasslands. They avoid competing with their nearest relative, the Hooded Vulture *Necrosyrtes monachus*, both by occupying more open ground (away from trees) and by exploiting different food-sources. In conclusion, the Palaearctic migrants maintain themselves in Africa by continually moving to exploit temporary food sources, by occupying areas largely free of competitors, or by a combination of both strategies.

Much less is known about the New World migration system, so direct comparisons with the Old World are difficult. However, the New World migrants meet some suitable habitat in Central America, instead of a large, mainly lifeless desert, which occupies the comparable position in Africa. The New World species may thus often be faced with a shorter and less arduous journey. They also seem to make more use of the forests than do the Palaearctic raptors in Africa, and only the Swainson's Hawks and some Peregrines are known to extend south of the tropic of Capricorn. The literature known to me contains nothing on how the migrants avoid competing with the residents, but in some species itinerancy is probably involved, and in others not, as in Africa.

MIGRATIONS OF AFRICAN SPECIES

Among the African raptors themselves, regular migrations are most apparent in seasonal environments, namely the savannahs and grasslands as opposed to the evergreen forests. Some species migrate so completely as to be totally absent from large parts of their range for several months each year, while others are partial migrants. The amount and timing of movement seems to be related to rainfall (and thus to prey availability) and, as in the temperate zone, the appearance or disappearance of a species in a given locality often fits closely with local conditions.

In West Africa, migrants generally stay in the southern woodlands during the local dry season, while food is plentiful and hunting conditions are good. When it rains heavily there and the grass grows rapidly to 1–3 m high, the migrants move north to the short grass areas, where the rains are later and lighter and produce less growth of vegetation (Figure 31). The birds thus manage to remain in a fairly favourable environment all year and, while in the north, they take advantage of a short seasonal surplus of food which is not fully utilised by the sparse resident population (Thiollay 1978). Different migrant species spend different periods in the north, depending on their particular needs, but the general northward passage occurs at the start of the rains in 'spring', and the southward one at the end of the rains in 'autumn' (Figure 31). Complete migrants include the African Black Kite, Grasshopper Buzzard Eagle *Butastur rufipennis*, Red-tailed Buzzard *Buteo auguralis* and Shikra; among partial migrants are several large vultures, Lanner Falcon, Chanting Goshawks *Melierax metabates* and *M. gabar*, Cuckoo Falcon *Aviceda cuculoides*, Harrier-hawk *Polyboroides typus*, Secretary Bird and many others (Elgood *et al* 1973, Thiollay 1978, P. Ward).

The Black Kite is scarce or absent between May/June and August/September from much of its African breeding range north and south of the equator; and Wahlberg's Eagle is absent in the same months from its extensive breeding range south of the equator (Brown 1970). The off-season ranges of these species lie well north of the equator, but are incompletely known. Several other species perform seasonal movements elsewhere in Africa, and yet others which are at present considered as resident may be found to be partly migratory once they are better known.

MIGRATION STREAMS

All raptors migrate by day but, while some species progress mainly

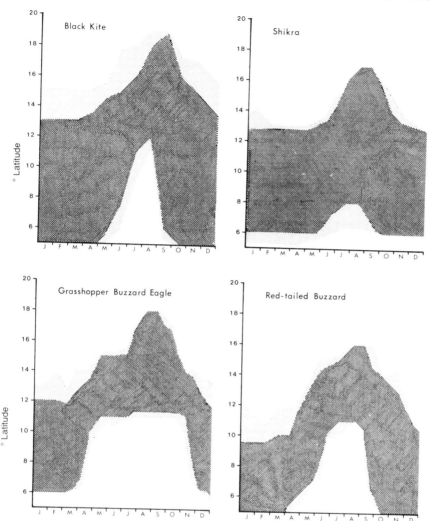

Figure 31. Seasonal changes in the distribution of four migrant raptors in West Africa. Dark shading – frequent sightings; light shading – infrequent sightings. Species differed in the extent of their migrations, and in the periods spent at different latitudes. Re-drawn from Thiollay 1978.

by active flapping flight, others progress mainly by soaring, climbing in successive updrafts or thermals and gliding with some loss of height to the next. The first group include the narrow-winged raptors, such as falcons and harriers. They expend considerable energy on migration, but can cross water with ease, and can thus migrate on a direct course. The second group include the broad-winged raptors, such as eagles, vultures and buteonine hawks. They perhaps expend less energy on migration but, being more dependent on updrafts or thermals, they are obliged to make as much of their journey as possible overland, on what is often an indirect route, taking advantage of land bridges or short sea-crossings. In the absence of mountain updrafts, they are also restricted to the middle part of the day, when thermals are best developed. In an intermediate category are the accipiters, which take advantage of updrafts when available, but rely less heavily on them than do the other broad-winged species. The latter provide some of the most striking examples in birds of narrowly channelled migration routes, and in Europe thousands of individuals circumvent the Mediterranean each year via the Straits of Gibraltar in the west, or the Bosphorus and Dardanelles in the east (Figure 32). Through the influence of land form and topography, the birds form concentrated migration streams.

The extent to which a species takes mainly the east or west route out of Europe seems to depend partly on whether it winters in East or West Africa. Honey Buzzards winter mainly in the woodlands of West Africa, and some individuals from far east in Europe take the Gibraltar crossing rather than the Bosphorus one. This is evident from the numbers involved, up to 117,000 over Gibraltar in autumn, compared with 26,000 over the Bosphorus (Table 46), even though much greater numbers breed in eastern than in western Europe. On the other hand, Lesser Spotted Eagles breed in Europe as far west as the Baltic, yet head southeast so exclusively in autumn that they are unknown at Gibraltar or in northern Africa west of the Nile (Moreau 1972). Their wintering areas lie in East Africa.

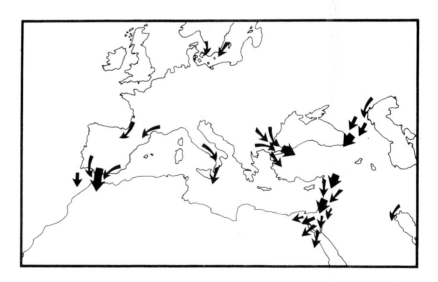

Figure 32. *Major migration routes for broad-winged raptors into Africa.*

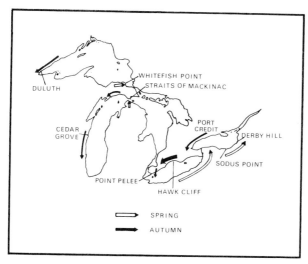

Figure 33. *Places of concentrated raptor migration around the Great Lakes.*

Another more eastern route for broad-winged raptors into Africa lies between the Caspian and the Black Seas, joining the Bosphorus stream in the Levant, and crossing near Suez. The sheer enormity of this movement only recently became apparent, when 380,000 raptors of 28 species were counted at the east end of the Black Sea between 17 August and 10 October, including 205,000 Steppe Buzzards and 138,000 Honey Buzzards (Beaman *et al* 1979). This compares with maximum totals recorded over several autumns at Gibraltar and the Bosphorus of about 194,000 and 77,000 raptors of all species (Table 46). A fourth route is probably from Asia across Arabia, and into Africa over the south end of the Red Sea, but there is uncertainty about the numbers of birds involved. Fifthly, a small passage occurs down Italy, over Sicily and Malta, and across to North Africa (Beaman & Galia 1974). On the return migration, no less than 605,000 raptors of about 30 species were counted in 1977 at the north end of the Red Sea in Israel (Table 46; Christensen *et al* 1979). Further north this stream splits into several smaller ones, including the Bosphorus and Black Sea contingents.

Less marked migration routes occur in North America where, in contrast to Europe, the main mountain chains run north-south. Crosswinds hitting the ridges create updrafts which are good for soaring, so major migration streams occur along the Rockies in the west and along the Appalachians in the east. They also occur round the edges of the Great Lakes (Figure 33) and along various coastal peninsulas, as at Cape May in New Jersey (Dunne & Clark 1977). Particularly well known as an observation post is Hawk Mountain in the Appalachian chain of Pennsylvania, where 10,000–20,000 raptors of 15 species have been counted in most autumns over the last 40 years (Broun 1949, Nagy

1977). For the journey into South America, the main routes for soaring species are down the Florida Peninsula across the West Indies and on to the mainland, and also through Central America over the land bridge at Panama. Soaring raptors form enormous roosts near concentration points and at Panama, Smith (1973) once counted 29,753 Broad-winged Hawks which spent the night in two square miles (518 ha) of forest. Over flat country, migration routes are less clearly defined, evidently because the passage is on a broad front, and often high above the ground.

Although much information has accumulated in recent years on the physiological state in which passerine and a few other birds make their journeys, and especially on the amount of fat they carry, little in this respect is known about raptors. Through most terrain, accipiters, harriers and falcons could normally pick up food (especially avian) on the way. Peregrines and Sharp-shinned Hawks tracked by radio while on migration through the United States were found to hunt for periods before or after each day's flight (Cochran 1972, 1975). These and other species that have been trapped in numbers while on migration almost invariably had food in their crops (F. P. Ward). On the other hand, the small Eastern Red-footed Falcons have such an immense sea crossing from India to Africa (minimum 3,000 km) that they must surely put on body fat for the journey – which fits the finding that in India in autumn they are 'very good eating' (Moreau 1972). The big soaring species probably have difficulty in feeding over much of their land route, but their normal lifestyle is such that they can go for days without food, and several casual observations suggest that fasting is usual on migration (Safriel 1968, Moreau 1972). This applies especially to those Palaearctic species which have long flights through deserts bereft of other life. To some extent fasting might be beneficial, through reducing the wing-loading and hence the energy needs of the bird.

Another problem concerns navigation. It is usually assumed that birds inherit whichever directional tendencies get them to their destination most economically. To judge from ring recoveries, the narrow-winged migrants keep more or less on a straight course all the way, as do many other birds (Newton 1972a). For broad-winged soaring raptors, the problems of navigation seem greater because of the indirect routes they often take. In North America, if such birds merely drifted in a general southwesterly direction, taking advantage of any local topographical features, they could reach their wintering grounds in South America. But in Europe the problem is less simple, for many populations have a strong easterly or westerly component in their migration, before turning south into Africa, and then east or west again within that continent. Do they then inherit several directional tendencies, which are changed at specific points in the journey, as seems to be true for other birds?

In the temperate zone, weather has a strong influence on the timing of the actual flights. The passage of a cold front to the north is often followed by a strong movement in autumn and the passage of a warm front to the south is often followed by a strong movement in spring. But whether the birds are responding to the temperature change as such, to its effects on food, or to the associated sky and wind conditions, is hard to tell. For like other birds, raptors prefer to migrate under clear skies and with following winds, so that at a given observation point a north wind brings the birds in autumn and a south wind in spring. Juvenile raptors are drifted off course by crosswinds more than adults are, and many of the big days at certain observation posts occur when the wind has a strong easterly or westerly component. This may in turn give a misleading impression of the proportion of juveniles in the flight as a whole, as may any tendency the juveniles have to fly closer to the observer, so that more are positively identified.

Winds are crucial to the volume of migration that can be seen by an observer on the ground. They influence not only the numbers of birds migrating, but also the proportion of migrants which pass over the observation point (as opposed to passing elsewhere or on a broader front), and the proportion of those over the observation point that pass within visual range. Compared with the picture of raptor migration revealed by radar, visual observations at Gibraltar give a false impression of the narrowness of the stream, and of the volume of movement, both on an hour-to-hour and on a day-to-day basis (Evans & Lathbury 1973). Certain wind conditions favour a concentrated low altitude passage (mostly visible through binoculars), and other wind conditions favour a broader or higher altitude passage (visible only by radar). Likewise in eastern North America, the numbers seen from the ground are strongly correlated with side winds, whereas those seen by radar are more correlated with following winds, when birds fly higher and on a broader front (Richardson 1975). The particular wind conditions that lead to good visual passage vary between localities, depending on local topography, and on the desired migration direction of the birds. In some places the largest counts are made in spring and in others in autumn, but to what extent this reflects differing conditions for concentrated visual passage, rather than differing migration routes at the two seasons, is hard to tell in the absence of enough ring recoveries.

Visual counts cannot therefore be expected to give anything but a minimum (and possibly misleading) estimate of the numbers of birds migrating. They do however give an idea of relative numbers and species composition at different places, of the migration seasons of particular species, and of sex or age differences. Repeated over enough years, such counts also help to reveal long-term population trends, though year-to-year fluctuations are usually great (Edelstam 1972, Ulfstrand et al 1974).

At the Bosphorus, Belon (1555) gave an early and vivid description of migration in Black Kites: 'if they had continued for a fortnight in the same strength as on that day, we could surely have said that they were in greater number than all the men living on the earth', and added that 'they are seen to pass in this way as thick as ants, and so continue for many days'. Later, more detailed, studies at the Bosphorus indicated

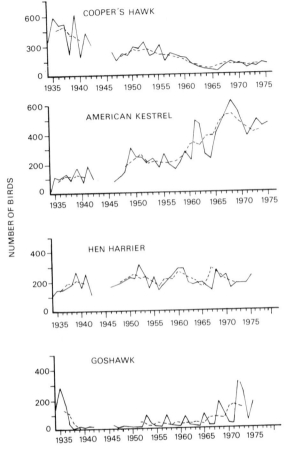

Figure 34. Population trends over 40 years as suggested by autumn counts of migrating raptors at Hawk Mountain in Pennsylvania. Continuous lines join the annual totals, broken lines show 5-year moving averages. From Nagy 1977.
  (a) Cooper's Hawk – long-term decline, associated initially with heavy persecution and latterly with organo-chlorine pesticide use.
  (b) American Kestral – long-term increase, cause not known.
  (c) Hen Harrier – no obvious long-term trend.
  (d) Goshawk – an increase in recent years, but with marked peaks every 3–5 years, presumably related to prey cycles.

massive declines in several species between 1870 and 1930, especially in Short-toed Eagles, Booted Eagles, Black Kites, Egyptian and Black Vultures, and in more recent times in other species as well (Nisbet & Smout 1957). In America, counts have been made at Hawk Mountain almost every year since 1934; and in some species have revealed long-term declines, in others increases, and in yet others more or less regular fluctuations or no clear trend (Figure 34). The changes are perhaps more meaningful if they occur simultaneously in the records from several observation points (Robbins 1975).

SUMMARY

Much of the diversity in the movement patterns of raptors can be explained in terms of food supply: (a) the seasonality of movement, such that raptors are absent from breeding areas at a time when food is scarce there; (b) the regional variations in extent of migration; (c) the annual variations in the migrations, especially of northern species dependent on cyclic prey; (d) the relationship between migration and diet, with those species that depend on cold-blooded prey withdrawing earlier and most completely from the temperate zone in winter; and (e) the precise timing of movements in relation to changes in prey numbers. Within species, the distribution of each population is related to the distribution of other populations, the east-west pattern in breeding areas being maintained in wintering areas by parallel migrations; but in their north-south distributions, leap-frog migrations are frequent, with the northernmost breeding populations wintering furthest south. Within species, sex and age differences also occur in the proportions of birds leaving, the timing of migrations and distances moved, all of which can be interpreted partly in terms of dominance and food relationships within populations. In large parts of Africa for half the year, Palaearctic migrants greatly outnumber the residents; some migrant species avoid competing with the residents by exploiting transient food supplies, a strategy which entails continual and extensive movements. Broad-winged soaring raptors provide some of the most striking examples of narrow migration routes, on which many thousands of birds can be seen in a day. Counts over many years at concentration points have revealed long-term population trends, as well as sex and age differences in migration.

Peregrine

CHAPTER 12

Mortality

Much less is known about the mortality of raptors than about most other aspects of their ecology. This is partly because the actual deaths are seldom witnessed (unless at the hand of man), and also because no known method of study gives unbiased information on the ages at death, or on the causes of death. In this chapter, I shall therefore be concerned as much with the problems of studying mortality, as with the knowledge gained. Mortality estimates are important in understanding population turnover, and in pinpointing particularly vulnerable age groups or populations.

The underlying trend is for large raptors to live longer than small ones, and for fewer to die each year (Chapter 7).* In the security of zoos, individuals of large vulture and eagle species have occasionally lived for 40–55 years, medium-sized buzzards and kites have lived for 20–40 years, and small falcons and accipiters for about 15 years (Table 47). In the wild, the large species have been insufficiently studied to offer

---

* This fits the general trend in birds which has been expressed as $t = 17 \cdot 6 \ M^{0 \cdot 20}$ for wild birds, as $t = 28 \cdot 3 \ M^{0 \cdot 19}$ for captive birds, and as $t = 16 \cdot 6 \ M^{0 \cdot 18}$ for wild non-passerines, where t is the maximum longevity in years, and M the body weight in kg (Lindstedt & Calder 1976).

Illustration: Eagle Owl with Kestrel

comparisons with zoo birds, but ringed individuals of other species have occasionally lived as long as the oldest captives (Table 47). And as more ringed birds are recovered, the longevity records of many species in the wild may well be extended. Nonetheless, the majority of individuals do not reach anywhere near their potential lifespan: in all raptor species so far studied in the wild, more than half the birds that fledged died in their first year.

As in other birds, the annual death rates are usually calculated from national ringing recoveries or from detailed studies of particular populations. The first source gives results which cover wide geographical areas and periods exceeding 50 years, but they are often biased in some way; the second source gives more accurate mortality estimates, but they are not necessarily applicable outside the study areas. For both types of estimate, it is helpful to know whether the population is numerically stable in the long-term or is changing, because mortality during a period of stability may differ from that during rise or fall.

MORTALITY ESTIMATES FROM RECOVERIES

This method entails comparing, for a given cohort of nestlings, the number recovered dead in each year after ringing. The rate at which the recoveries decline year by year reflects the mortality. Usually there are too few recoveries for any one year-class to analyse them separately, so the records of individuals born in different years are combined into a 'composite life-table', and treated as though they were members of the same age group (Table 48). In the method most commonly used, enough time has to be left after the last year of ringing to allow all the recoveries to come in, a period which depends on the lifespan of the species concerned.* Mortality can also be examined separately for different sex and age groups, as can its seasonal distribution.

This method gives a good estimate of mortality only if the deaths of birds reported are typical (with respect to age) of all deaths in the population. This condition will hold only (a) if the birds that are ringed are representative of the population as a whole, (b) if the process of ringing does not itself influence survival, (c) if there is no loss of rings with age, and (d) if the reporting rate is independent of age, cause of death, date and location. The degree of error in mortality estimates depends on the extent to which these various conditions are met, and in practice it is the last condition that gives the most trouble in raptor studies.

Shooting and other killing by humans tend to concentrate on the younger age groups of a population, so that these are over-represented in the sample, and give rise to an inflated estimate of the overall mortality. The Goshawk in northern Europe is an extreme example, for more than 90% of the recoveries in one analysis were of birds killed by

---

* Haldane's 'incomplete data' method enables all recoveries up to the present to be used, but entails more difficult computing.

man (Haukioja & Haukioja 1970). The authors calculated mortality separately for killed birds and for birds found dead (this last category might have included a few killed birds). In the first three years of life, the estimates were 70%, 57% and 37% for killed birds, compared with 52%, 11% and 11% for the much smaller numbers found dead. Taking both sets of figures into account, and estimating the proportions of birds in the whole population that were shot, the authors produced 'corrected' mortality estimates of 63%, 33% and 19%.

The shooting pressure on north European Goshawks affected other aspects of the recoveries, for it transpired that birds which were on the move were especially likely to fall prey to hunters. Thus most recoveries came in autumn and spring, which were the two main periods of wandering. Autumn records were more numerous than the spring ones partly because the population was larger in autumn and partly because, being the game season, more hunters were in the field then. Recoveries were also heavily biased towards those sectors of the population which were most on the move, namely first-year birds of both sexes and second-year males. The adults tended to remain near their breeding areas, and the under-representation of second-year females was because by that age more females than males had become resident on breeding areas. On the other hand, adults were more likely than yearlings to be killed in summer, especially females, which were easily shot off nests.

I have discussed this species in some detail in order to indicate the kind of problems encountered in using recoveries to estimate mortality of heavily persecuted raptors, and also to show how the pattern of recoveries is influenced by the behaviour of the birds themselves and of their human hunters. The Goshawk was perhaps extreme in the bias towards young birds caused by shooting and trapping, but in several other accipiter and harrier populations, mortality estimates in the literature are so high that, on known breeding rates, the population concerned would have soon died out. The fact that they persist implies that the mortality of the recovered birds was not the same as that of the whole population.

In Table 49, which summarises the mortality estimates of different raptors from the literature, I have also given the percentage of killed birds (where known), so as to indicate how reliable the estimates are likely to be. When this percentage is as low as 40%, as in North American Ospreys, the bias in adult mortality is probably negligible (Henny & Wight 1969).

The Osprey suffers from another source of bias, which applies in some degree to all migrant populations, namely that the chance of being recovered differs between geographical regions. Thus Ospreys are more likely to be reported at certain seasons than at others, and more important, at certain times in their lives. Yearling and many second-year birds remain on 'winter quarters' for the summer (the North American birds in Central and South America, and the European ones in Africa), so both groups are less likely to be reported in summer

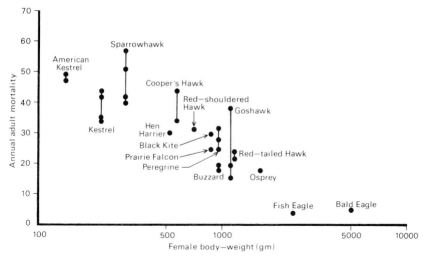

Figure 35. Annual mortality of adult raptors in relation to female body weight. Different estimates for the same species are joined by lines. The scale for body weight is logarithmic. Estimates for eagles are based on plumage ratios (see text), those for the remaining species on ring recoveries.

than the older birds in their northern breeding areas. However, by restricting analysis to older birds, a reasonable estimate of their mortality can be gained (calculated at 18% per year). In other species, differences in recovery rates occur even within Europe or North America, when the young migrate more than the adults, but in these cases the bias is probably small.

In other long-lived birds, the rings become worn and corroded, and may drop off in later life. This means that old individuals are under-represented in the sample, which again inflates the estimate of overall mortality. Ring loss is not generally a problem in raptors (Henny 1972, Österlöf 1977), except perhaps in certain eagles and vultures, which may sometimes manage to remove their rings. However, no-one has yet seriously tried to estimate the adult mortality of large raptors from ring recoveries, and in the meantime changes in ring construction have done much to lessen the loss.

These various sampling biases mean that mortality estimates derived from ring recoveries should be used with caution. They are nonetheless accurate enough for some purposes, and reveal important trends. The first is for mortality to vary with body size, as already indicated (Figure 35, Table 49). The adult death rate has been calculated at 19–31% per year in different species of buteonine hawks, at 25–32% in large falcons and harriers, at 34–49% in kestrels, and at 49–57% in European Sparrowhawks. Second, within each population, the young birds showed greater annual mortality than the older ones. This difference may have been exaggerated by the greater vulnerability of young birds

to persecution, but even when analysis was restricted to birds foun
dead, the difference still held. In the smaller raptors, the difference wa
apparent for the first 1–2 years, but in larger raptors up to the 3rd to 5
year, depending on species. After this stage, the estimated mortali
was more or less constant until the age at which extremely fe
individuals were left alive. Hence, no evidence was obtained fe
senescence, because too few birds reached an age at which it mig
show. In all these respects, raptors reveal the same trends as oth
birds, and the greater mortality of young individuals is readi
explained in terms of their lesser experience and social status (Lac
1954, 1966).

For some species, significantly different estimates of mortality wei
obtained from one region to another (Table 49). Such differences coul
have been genuine, or they could have been due to differential bia
between samples, or to differences in analysis (some authors started th
first year from the date of ringing and others from a later date). In th
North American Red-tailed Hawk, however, data from north and sout
regions were analysed in the same way, and the significantly greate
mortality in the south was associated with a greater production c
young there, a finding that would be expected in stable populatior
(Henny & Wight 1970).

All the mortality figures in Table 49 are necessarily averages fron
many years, so any annual or other variations are masked. Amon
Kestrels in Britain, first-year mortality was slightly less over perioc
when voles were increasing than over periods when voles wei
declining (Snow 1968). In the same species in Holland, young hatche
early in the season suffered significantly less mortality in their fir
year than did those hatched late (Cavé 1968). In no species wa
mortality examined separately for the sexes, but in some a differenc
would be expected from what is known of sex ratios (Chapter 1).

MORTALITY ESTIMATES FROM POPULATION STUDIES

The second main way of studying age and mortality in adults is t
catch and mark all the individuals of a species breeding in a particula
area, and find what proportion is present in each succeeding year. Suc
studies have been done for many years on songbirds and seabirds, bu
have only recently been started on raptors. In some raptor species ther
is the problem that many individuals nest in widely separated tex
ritories in different years, so that a missing bird may have died or i
may have moved out of the study area. Movements of more than 20 kr
between the nesting territories of successive years have been noted i
European Sparrowhawks and of more than 100 km in Canadian Mer
lins (Chapter 10). So far, only in Buzzards in northern Scotland wer
the movements so rare and so short that mortality could be found i
this way (D. N. Weir). The figure was 20%, about the same as thos
from ringing schemes in central and northern Europe (Table 49).

In the absence of marked birds, the mean mortality in certain populations can be calculated from a knowledge of the adult numbers over several years, and of the production of young. The method depends on the fact that, in any stable population, the birth rate must equal the death rate. Imagine that a population at the start of one breeding season was 20 birds, which altogether produced 20 young, and that the population at the start of the next breeding season was 24 birds, then 16 must have died in the meantime, an annual mortality of 40%. The method tells us nothing of how mortality is distributed between different age classes, and becomes complicated when account must be taken of movements. It has been used successfully on the isolated population of Red Kites in Wales, in which the mean annual mortality over 22 years was estimated at 17% (Davies & Davis 1973).

Another method which has been used to learn something of mortality in the absence of marked birds entails finding the age structure of populations in which the immature year classes can be distinguished in the field. In some large raptors the definitive plumage is not acquired until about the fifth (or a later) year, so if a representative cross-section of birds is examined, the age structure of the population until the fifth year can be obtained. Such counts of age-groups have been made in Golden Eagles, Lammergeiers, Bateleurs, Bald Eagles and Fish Eagles (Brown & Watson 1964, Brown & Cade 1972, Brown & Hopcraft 1973, Brown 1977, Thiollay & Meyer 1978). In the Fish Eagle, the sub-adult plumage class just prior to adult formed about 3–4% of the population; so if the annual adult mortality was equal to this sub-adult class, then it in turn was about 3–4% per year. Using similar methods, mortality of Bald Eagles in the Aleutians was estimated at around 95% up to the acquisition of adult plumage in the fourth to sixth year, and at around 5% per year thereafter (Sherrod et al 1977).

So far, these are the only mortality estimates made for large, long-lived raptors, but their accuracy again depends on how representative a cross-section of the population was examined. In all the species concerned, the counts showed that immatures concentrated in partly different areas from adults; hence, however many birds were seen, the mortality estimate could be in considerable error, depending on where the counts were made. It is also possible that individuals vary in their moults and acquire their adult plumage at different ages, but I know of no firm evidence for this.

CAUSES OF DEATH

Starvation, disease, predation, electrocution, shooting, trapping, poisoning, collisions and other accidents all kill birds-of-prey, but for any one population it is hard to tell the relative importance of each cause. This is partly because most mortality goes unobserved in the wild, and the watcher can usually record only the disappearance, not the death, of his birds. And even when the manner of death is seen, its underlying cause may not be obvious. Thus a bird may be excluded from suitable habitat by territorial behaviour, so that it becomes short of food, and thus weakened, it may then become diseased or caught by a predator. In this case, the ultimate cause of death is socially-induced starvation, but the proximate cause is disease or predation. Most information on proximate causes comes from autopsies of birds found dead, and from the recoveries of ringed birds in which the cause of death is given. Both sources are inevitably biased towards mortality associated with human activity and habitation. To pick an extreme contrast, a bird which crashes through a kitchen window is almost certain to be noticed, but one which is eaten by a predator in some remote forest is not.

The United States Bureau of Fish and Wildlife operated a scheme in which refuge managers and others were asked to send in any Bald Eagles found dead. These birds were then subjected to detailed post-mortem examinations. Of 231 birds inspected in the period 1966–74, 43% had died from direct persecution (mainly shooting), 10% from pesticide poisoning, 21% from accidents, 12% from natural causes and 14% from unknown causes (Table 50). Another American programme involved moribund birds of prey found by the public and offered for veterinary treatment (Redig 1978). Among 850 birds of 30 species examined during 1972–77, 23% had been shot, 15% trapped, 34% were accident victims, 8% were diseased, and the remaining 20% were brought in for various other reasons. Perhaps not all these birds would have died if left in the wild, but they gave a further indication of the role of different mortality factors among birds found by people. Other similar schemes were either concerned with a wider range of birds and involved very few raptors (Jennings 1961), or were concerned only with a particular cause of death, such as pesticide poisoning. So they tell us little of value in the present context.

Ringing recoveries inevitably provide less information than post mortems, and in some countries they may have become less reliable if raptors have received legal protection. People who now kill these birds are likely not to report the fact, or they may falsify the cause of death. Nonetheless, information gained by each method has proved useful in confirming certain kinds of mortality, and in making comparisons between species. Some data on natural and accidental mortality are summarised below and those on persecution and pesticides are given later (Chapters 13–15).

1. Starvation

Ring recoveries include reports of birds in a weakened and starving condition, and emaciated corpses are also occasionally found by biologists during field studies (Snow 1968, Sherrod et al 1977). The striking fact is that starving raptors are found so seldom. This is true of many other birds, and may be due to the fact that very few individuals are starving at any one time, while those that are soon fall victim to predation or disease, or are eaten promptly after death. It is a matter of

(Upper) Wahlberg's Eagle at nest with young in Zimbabwe Rhodesia. One of the most obvious migrants ng the African raptors, the species arrives in the southern part of the continent in August/September each breeds, and then moves north again in April/May. Photo: Peter Steyn. (Lower) Booted Eagle female at nest ain. The nest surface is decked with fresh green sprays. Photo: M. D. England.

24 Broad-winged Hawks migrating over Ancon Hill, Panama. On their way between North and South Am
these birds form a concentrated stream, passing each October/November within 5 km of Panama City.
Broad-winged Hawks pass this way, as do Swainson's Hawks; totals of 395,000 and 344,000 were coun
1972. Photos and details: N. J. Smith.

observation that small carcasses in the countryside usually disappear within a day or two, having been removed and eaten by scavengers.

2. Disease

More is known about disease in raptors than about any other form of natural mortality, for there have been many studies of wild and captive birds (Trainer 1969, Keymer 1973, Cooper 1972, 1973, 1975, 1976, Greenwood 1977). This is a good thing because, in view of their close contact with prey and their supposed tendency to select diseased individuals, raptors have unusual opportunities to become infected.

Among the bacterial diseases, tuberculosis was the most often recorded in wild raptors, while others included erysipelas, listeriosis, salmonellosis and pseudotuberculosis. Most of the relevant bacteria occur in the environment but the last two can also be acquired from other birds and from rodents. The fungal disease, aspergillosis, affects the bronchial system, and was especially frequent in captive birds, but much less common in wild ones (Keymer 1973). Among the viral diseases, Newcastle disease (fowl pest) was found in virulent form in several raptor species, while others included avian leucosis (causes tumours), avian pox (a proliferative skin disease), herpesvirus hepatitis (affects mainly the liver), ornithosis (general) and encephalitis (affects the brain) (Trainer 1969, Greenwood 1977).

All the major groups of endoparasites have been found in raptors, as have most of the blood protozoa and the ectoparasites (Trainer 1969). Parasitism has seldom been found to be serious, but in one United States survey of 216 raptors, 39% were infected, and 9% had serious or fatal disease (Ward 1973). Among the protozoa, two types are well known in birds-of-prey. Coccidia live in the intestines, can spread from bird to bird, and have caused some deaths among raptors (Mathey 1966, Coon et al 1969). Trichonomonas are common in pigeons which, when eaten by raptors, cause an often-lethal disease, known as 'frounce', which is especially frequent in captive birds (Stabler 1969). Among the parasitic worms, *Capillaria* nematodes, which live in the gut, have seriously affected Gyr Falcons (Trainer et al 1968), while *Serratospiculum* nematodes, which live in the air sacs, have killed Prairie Falcons (Ward & Fairchild 1972), and have also been found in other falcon species. Myiasis (infestation with fly maggots) has several times been found in North American raptors, affecting nestlings in wet summers, but not always killing them (Sargent 1938, White 1963). Among captive raptors, disorders of the digestive and urinary systems are frequent; cardiovascular disease is common in old birds in zoos, and nutritional bone disease in pet birds fed exclusively on plain meat.

Botulism is a form of food poisoning caused by a toxin from the bacterium *Clostridium botulinum*, which develops in stagnant water, and over the years has killed millions of waterfowl in North America (Sciple 1953). Affected birds are eaten by raptors, which may in turn be affected, though Turkey Vultures have shown unusual resistance (Kalmbach 1939). Raptors may also become exposed through drinking or bathing in water containing concentrated toxin (White 1963).

Diseases may affect wild populations either as widespread or local outbreaks, as sporadic individual mortality, or in mild form which may lead to reduced breeding success or lifespan. This last type is probably the commonest, but is hard to detect (Trainer 1969). Disease may be further considered as 'population dependent' or 'population independent' (Hanson 1969). The first type depends for its spread on contact between individuals of a species, and in raptors is unlikely to be important, except perhaps among colonial breeders. An example would be a host-specific parasite, such as the *Serratospiculum* mentioned above. In contrast, population independent diseases are contracted from the environment, food or water. Raptors could thus become infected from their prey, irrespective of the size of the raptor population and the amount of contact between individuals. Apart from host-specific parasites, no known disease of raptors is specific to raptors alone, and so population independent diseases are the most common. An example is the fowl cholera (caused by *Pasteurella* bacteria), which passed through dead waterfowl and rodents, and affected large numbers of Hen Harriers and Short-eared Owls in California (Rosen & Morse 1959). Other predatory birds in zoos have developed erysipelas and anthrax from feeding on diseased mammal carcasses (Hamerton 1942).

Although a wide variety of diseases and parasites has been found in raptors, with effects from slight to fatal, their significance to wild populations is far from clear. The view is now generally accepted that most disease organisms and parasites evolve in such a way as to become less harmful to their host with time, since the organisms have a better chance of surviving if they do not destroy their habitat. Likewise, the host species tends to evolve resistance to its parasites and to the toxins that they produce, but it may be vulnerable when weakened by other factors or when exposed to a new disease. Almost certainly, disease plays an insignificant role in the control of raptor populations, and accounts for only a small part of the total mortality. But this negative statement cannot be considered established without more research. In particular, it is possible that diseased birds are rare, not because disease is rare, but because sick birds are usually eliminated at once by predators or starvation. On the other hand, mortality attributed to disease may have really been due to food-shortage, with disease as a secondary consequence.

3. Predation

In the extensive prey lists of Uttendörfer (1952) from central Europe, certain raptors figured prominently in the diets of other raptors and of owls, especially of the Goshawk and Eagle Owl *Bubo bubo* (Table 51). The smaller species fell victim more often than the larger ones, partly because the small were vulnerable to a greater range of enemies. Some raptors also preyed on others of their own species, especially the accipiters in which the females took the smaller males (I. Newton & M.

Marquiss). Mikkola (1976) listed 231 cases from the European literature of diurnal raptors taken by owls (he included Uttendörfer's data). At least 207 individuals of 13 species were recorded in the food of the Eagle Owl, including 65 Buzzards, 55 Kestrels, 26 Goshawks and 19 Peregrines. Raptors formed 3–5% of the Eagle Owl's diet, and 23–26% of its bird prey. Smaller raptors were also taken by the Tawny Owl *Strix aluco*, and the 22 records included 11 Kestrels and nine Sparrow-hawks. Not all this killing may have been for food, as some may have been in nest defence. Both these owls seem to dislike raptors nesting nearby, and prey on any species no larger than themselves, mainly breeding adults and feathered young. The same is true of the Great Horned Owl in North America, which kills raptors up to the size of the Red-tailed Hawk (Chapter 8). Data such as these confirm the prevalence of predation on small and medium sized raptors, but again they tell us nothing about the contribution of predation to the total mortality of a species, nor of its role in population control.

4. Accidental deaths

The ones that come to notice usually involve collisions of birds with vehicles, buildings and other structures. Nearly one-third of all ringing recoveries of British Sparrowhawks and Kestrels in the period 1959–69 were obtained in this way. This proportion had increased relative to an earlier period, perhaps partly due to the increased speed and density of traffic and the greater spread of wires and buildings over the landscape (Table 52). In other countries, the proliferation of power poles and wires has led to the deaths of many raptors. Electrocution is especially frequent in large species, whose wings can touch two wires at once. The Cape Vulture suffers a large mortality in this way, and in one survey 148 birds were found under wires in the southwest Transvaal over 26 months (Markus 1972). With 67,000 km of wires in South Africa as a whole, this number is probably only the tip of an iceberg. Until recently, several hundreds of Golden Eagles died in this way each year in the western United States, and in some localities the inves-tigators found more dead birds than poles (Chapter 16). In both species most losses were of juveniles.

Road deaths are most frequent among species that eat carrion on roads. In the western United States migrant Rough-legged Buzzards, still tame from the Arctic, are often hit by traffic while feeding on road-killed rabbits. Along one 240 km highway in Utah, White (1969a) found 19 freshly dead Rough-legs and eight Horned Owls in only one week, which he presumed was an underestimate of the total killed. These various man-made causes of mortality may not seem overwhelm-ing when considered singly, but their combined impact is increasing. For certain populations, they already form a major part of the total mortality, and in extreme cases they may be contributing to decline.

5. Influence of weather

Hard weather can accentuate most kinds of mortality. It can increase the food needs of the birds or make food harder to obtain; it can make birds more susceptible to disease; or it can force them nearer to human settlements, so that they are vulnerable to shooting and accidents. In the hard winter of 1939/40, for example, starved Buzzards were found all over Germany, yet very few were found in normal winters (Drost & Schüz 1940). Nonetheless, hard weather has less effect on raptors than on most other species; 15,000 birds found dead in Britain in the severe winter of 1962/63 included only nine raptors, all mammal feeders (Dobinson & Richards 1964).

Of the various kinds of mortality discussed, starvation, predation and population-dependent diseases are the ones most likely to act in a 'density-dependent' way, claiming an increasing proportion of a population as its numbers rise. They are therefore the only killing factors that are likely to regulate populations in the strict sense. Population independent diseases and at least some accidental forms of death are 'density-independent' in that the proportion of birds they claim is for the most part unrelated to population density. In the early days of bird population studies, a lot of effort was devoted to studying mortality causes, in the hope that these would reveal what determined population levels. But some raptor populations are clearly limited by the carrying capacity of their environment, and the various mortality factors act mainly to trim down numbers each year to a level that the environment will support. This suggests the over-riding influence of density-dependent factors. Other populations are well below the environmental carrying capacity, as a result of overkill from persecution or pesticide poisoning, as discussed in the following chapters.

SUMMARY

The study of mortality in raptors is much less advanced than the study of breeding. Accurate estimates are hard to obtain, and practically nothing is known of how mortality might vary from year to year, or from area to area, according to food and other conditions. The trend within the raptors for annual mortality to decline with increasing body size is well established, as is the trend within species for greater mortality in immatures than in older birds. More study is needed of the mortality of marked adults in local breeding populations, but this is likely to give good estimates only where individuals remain faithful to their territories from year to year. The importance of different mortality causes is also poorly understood, partly because it is hard to find a sample that is representative of the whole population, and partly because of the operation of pre-disposing causes. Starvation, predation and disease are all recorded as causing deaths of raptors, as are various accidents and collisions, electrocution, shooting, trapping and poisoning. The ringing recoveries and post-mortem analyses which provide most information are inevitably biased towards deaths that occur from human action or around human habitation.

CHAPTER 13

Human persecution

Because some raptors eat lambs, poultry or game birds, they have been
slaughtered in millions over the last 150 years, especially in Europe.
This chapter is concerned with the extent of this killing and with its
effects on populations. It is not concerned with whether the killing is
justified, for this question involves value judgements, based on per-
sonal preferences. The tremendous effects of this persecution on
populations are seldom doubted, yet little scientific documentation is
available. This is partly because most reduction of numbers occurred
between 1850 and 1900, before biologists were interested in recording
it, and in recent years, killing has become illegal in many countries so,
practised subversively, it has proved hard to study. Moreover, in their
efforts to understand the birds, biologists have generally avoided
studying populations that they knew were being heavily shot.

In an attempt to protect domestic stock, the killing of the larger
raptors was officially encouraged in parts of Europe as early as the 16th
century, by payment of bounties. This seems to have been sporadic,
however, and to have had no marked or long-term effects on populations. It was with the rise in small game management in the 19th
century that persecution reached its peak, and spread to the smaller
species. Game shooting increased in popularity with the introduction
of Pheasant *Phasianus colchicus* rearing, and increased again with the
improvement in the shotgun, from muzzle loader to breech loader. The

Illustration: Kestrel in pole trap

213

objective of total elimination of raptor populations was soon achieved for some species over large areas.

In many countries the destruction of all raptors became an accepted practice, and no one in his right mind was expected to pass up the chance to kill a hawk. This attitude was reflected not only in the lack of protective legislation, but also in the widespread payment of premiums for birds killed, and in the employment of 'game-keepers', with the specific task of destroying predators. In Britain every sizeable estate had at least one keeper, and some idea of their total numbers can be gained from *Castle's Fishing and Allied Trades Directory* (1910) which lists 1600 registered gamekeepers, a total that excludes under-keepers and others concerned with 'vermin' control. The same attitude to raptors persisted throughout Europe well into the 20th century but, with increasing education and the rise of a conservation movement, public opinion is gradually changing. It has so far been reflected in the abolishment of many bounty schemes, and in the introduction in one country after another of protective legislation. In some countries, however, such as Britain and France, this legislation is still resisted or largely ignored by the shooting fraternity.

In North America, persecution seems never to have reached the levels that it did in Europe, but in some parts it was officially encouraged with bounty schemes. The State of Pennsylvania enacted the now famous hawk-and-owl bounty law in 1885, by which the sum of 50 cents was paid for the 'scalp' of any hawk or owl. Immediately the slaughter began, and in two years 180,000 scalps were presented for payment (Hornaday 1913). Fairly heavy persecution continued over much of the country into the nineteen-forties, but following the introduction of protection laws, both organised and casual shooting of raptors seems to have much declined (see later). The legislation has been more effective than in Europe, perhaps partly because the hunting public were already well used to the rigid enforcement of regulations and bag limits for waterfowl and other game species. Also, the destruction of predators has never figured as prominently in small game management in America as it has in Europe. Persecution was nonetheless considered a major factor in the decline of the California Condor to its present population of less than 60 individuals. This persecution included the collecting of skins and eggs for museums (Miller et al 1965, Snyder & Snyder 1975).

SOME GENERALISATIONS

The only permanent way to reduce a bird's population is to reduce its habitat and food supply. The alternative entails holding the numbers of the bird below the level that the environment will support and the removal of birds year after year to counter their breeding.

Whether sustained killing leads to long-term population decline depends on whether it replaces the natural mortality, or adds to it.

Thus if the increased mortality from shooting is offset by reduced mortality from natural causes, so that the number of birds which die each year is about the same, then the population will not decline. But if the mortality from shooting, or from a combination of shooting and natural causes, exceeds that which would otherwise occur from natural causes alone, then the population will decline. In practice, much depends on when the killing occurs. Its effect is likely to be minimal in the months following breeding, for the population is then at its seasonal peak, and contains many juveniles which would die anyway or disperse before the next breeding season. In such cases, the shooting has to be exceptionally heavy, if it is to do more than merely crop an expendable surplus. The effect of killing is greatest if it is done at the start of a breeding season, for the population is then at its seasonal low, after most natural mortality has occurred. Then shooting not only adds to the natural mortality, but also concentrates on the breeding adults, the most valuable sector of the population, so that decline is rapid. It was through the annual wholesale destruction of breeding pairs that the populations of several species were wiped out over much of Britain before 1900. The traditional nesting places of the birds were well known to the landowners and their keepers, and even today many Peregrine cliffs in the Scottish Highlands show the remains of stone shelters, built by the keepers for use as shooting hides. My friend Doug Weir assures me that some such shelters are still used for this purpose today.

In practice, the vulnerability of any species depends partly on how easily it can be killed. Some species are fairly tame and easy to shoot; others use conspicuous perches and are easy to catch in leg traps; while yet others eat carrion, so are easy to poison. Throughout Europe, it is the carrion-feeding species that have suffered most, simply because they can be killed in large numbers with minimum effort. Secondly, large species are inevitably more susceptible to the effects of persecution than small ones. This is partly because they live at lower densities, but mainly because they have much lower breeding rates and take much longer to reach breeding age (Chapter 7). Following a 50% kill, a slow-breeding condor population could take decades to recover, whereas a fast-breeding kestrel population could be back in a year or two. In the long-term, therefore, it is the small fast-breeding species that are most resistent to sustained killing. A third factor influencing vulnerability is the size and distribution of the population to begin with. Any small population which is localised in a restricted habitat is more easily eliminated than a large population that extends into wild country where it is hard to reach. Events over the last 150 years have lead to the fragmentation of many formerly widespread populations. Such isolated remnants are vulnerable for another reason, namely, the reduced chance of immigration which might otherwise serve to counter the effects of, say, local persecution.

LONG-TERM TRENDS IN BRITISH POPULATIONS

These various generalisations can be illustrated by reference to the British raptors, whose history over the last 150 years is well known. Early in the 20th century, five species were apparently eliminated completely for a period as breeders. These species (and their approximate dates of disappearance) were Marsh Harrier (1898), Honey Buzzard (1911), Goshawk (1889), Osprey (1908) and White-tailed Eagle (1916). The first three had anyway been restricted to small areas by habitat destruction, so their tiny populations would have been easy to find and eliminate. But the White-tailed Eagle was widespread, and probably numbered more than 200 pairs; its extirpation would have been facilitated by carrion feeding (and poisoning) and by a low breeding rate. It is not certain to what extent collectors of skins and eggs were involved in the final demise of these species, but there can be no doubt that it was the game-keepers who brought them to a low point to begin with. Only they had the guns, traps and poison to do it. Four of the species concerned later re-colonised from the European mainland and have small British populations at the present time.

Several other previously widespread species were much restricted in range, the Buzzard to some western districts, the Hen Harrier to the northern and western Isles, and the Red Kite to a tiny area in central Wales, where game preservation did not take hold. People often wrote of these species as having 'retreated' to remote areas – perhaps from analogy with human behaviour under persecution – but with the birds retreat was not involved. Populations were wiped out from all but remote areas where birds survived in no greater numbers than previously. Under reduced persecution, the Buzzard has since re-colonised large parts of the country; the Hen Harrier has re-occupied many mainland areas; but the Kite has taken an extremely long time of dedicated protection to reach its 1975 level of 33 pairs; it was hampered by an exceptionally low breeding rate and the continued use of strychnine baits in the breeding areas (Moore 1957, Watson 1977, Davies & Davis 1973, Newton 1972).

The Golden Eagle also suffered a considerable diminution in range, and its survival through the worst period, in contrast to the White-tailed Eagle, could be attributed to its occupation of some high, inland areas which, at that time, were extremely remote and hard of access. The Peregrine was eliminated from a few areas (eg, the southern Pennines) but had large reservoir populations on coasts and islands where it could breed free from persecution, and produce recruits to offset the losses in other areas. The species least affected were the Merlin, Kestrel and Sparrowhawk. These were the three smallest species, breeding at highest densities with the highest breeding rates, and thus having the best ability to recover year after year from persistent killing.

Summarising, marked reductions in numbers and range were associated with low and localised populations at the start, with carrion feeding, with slow breeding rates, or with a combination of these features. Lesser reductions were associated with large populations living partly away from game preserving areas, little or no carrion feeding, and high breeding rates. The Hen Harrier was the only species with a high breeding rate that was markedly restricted by shooting, but practically the whole population nested in grouse-preserving areas, and with the fearless nest defence, the species would have been especially easy to shoot at the nest.

Persecution remains a threat to the British raptors, and is clearly the main factor restricting the present range of at least the Buzzard, Hen Harrier and Golden Eagle, none of which occupy more than about half of their potential range in the British Isles, including Ireland. The Red Kite and others are low because, although perhaps no longer restricted by persecution, they were reduced by it in the first place. Some species are unlikely to achieve their former numbers in the foreseeable future because the habitat is no longer sufficiently widespread.

To end this section, it is worth emphasising that game preservation – the main practice leading to the destruction of raptors – has also had its positive side, in the provision of habitat. The woodland and other cover which, in cultivated areas, is often left primarily for game, incidentally provides the habitat on which many raptors and other wildlife depend.

EFFECTS ON POPULATIONS

Evidence for the effects of persecution on populations comes from (a) records of numbers killed; (b) correlations between changes in killing and changes in population; (c) recoveries of ringed birds; and (d) studies of birds found dead.

(a) Numbers killed

The payment of premiums for dead raptors has often meant that good records have been kept of the totals killed. The numbers can be extremely impressive, as the following examples show:

In Alaska in 1917–52, rewards were paid for 128,273 Bald Eagles (White 1974).

In Western Australia in 1928–68, rewards were paid on 147,237 Wedge-tailed Eagles, an average of 3,591 per year. Bounty payments were stopped in 1968, but birds continue to be shot in many sheep-rearing areas. The kill for the whole continent may still exceed 30,000 per year (Serventy & Whittell 1976).

In Norway in 1846–1900, rewards were paid for 223,487 birds-of-prey, which included 61,157 Golden and White-tailed Eagles up to 1869, dropping to 27,319 eagles in 1870–99 (Johnsen 1929). As late as 1963, bounties were paid on 168 eagles.

In the Netherlands in 1852–57, rewards were paid for 219 'eagles', 12,787 'falcons', 2,828 'goshawks', 16,626 'sparrowhawks', 1,756 'buzzards' and 5,017 'harriers', making a total of 39,233 birds-of-prey, probably largely migrants (Braaksma et al 1959).

In Germany, in the Nordrhein-Westfalen districts in 1951–68, a total of 210,520 raptors was recorded killed; in Lower Saxony in 1959–63, a total of 38,432; in Schleswig-Holstein in 1960–68, a total of 37,793; and at Hessen in 1951–67, a total of 61,353 (Bijleveld 1974).

From a single Scottish estate at Glengarry in 1837–40, the kills included 98 Peregrines, 78 Merlins, 462 Kestrels, 285 Buzzards, 3 Honey Buzzards, 15 Golden Eagles, 27 White-tailed Eagles, 18 Ospreys, 63 Goshawks, 275 Kites and 68 harriers, making a total of 1,372 birds-of-prey (Richmond 1959).

From one 1,200 ha hunting preserve in southern England in 1952–59, 344 Sparrowhawks (Ash 1960).

Bijleveld (1974) has recently assembled from official statistics totals such as these for many European countries. He estimated that in the 20 years up to 1970 several millions of raptors had been killed on the continent by game-bird hunters alone, with especially large numbers in France and Germany. The sheer magnitude of such figures has led some people to doubt them, but they are repeated in similar order in region after region, and in each case feet or beak were required as proof of killing. (Moreover, in the nineteen-sixties, similar numbers to those killed by hunters also died in several areas in seed-dressing incidents, Chapter 14.) Confusions of species in bounty schemes were probably common, however. The annual figures for particular estates or districts often included many more raptors than could have lived there at one time, a testimony to the effects of movements or to the existence of neighbouring less disturbed populations, from which new recruits continually came.

When culling occurred on migration routes, the totals were often extremely large, but drawn from populations covering a wide area. Each autumn in southwest France, an estimated 30,000 to 50,000 small raptors fall victim in the nets of birdcatchers, in addition to the many killed in Pyrenean passes by 'pigeon shooters'. Among the ringed raptors reported from southwest France, 21% came from Scandinavia, 12% from Poland and Russia, 12% from central Europe, 7% from south Germany and Switzerland, 7% from north Germany, 11% from England, 23% from the Low Countries and 7% from France (Yeatman, in Bijleveld 1974).

The numbers alone tell us little about the effect of this slaughter on populations, except that in some cases they must have represented at least the bulk of the local stock. Comparing the figures of the present century with those of the previous, the main difference is in the reduced representation of eagles and other large species in many lists, and their complete disappearance from others. That this was in some regions due to the culling itself is suggested by the large initial kills, followed by a swift decline, as the scheme continued. For example, in 1676 at Tenterden in Kent, 'an intensive campaign for the thinning out of vermin' began, and in the next ten years payments were made for 380 Red Kites, after which numbers dropped away rapidly, with annual totals of 35, 13, two and two (Ticehurst 1920). Likewise, the Scottish Glengarry figures included at least four species which were no longer present a century later, but for which the habitat still seemed suitable (two have since returned). Evidently there have been long periods in recent history when raptor numbers in many European areas were well below what habitats would support.

In other lists, there was no obvious decline in the totals over many years, which suggests that in these areas the hunters were merely cropping the populations concerned, and causing no long-term decline. This is indicated in some official statistics from Austria, which show that between 1948 and 1968, premiums were paid annually on about 12,000 to 20,000 birds (Table 53). In this and other parts of Europe, the cull by hunters was especially great in the severe winter of 1962–63, when the birds were more than usually vulnerable. Likewise, the 6,000 Goshawks destroyed annually by Finland's 170,000 hunters is also thought to be causing no long-term decline in the Goshawk breeding population, but most of these birds are juveniles killed in the few months following breeding (Moilanen 1976, Saurola 1976). Care is needed in using only the records of recent years, however, because in any long-running bounty scheme covering several species, one might expect there to be less change as the years go by, as the larger species are eliminated to leave the smaller, more resilient ones.

(b) Changes in persecution and population status
The evidence is of two kinds: first, where the distribution of a species over a wide area fits with variations in persecution; and second, when some marked improvement in the status of a species follows a known decline in killing. The Buzzard in Britain provides an example of both kinds of correlation. In 1800 the species bred throughout the country; by 1860 it had been eliminated from all but a few western districts; by 1954 it had spread considerably; and by 1970 it had spread even further (Moore 1957, Sharrock 1976). These changes correlate with changes in the intensity of game preserving, helped in later years by a change in attitude. A particularly detailed survey in 1954 showed that the distribution of Buzzards at that time closely mirrored the distribution of game-keepers (Figure 36). The bird was commonest in districts where game-keepers were scarcest and absent altogether from districts where keepers were numerous (nesting habitat was available throughout).

Further evidence for the influence of game-keeping on British raptor populations came during the 1914–18 and 1939–45 wars, when many keepers were employed on other things. At these times all raptors increased and extended their range, and for the commoner species the changes were reflected in the numbers of nestlings ringed each year by amateur bird-ringers (Newton 1972). There was a big increase in the numbers of Sparrowhawks and other raptors ringed within two

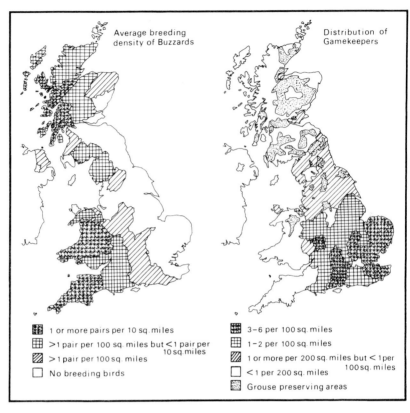

Figure 36. *Distribution of breeding Buzzards in Britain in 1954 (left)
compared to the contemporary distribution of gamekeepers (right).
Re-drawn from Moore 1957.*

years of the war starting and a rapid drop to former levels within two
years of the war ending (Figure 37). The numbers of nestlings ringed
must to some extent have reflected the numbers available for ringing,
the increase representing the combined effects of improved population
and breeding success, under lessened game-keeping. It was also during
this war that the Hen Harrier became properly re-established on the
Scottish mainland, nesting largely undisturbed in the young forestry
plantations which had appeared since it was here before. Early naturalists wrote about a similar increase in raptors during the 1914–18 war,
but ringing was not sufficiently developed to document it; and in both
wars increases were not confined to Britain, but occurred throughout
Europe. Wolves and other mammal predators gained a similar respite
and also spread.

It is curious that, with such widespread persecution, only one
species seems to have been exterminated altogether by it. This was the
caracara *Polyborus lutosus*, which was particularly vulnerable because

it occupied a small area, namely Guadalupe Island, off western Mexico. Although 'abundant on every part of the island' in 1876, not one could be found in 1906 when a collecting expedition scoured the island from end to end. In the interim, goat-herders had initiated a campaign against them, and wiped out the lot. Not much later, the people themselves left.

Where persecution was insufficient to eliminate populations, it sometimes affected their age structure and breeding success, as was apparent among Golden Eagles in Scotland. Sandeman (1957) compared the breeding in deer areas, where eagles were not persecuted, with that in grouse and sheep areas, where they were persecuted (Table 54). In deer areas, there was no instance of an eagle lacking a mate, but in sheep and grouse areas eight such instances were recorded. In deer areas there was no instance of an adult eagle paired to an immature partner, but in grouse and sheep areas there were four such instances. Both these features were symptoms of excessive killing. An immature partner in a pair meant either that the pair did not lay or that they produced infertile eggs. The mean size of successful broods was the same throughout, but the overall brood size, when pairs that raised no young were taken into account, was 0·6 in deer areas and 0·3 in grouse and sheep areas. In these latter areas, killing was suppressing the breeding output so much that the population could not have been sustained without continued immigration. In populations subjected to less persecution, the removal of breeding birds does little more than create temporary gaps, which are soon filled by new recruits. Or it may reduce breeding rate, but not enough to cause population decline.

Widespread shooting over decades seems also to have affected the

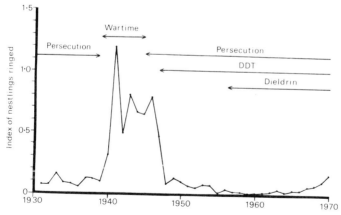

Figure 37. An index, based on ringing, of the output of young Sparrow-hawks in Britain showing the temporary increase during the war, associated with the decline in gamekeeping. The index is the percentage that nestling Sparrowhawks formed of all nestling birds ringed in Britain each year. Re-drawn from Newton 1972.

behaviour of individual raptors and their reactions to man, perhaps partly through the selective removal of the tamer individuals. This is apparent from comparison of, say, the African populations with the European ones, or even of the western with the eastern North American ones. The African ones generally show themselves more, nest in closer association with man, and allow a much closer approach before taking flight than do their European equivalents. Similarly, Snyder & Snyder (1975) wrote that 'one need only visit a country such as Guyana in South America, which has rigidly enforced laws prohibiting firearms, to realise how abundant and relatively tame raptors can become in the absence of shooting'. Perhaps the extreme in tameness is found in parts of Hindu India, where several species nest at extremely high density within cities; the food supply is there, but these birds could not remain without a tolerance from the human population (Galushin 1971). The Galapagos Hawk is also extremely tame, and will allow observers close enough to read the colour rings. Shooting seems to have affected the nest defence behaviour of raptors, as discussed in Chapter 5.

(c) Ring recoveries

Compared with most other birds, not only are more ringed raptors recovered, but very many of the recovered birds are reported as shot or trapped. For the common British species, the percentage of ringed birds that were later recovered varied from 7% to 14%, and the proportions of these reported as killed up to 1954 was as high as 68%, depending on species (Tables 55 and 56). The proportions recovered are greater than those for some waterfowl and game birds exposed to proper hunting seasons, and for recognised pest species. Only large waterfowl and cormorants showed a higher recovery rate, the former being legally hunted or specially studied and the latter killed as pests. After 1954, when protective legislation was introduced for raptors, the proportions reported as killed declined (Table 56). This may have been genuine, or it may also have been due to many people omitting to report the birds they had killed, or falsifying the cause of death. In both periods, birds reported as 'found dead' may have included some killed by people.

Similar analyses of European recoveries also indicated the significance of persecution, the Buzzard and Goshawk being particularly susceptable (Table 49 and 57). They also reflected the regional variations in shooting pressure. Among Kestrels ringed as nestlings in Holland, intentionally killed birds formed 82% of all recoveries from Belgium and France, but only 10% of those from other west European countries. The mean annual mortality calculated from the two sets of recoveries was significantly different, at 59% and 44% (Cavé 1968). In some species, the recoveries implied a difference in wariness between young and old birds, for more of the birds recovered in their first year had been shot or trapped than of those recovered in later years. Among Peregrines in North America, the figures were 50% and 33% for first-year and older birds, among Prairie Falcons they were 67% and

45%, among Goshawks in Fennoscandia they were 87% and 78% and among Kestrels in Holland they were 34% and 19% (Enderson 1969, Höglund 1964, Cavé 1968).

In eastern North America, recoveries from three species suggested a reduction in shooting in recent years (Table 58), and in two of these, a lessening in the overall mortality. The Cooper's Hawk is commonly known as the 'chicken hawk' and has long been a favourite target of hunters (Henny & Wight 1972). Ringed nestlings that were recovered in their first year dropped from 21% in 1929–40 to 10% in 1941–57, but the proportion of these reported as shot remained about 74% throughout (Table 58). These recoveries gave estimates of the annual mortality for first-year and older birds of 83% and 44 ± 7% for the years up to 1940, and of 78% and 34 ± 4% for the years after 1940. In the early period the heavy hunting pressure could only have caused population decline, a trend which was confirmed from long-term counts at migration stations.

In the American Kestrel, diminished shooting pressure was associated with a significant drop in the overall juvenile mortality from 69% in 1935–45 to 61% in 1946–65, but not in the overall adult mortality, which remained the same throughout (Henny 1972). In the Red-tailed Hawk, shooting pressure had always been slight, and the apparent drop in recent years had no effect on the overall mortality of either juveniles or adults, as calculated from the recoveries (Henny & Wight 1972). So far as I know, these American studies are the only attempts to find the effects of shooting on population trends, but similar studies could easily be done for some European species on the recoveries available.

Summarising, ringing recoveries have shown the importance of persecution in the overall mortality of raptors compared to that of other birds, as well as differences in the importance of shooting between particular species, regions and age groups. For some regions, they have also suggested that shooting pressure has dropped in recent decades and that, in at least two species, diminished shooting has led to reductions in overall mortality.

(d) Studies of carcasses

The significance of human persecution in the overall mortality of species is also shown from an examination of birds found dead or dying. In the American scheme mentioned in Chapter 12, 231 Bald Eagle carcasses were received for autopsy during the period 1966–74. A minimum of 98 (43%) had been shot, poisoned or trapped, making direct persecution the major single cause of death. Other birds died from organo-chlorine contamination or from natural factors. In an earlier survey, Sprunt (1963) reported that 91 (77%) of the 118 Bald Eagles known to have died in 1962 had been shot. In the other American scheme described in Chapter 12, 23% of 850 birds of prey found moribund had been shot and another 15% had been trapped (Redig 1978).

Further evidence is available from individual field studies. Among

35 raptors of various species found dead or disabled in northeast Scotland in 1964–69, 20% had been killed by man (Weir 1971). In a study of Buzzards in the same area, Picozzi & Weir (1976) used a trained dog in regular searches for poisoned baits and for dead birds. They found 52 dead Buzzards in the period 1964–72, and ascertained the cause of death in all but five: 29 (54%) were poisoned and nine (15%) were shot or trapped, making a minimum of 69% killed by man. Of the 42 birds aged, 27 (64%) were in their first year. Between 1968 and 1972, the authors found poisoned baits on 12 of 15 estates within 30 km of the study area, together with 28 Buzzard carcasses. In four years before poisoning started, they found six adult pairs each spring on two of these estates, with 2·3 pairs on average producing fledged young annually. After poisoning started there were only four pairs, with 0·5 pairs producing fledged young annually. Little wonder that the distribution of this species in Britain inter-digitates so closely with that of gamekeepers. Like all the other species discussed in this section, the Buzzard was legally 'protected' during the period concerned.

Such figures show that a high proportion of the deaths of birds examined by biologists can be attributed to direct killing by man. Whether the proportion is representative of all deaths depends on how typical a sample was found. One can easily imagine that in the same areas some birds might die in ways which would make them unlikely to be found (for example killed and eaten by predators). This would mean that the role of human persecution in the overall mortality would be exaggerated. Ringing recoveries suffer from the same drawbacks when used to apportion causes of death; but both methods give useful comparisons with results from other birds and show the prevalence of persecution on sparse protected populations.

Studies of local populations could not be expected to reflect the general levels of persecution, because biologists normally select study areas where it is low. Nonetheless, human interference was the commonest cause of adult mortality and nest failure recorded in many studies (Chapter 8). In some early British work, it accounted for every nest over a several year period (Owen 1916–22, Rowan 1921–22).

METHODS OF KILLING

When the gun is used, the birds are often shot at the nest, or whenever they fly within range. In parts of Europe it was common to set out a live Eagle Owl and shoot from a hide any raptors or crows that came to mob it; and in both Europe and in North America, it was formerly common to shoot large numbers of raptors at concentration points on migration. As for traps, the commonest types are leg traps of various kinds; these have sprung jaws which snap together when the birds steps on a central treadle, holding firm until the bird is removed or dies. They are placed on nests, around carcasses or on natural or artificial perching places, as in the pole trap. For eagles and other large

25 Merlin females at nests in Wales with eggs and young. Photos: D. Green.

26 Merlins in Britain nest either on the ground among rank vegetation, or in trees in the old stick nests of oth
birds. Photo: D. Green. (Upper) Male incubating at ground nest; young in old nest of crow (lower). Photo:
Papke.

species in open country, it is usual to build a small mound of stones on which to place the trap. Another type is the cage trap with two compartments, in one of which live pigeons or other animals are placed to act as decoys; in the other the raptor is caught alive when it steps on a treadle which closes the lid. Situated near woodland, such traps are especially effective against accipiters, but need daily attention to keep the decoys fed and watered.

Regarding poisoning, raptors are sometimes killed deliberately in this way, and sometimes incidentally during attempts to get rid of other animals, such as wolves and foxes. Widespread poisoning alone has caused serious declines, as shown from (a) the coincidence between the periods of poisoning and decline, (b) the finding of corpses at bait, sometimes in numbers large enough to form the bulk of a local stock, and (c) the presence of poisons at lethal levels in tissues. Several cases have been documented in recent years, involving local populations of eagles, vultures and others (Bijleveld 1974). One striking instance was the virtual disappearance in recent decades of the once-common Griffons from Roumania and Bulgaria, linked with the widespread use of strychnine for wolf control. In one Roumanian area, 60 White-tailed Eagle carcasses were picked up in one week, and in another area ten Egyptian Vultures were found dead round a single bait (Bijleveld 1974).

The main poisons used in different areas are shown in Table 59, and their toxicities are compared with those of some common pesticides (some chemicals are used for both purposes). In Britain, the most-used poisons include the traditional strychnine, the more recent organo-phosphorous pesticide known as phosdrin (or mevinphos) and the narcotic alpha-chlorolose. Instances of secondary poisoning (of a second animal being killed by eating the first) are known from all these compounds. When used in eggs they kill a few raptor species (mainly harriers), but are very effective against corvids, whereas on meat baits they kill many raptors. The advent of alpha-chlorolose has led to a great increase in the illegal persecution of British raptors because, compared to other poisons, it is relatively safe to use. Instances came to light in 1971–76 from analyses performed by the Government Agricultural Departments of carcasses found by amateur naturalists (Brown et al 1977). Most carcasses were found during March–May each year, which is when gamekeepers and shepherds have a blitz on 'vermin'. Different poisons were favoured in different regions, depending on local availability.

In North America, the frequently used 'ten-eighty' poison (sodium monofluoracetate) has now been outlawed. It was used mainly against coyotes and wolves, but killed many raptors and other birds incidentally and was also a source of secondary poisoning. Progress towards the selective killing of the target species was made with the development of the 'coyote-gitter'. This consists of a cartridge which is buried below ground, and an odorous wick which is left above. The wick is attractive only to coyotes and other canids, and when it is tugged, the cartridge shoots cyanide into the animal's face. Use of this method almost eliminates the chance of killing raptors.

THE SHOOTING OF EAGLES FROM AIRPLANES

Wherever eagles live alongside sheep, they feed from dead sheep and lambs, and also kill some live lambs. This is true of the Golden Eagle in parts of Europe and North America, the White-tailed Eagle in Norway and Greenland, the Wedge-tailed Eagle in Australia, and the Black and Martial Eagles in southern Africa. The killing of birds that follows reached a considerable scale in western Texas and southeast New Mexico after the discovery that Golden Eagles could be shot down from airplanes with complete success. Over a period of 20 years until it was banned in 1962, 1,000–2,000 birds were shot annually in sheep ranching areas (Spofford 1964).

The ranchers clubbed together to finance the operations which were performed by a small number of pilots, based at airports throughout the area. Lambing occurred at different dates in different regions, and a few days before a particular flock was scheduled to lamb, the rancher would arrange for a 'shoot-off'. This would continue over several days until virtually all the eagles had been removed from a large area, extending beyond the ranch into surrounding mountains and rough country, up to 150 miles from an airport. As the eagles were removed, others would begin drifting in from surrounding areas, so it was usual to conduct a second shoot-off 2–3 weeks later. Again the total numbers killed in an area greatly exceeded those present at any one time.

Since these birds were winter visitors, it was not possible to assess the effect of the killing on breeding populations but, with such large numbers involved, it could clearly have drained the populations of a considerable area.

ROADSIDE SHOOTING

Sometimes killing of raptors has been done as a sport in its own right, such as the shooting of migrants mentioned earlier. In addition, roadside shooting was frequent in parts of the western United States. Ellis *et al* (1969) studied the mortality along an 18 km road paralleling a power line in Utah. The poles were used as perches by raptors, and in the 18 month study the authors found 38 carcasses on the ground below, an average of 2 per km, mainly Golden Eagles, with some Bald Eagles and others. Mortality was heaviest in autumn/winter, which was the season of heaviest game hunting in the area. Men were seen to drive back and forth along the road searching for tempting targets; and at night they used spotlights to locate the roosting birds. During the same period, only one carcass was found along another segment of power line, not parallelled by a road. Roadside shooting could of course be practised only where raptors had not been exposed to much previous persecution, and where they were still tame enough to allow a close approach.

OTHER HUMAN DISTURBANCE

Compared to killing, the effects of other human interference on raptors may seem slight, but in fact human presence has become increasingly important in rendering suitable habitat unattractive to raptors. This presence is often tourism or recreation, as an increasingly mobile public intrude on wilderness and other areas that were previously undisturbed. One of the best studies of the effect of continued human presence was in a forest which overlapped the Dutch-German border, roughly half and half (Table 60). Forest management was similarly intense on each side, but there was much greater human activity on the Dutch side, with more roads, more houses, more holiday cottages, and more recreation generally. Correspondingly, only four pairs of large raptors bred on the Dutch side in 1969, compared with 37 on the German side.

Among Bald Eagles in North America and White-tailed Eagles in Europe, many instances are known of territories being abandoned following the opening of lakes for recreation and the building of holiday chalets (Sprunt 1969, Bijleveld 1974). In Washington State wintering Bald Eagles were similarly thought to be confined by human presence to an area smaller than they would otherwise use. Old birds were more sensitive to disturbance than were immatures, the mean take-off distance on approach by a man being 196 and 99 metres respectively (Stalmaster 1976). Similar age-linked differences were noted among Golden Eagles in Utah, and in both species they might have accounted for more immatures being shot (Ellis *et al* 1969).

Another worry to conservationists is provided by low-flying aircraft, which are being increasingly used on construction and survey work in remote areas. In tests of the effects of helicopters on nesting Gyr Falcons in Alaska, Platt (1977) made 51 flights over 23 nests. The birds flew from the cliffs each time, but did not fail in their breeding; the next year, however, they had shifted to another site significantly more often than birds which were not disturbed in this way. Helicopters have been much used in surveys of breeding Peregrines in Alaska, and on this species there is no indication of any adverse effect (C. M. White). The levels of disturbance that raptors will take are likely to vary regionally, with the birds more tolerant of human presence where they are exposed to people but not harassed or shot at.

The problem created by tourism and recreation is familiar to every naturalist. One solution is to try to restrict such development as much as possible to the places where it will do least harm, and prevent its uncontrolled spread through all wild places.

SUMMARY AND CONCLUDING REMARKS

Persecution causes population declines only if it adds to the natural mortality, rather than replaces it. Large species with slow breeding rates are less able to withstand heavy persecution than are small species with fast breeding rates. Over the last 150 years, persecution has eliminated some of the bigger species from large parts of Europe, and is still responsible for restricting the distribution of others. In Britain over this period, the ranges of several species have contracted and expanded again with the rise and partial decline in game preservation, with temporary expansions during the two wars when many gamekeepers were otherwise employed. In some lists of bounty payments, certain species declined or disappeared in the records during the operation of the scheme, suggesting that the killing reduced or exterminated them. But in other lists no decline in the numbers occurred over a long period, suggesting that hunters were merely cropping the population, and causing no long-term decline. The importance of deliberate killing of raptors is shown by the large proportions of ringed birds that were later recovered, and by the large proportion of these recovered birds that were reported as shot. Recovery rates were higher for European raptors than for many game-bird and waterfowl populations exposed to regular hunting seasons. Widespread use of poison on meat baits has had the most damaging effects on populations in Europe, often where it was used primarily against wolves or foxes. In recent years, human presence through tourism and recreation has increasingly rendered suitable habitat unattractive to raptors, thus further reducing their breeding numbers.

Throughout this chapter, I have avoided any discussion of whether the killing of raptors is warranted as a means of protecting livestock or game birds. Several studies are now available which show that the effect of such predators on these animals is negligible or small, though in occasional cases (as when Goshawks concentrate around Pheasant release areas) the losses can be greater (Lockie 1964, Leopold & Wolfe 1970, Newton 1970, Kenward 1977a). Whether you regard such damage as justifying the widespread carnage of birds of prey depends on what value you place on the presence of these birds in the environment. Many people appreciate the great beauty of raptors and enjoy having them around. They regard them as a natural and proper part of any landscape, and as epitomising all that is wild and free. On the other side, I can more easily sympathise with the shepherd earning a living than with the game hunter pursuing a hobby, and for whom the slaughter of raptors usually makes little difference to the final bag. The obliteration of whole populations from large areas of Britain and Europe seems an extraordinary selfish act by a small sectional interest. Perhaps it is this knowledge that makes sympathy for the game preserver's case hard to find, together with the total disregard that many landowners, gamekeepers and hunters still show for the protection laws.

CHAPTER 14

DDT and other organo-chlorines

Of all the insecticides ever used, it is the organo-chlorine compounds
that have done most harm to raptors and to other wildlife. These
compounds include DDT, and cyclodienes such as aldrin, dieldrin,
endrin and heptachlor. Besides being toxic, they have three main
properties harmful to wildlife. First, they are chemically extremely
stable, which means that they persist more or less unchanged in the
environment for many years. Second, they dissolve in fat, which means
that they can accumulate in animal bodies, and pass from prey to
predator, concentrating at successive steps in a food-chain. Birds of
prey are near the tops of food chains and are especially liable to
accumulate large amounts. Thirdly, organo-chlorines become dispersed over wide areas in the bodies of migrant animals, and in air and
water currents, so can affect populations far removed from areas of
usage. Moreover, at sub-lethal levels of only a few parts per million
(ppm) in tissues, they can so disrupt the breeding of raptorial birds as
to lead to population decline. Some other pesticides may cause locally

Illustration: Peregrine and Wood Pigeon

229

heavy mortality among wildlife, but because these pesticides break down more quickly, they neither have lasting effects, nor do they affect organisms in areas remote from places of application.

Dichlor-diphenyl-trichlor-ethane (later known as DDT) was first synthesised during the last century, but its remarkable properties as an insecticide were not discovered until early in World War II. It contributed to the outcome of the war through suppressing epidemics of insect-borne diseases, such as typhus, which had killed thousands of troops in World War I. But not until after 1946 was DDT used on a wide scale against agricultural pests and against malarial mosquitos. The world over, it has remained the most effective chemical against many insect pests, and today its use mainly helps to produce cotton economically in some poor countries. To man, its direct toxicity is said to be about the same as that of aspirin. Little wonder that for years DDT was considered to be the ideal insecticide, until its insidious side-effects on fishery and wildlife resources became apparent.

Inside the animal body, most DDT is quickly changed to a more stable metabolite, DDE, which is also of low direct toxicity; but it is DDE which is responsible for most of the adverse effects on bird breeding. Its main effect is in causing shell-thinning and egg breakage, but it also increases mortality of embryos in unbroken eggs, and in these ways lowers the breeding rate. The phenomenon of shell-thinning was first discovered by Ratcliffe (1967) on eggs in British collections.

Another organo-chlorine that came into agricultural use about the same time as DDT was sold under the names of gamma-BHC, gammexane or lindane (Table 61). Both compounds were used chiefly as insecticide sprays. After 1955, the much more toxic dieldrin and other cyclodiene compounds came into use, chiefly as seed-dressings, the thin coating on individual grains protecting against insect attack but also poisoning any other animals which ate the grains. These compounds were important to wildlife chiefly because of the large scale direct mortality they caused, the raptors dying as a result of taking contaminated seed-eating birds and mammals.

As the side effects of these various compounds became known in the nineteen-sixties, nation after nation placed restrictions on their use. Thus the general trend over much of Europe and North America was of increasing use of organo-chlorines from the late nineteen-forties to the early 'sixties, followed by decline, so that by 1975 relatively small amounts were used. However, manufacture continued and the main market shifted southwards. Areas of heaviest use now lie in a broad band across Central America, northern Africa, the Middle East, India and southeast Asia (Goldberg 1975). Particular black spots include Guatemala, Cyprus, Turkey, Lebanon, Pakistan and Ceylon. American and European factories still make most of this DDT, but many Asian and African countries are now producing their own. Thus at the time of writing, organo-chlorine use on a world scale is still increasing, especially in the developing nations.

Another group of organo-chlorine compounds which affect raptors are the polychlorinated biphenyls (PCBs). These are not insecticides, but industrial products used widely as plasticizers in paints, insulants, lubricants and other materials, from which they reach the environment (Table 61). PCBs are chemically similar to the insecticides, but even more persistent, with some similar effects on animals. They have been produced since 1929, and in steadily increasing amounts, so that by 1970 the annual production was estimated to have reached about half the maximum production of DDT. For a long time they were difficult to separate in animal tissues from the DDT-type compounds, and not until 1966 were they positively identified at great concentration in a Swedish White-tailed Eagle (Jensen 1966).

PCBs have since been found in almost all organisms examined, often at high levels in marine animals close to industrial cities (Riseborough *et al* 1968). In 1970 the main manufacturer for Britain and America (Monsanto) withdrew their use from those purposes most likely to lead to pollution. This was less effective in America than was hoped, partly because of importations from other countries (Stickel 1975), and in both Britain and America PCB levels in some animals continued to rise through the nineteen-seventies. Over most of the world, the production of these compounds continued unabated. Compared to the insecticides, PCBs have less obvious effects on animals, but in birds they cause some embryo deaths.

All these various organo-chlorines affect raptor populations through a well established sequence of events: (1) application of insecticide (or release of PCB); (2) contamination through the food-chain of prey species; (3) concentration of organo-chlorine as it passes from prey to raptor; (4) at sub-lethal level, breeding failure in the raptor, partly through shell-thinning and egg breakage; leading in turn to (5) population decline. An increased mortality from acute poisoning may also have contributed to the decline of certain populations. In the sections below, the evidence that has accrued for these various steps is examined in detail.

But it is first appropriate to mention sampling problems, for much depends on the results of chemical analyses. By the time the impact of organo-chlorines had been realised, some raptors had become extremely rare, and analyses of residue had to be restricted to specimens found dead or to unhatched eggs. In consequence, samples were often small and collected less randomly than biologists would have liked. Only in later studies on commoner species was it possible to obtain large random samples for chemical analysis and to collect eggs fresh before their fate had become apparent. Residue levels were expressed either as ppm in fresh (wet) weight, ppm in oven dry weight, or ppm in lipid (fat) weight. So when comparing results from different studies, check what units they are expressed in.

FROM ENVIRONMENT TO EGG

General occurrence of organo-chlorines

There can be no doubt that organo-chlorines soon became sufficiently widespread in the environment to affect raptors on the extensive scale observed. In all major regions of the world, they were found in air and water, in soils and in mud sediments on lake and sea beds. In Europe and North America, thousands of birds of many species were deliberately collected from wide areas for chemical analysis (Prestt *et al* 1970, Keith & Gruchy 1972, Prestt & Ratcliffe 1972, Peakall 1975). Organo-chlorines proved to be present in almost all these birds and, within each species, levels were generally greatest in areas of usage. The DDE and PCB compounds were the most widely distributed; the cyclodienes were present at lower levels and, at least in America, were more patchy in occurrence. Organo-chlorines were also universally present in seabirds, and again at greatest levels near to sources of pollution (Moore 1965, Risebrough *et al* 1967, 1968). They were at especially high levels in the enclosed waters of the Baltic and Mediterranean, but DDE was found even in such remote regions as Antarctica (George & Frear 1966).

In any one bird population, individuals varied greatly in the amount of organo-chlorine they contained, and usually a few had very much more than the rest (Figure 38). The proportions of different compounds varied between populations in different areas, partly reflecting levels in the local environment. In some populations DDE exceeded PCB and in others the reverse occurred. But whatever the overall ratio, among the birds in any one population the levels of these two groups of compounds tended to vary in parallel, so that individuals high in one group were also high in the other (Risebrough *et al* 1968, Newton & Bogan 1974). This made it hard to separate the effects on the birds of the two types of compounds. Only by comparing populations with

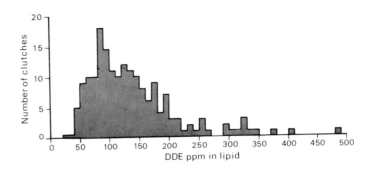

Figure 38. *Distribution of DDE among Sparrowhawk clutches from southern Scotland in 1971–74. As is usual in birds, the distribution is skewed, with a few clutches containing exceptionally large amounts. From details in Newton & Bogan 1978.*

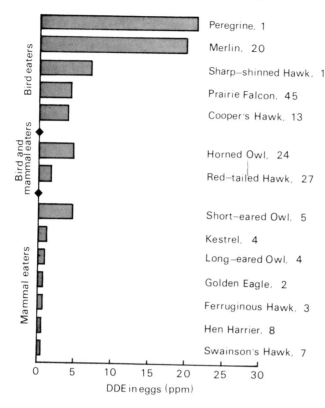

Figure 39. Mean DDE levels in the eggs of various predatory birds on the Canadian Prairies in the nineteen-sixties. Bird-eaters contain the highest levels. Numbers are specimens examined; ppm in wet weight. Re-drawn from Keith & Gruchy 1972.

different DDE: PCB ratios, by using sophisticated statistical techniques, and by experimental work in laboratories was this problem overcome (see later).

Magnification of residue levels in food chains

Concentration at successive trophic levels was evident in all studies in which different kinds of organisms from the same place were examined (Moore 1966, Walker *et al* 1967, Keith 1966, Vermeer & Reynolds 1970, Peakall 1975). Within areas, concentrations were lowest in plants (1st trophic level), higher in herbivorous animals (2nd trophic level), and higher still in carnivores (3rd trophic level), and so on up the food chain. The rates of concentration were usually greater in aquatic than in terrestrial systems, probably because many aquatic animals absorb organo-chlorines directly through their gills, as well as from their food. Fishes rapidly pick up pollutants from the water, and concentration factors of 1,000 or 10,000 times between water and fish are not uncommon (Stickel 1975).

Among the raptors in any given area, mammal-eaters invariably contained lower organo-chlorine levels than did bird-eaters or fish-eaters (eg Conrad 1977, Henny 1977). This was presumably because the mammal-eaters were living on herbivorous prey (a food chain with two steps), whereas the bird-eaters and fish-eaters were living largely on carnivorous prey (a food chain with at least three steps). Differences in organo-chlorine levels between raptor species are shown in Figure 39, based on data from the Canadian prairies (Keith & Gruchy 1972). Among three accipiter species in western North America, residue levels in eggs were related to the proportion of insectivorous birds in the diet (these being more contaminated than herbivorous or omnivorous prey, Snyder et al 1973). The Sharp-shinned Hawk fed almost entirely on insectivorous birds and had the highest DDE levels in its eggs, the Goshawk fed mainly on herbivorous mammals and birds and had the lowest DDE levels, while the Cooper's Hawk was intermediate in both respects. Moreover, in this last species a relationship existed between the diets and egg DDE levels of particular pairs ($P<0.05$, on a rank correlation test). A similar difference between these three accipiter species was noted in a later, more detailed study in another region (Henny 1977).

Mammal-eating raptors also tended to be affected chiefly in areas where organo-chlorines were applied directly, whereas the bird-eaters and fish-eaters were affected over much wider areas. This was partly because organo-chlorines were transported over shorter distances in the bodies of small mammals than they were in the bodies of migrant birds and fish or in water currents.

Perhaps the most organo-chlorine recovered from any bird was from some White-tailed Eagles found dead in south Sweden in 1965–68 (Jensen et al 1972). In one individual, DDT and PCB compounds comprised as much as 3·6% and 1·6% respectively of its body fat. The mean figures for four birds from this area were 25,000 ppm DDT/DDE in fat and 13,000 ppm PCB. Undeveloped eggs of White-tailed Eagles from southern Sweden contained an average of 1,000 ppm DDE and 600 ppm PCB in lipid, the highest levels recorded in raptor eggs.

Relationship between dietary and tissue levels

Raptors do not get their high organo-chlorine contents simply by storing the accumulated residue from all their prey, but rather because their average rates of intake are higher than in most other types of bird (Moriarty 1972). Organo-chlorines behave in the same way as various drugs and, over a certain range, settle in body tissues in direct proportion to the concentration in the diet (Figure 40). If an animal is put on a contaminated diet, the concentration of organo-chlorine in its body tissue builds up steadily (Lincer & Peakall 1973). If the rate of intake is constant and continues for long enough, the concentration in each tissue eventually reaches a plateau level. At this point the

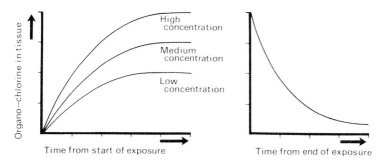

Figure 40. Rise and fall of organo-chlorine residues in tissues in relation
to levels in diet. Left – rates of increase in tissues/ at different dietary
concentrations; right – rate of decline in tissues after exposure ceases.
Partly after Moriarty 1975.

substance is being excreted (or broken down) at the same rate as it is
being absorbed. If exposure to the organo-chlorine then stops, the
concentration in tissues falls off, but it may take many months to reach
negligible levels again.

It is largely because of differing rates of intake that animals at
different trophic levels show different tissue concentrations. A carni-
vore with a high intake rate has a higher plateau level in its tissues than
does a herbivore with a low intake rate. In addition, species differ in
their ability to rid themselves of particular compounds, and in their
sensitivity to them. Death occurs if the rate of intake is so great that the
plateau level is lethal for that tissue (or egg). Nervous tissue is the most
sensitive, so the level in the brain gives the best indication of whether a
bird has died from organo-chlorine poisoning.

The constant relationship between DDE in diet and tissues (eggs) is
shown in Figure 41 for captive American Kestrels (Lincer 1975). Over
the dietary range 0·3–10 ppm, levels in eggs were consistently about
seven times greater (on a dry weight basis) than in food. From this and
other studies, the ratio of dietary to tissue (or egg) concentrations
varied with the tissue and with the species, but was always greater than
one, showing that some magnification of residue levels had always
occurred. Departures from linear relationships between diet and tissue
levels occurred when tissues changed in weight or composition, for
example when fat was lost.

Such precise studies were not possible in the field, but the results of
chemical analyses of the eggs and prey of wild raptors in particular
places were consistent with laboratory findings. Thus a 20-fold difference in DDE levels between the resident mammal-prey of the Rough-
legged Buzzard and the migrant bird-prey of the Peregrine in Alaska
was associated with a similar or greater difference in the DDE levels in
the eggs of these two raptors (Lincer et al 1970). Resident seabirds in
the Aleutian Islands contained very low levels of pesticide, and the

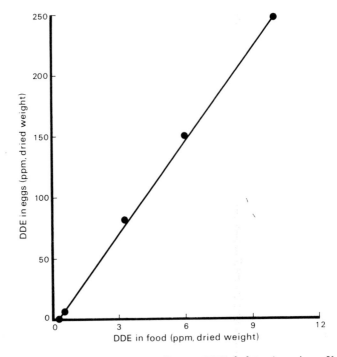

Figure 41. *Relationship between dietary DDE fed to American Kestrels and resulting DDE residues in eggs produced. Re-drawn from Lincer 1975.*

Peregrines that fed on them laid eggs much less contaminated than the Peregrines in interior Alaska (White *et al* 1973). Likewise, differences in the egg DDE levels between two Osprey populations in the eastern United States were linked with differences in the DDE levels in their respective food-fishes (Ames 1966, Wiemeyer *et al* 1975). As both Osprey populations wintered in the same area, it was the food difference on the breeding areas which caused the difference in egg levels.

In general in these studies, at least a seven-fold differential was found between body residues of prey and egg residues of raptor, but in some studies the difference was more than 20-fold (Cade *et al* 1968, Enderson & Berger 1968). It probably depended partly on the species, and partly on how long the birds concerned had been on that particular diet. So both field and laboratory studies show that residues concentrate at successive trophic levels, while laboratory studies also show that, over a certain range, the levels in the tissues of an animal bear a direct linear relationship to the levels in its diet.

Relationship between tissue and egg levels

Because organo-chlorines dissolve in fat, the level they reach in any one tissue depends on how much fat that tissue contains. The brain contains less fat (and so less organo-chlorine) than muscle, liver, ovary and eggs, and much less than adipose tissue, which is almost pure fat (Cade *et al* 1968, Vermeer & Reynolds 1970). Within the adipose tissue, the organo-chlorine is fairly harmless, but if for some reason the fat is suddenly metabolised, the residue shifts to other tissues and may cause death (Figure 42). Thus a bird in good fat condition can carry more organo-chlorine without hazard to itself than can a bird in poor condition, and deaths from direct poisoning are most likely to occur at periods of food shortage or migration, when fat is metabolised (Parslow & Jefferies 1973, Lincer & Peakall 1973, Bogan & Newton 1977).

The concentration in the female at the time of lay also influences the concentration in her eggs. This is expected because, until the shell is added, the egg is merely another body tissue. (The organo-chlorine is in the yolk, for there is practically no fat in the rest of an egg.) Thus strong correlations were found in the levels of DDE, PCB and dieldrin between the eggs and various body tissues of ten female gulls, and in dieldrin levels between the egg lipid and body lipid of 44 Prairie Falcons (Vermeer & Reynolds 1970, Enderson & Berger 1970). In a later study, DDE residues in the eggs of 47 American Kestrels were found to correlate closely with the DDE levels in the blood plasma of the birds laying the eggs (Henny 1977).

In the few species examined, the variation in levels between the eggs of a clutch was usually small. Among 20 Sparrowhawk clutches, two-thirds of all eggs had DDE levels within 12% of the mean for their clutch and PCB and dieldrin levels within 20%. The variations within these clutches were small compared with the big differences between clutches (Newton & Bogan 1978). Similar findings held for other

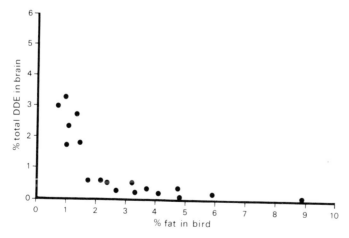

Figure 42. The proportion of the total body-load of DDE which was present in the brain of Sparrowhawks with different amounts of body fat. Loss of fat results in an increase of DDE in the brain. From Bogan & Newton 1977.

raptors and for other birds (Lincer 1975, Potts 1968, Vermeer & Reynolds 1970, Blus et al 1974). Hence, the levels in individual eggs could be taken as reflecting levels in the female at the time of lay, and as similar to levels in other eggs from the same clutch.

By laying eggs, female raptors rid themselves of organo-chlorine, an avenue of excretion not open to males. A female Sparrowhawk killed after laying had put more than half her total body load of DDE into her six eggs (Bogan & Newton 1977). Among several Sparrowhawk clutches, DDE levels did not necessarily decline from one egg to the next, but the eggs in two repeat clutches had lower DDE levels than their predecessors. Eggs from yearling females contained lower levels, on average, than eggs from older females in the same population, but among the older females there was no consistent trend of increasing or decreasing levels among the clutches from successive years (Newton & Bogan 1978). The lack of a trend within clutches, or between clutches from the same territories in successive years, was also noted among Cooper's Hawks in North America (Snyder et al 1973).

Problems of migrants

Although many raptors breed in remote regions, they cannot avoid serious contamination if they, or their prey, migrate to winter in areas where organo-chlorines are used. The marked declines of Peregrines that breed across the Arctic are linked with pesticide use on wintering areas in Western Europe and in Central and South America. Some summer prey of these Peregrines also winter in these areas, so the falcons are contaminated at both ends of their migration (Cade et al 1968). From surveys on breeding areas, it was estimated that by 1975, the Peregrines of northern Fennoscandia had declined by more than 90% and those of northern North America by about 50% (Lindberg 1977, Salminen & Wikman 1977, Fyfe et al 1976). On both continents, the only Peregrine populations that were not seriously affected were those resident in areas where pesticides were not used and which also fed on resident (pesticide reduced) prey. In Europe these conditions applied to the Scottish Highlands and other restricted areas, and in North America to the Aleutian and Queen Charlotte Islands (Fyfe et al 1976). Marked differences were also apparent between neighbouring Peregrine populations in interior Alaska: the birds on the Tanana River showed most DDE, shell thinning and population decline, followed in order by those on the Sagavanirktok, Colville and Yukon Rivers (White & Cade 1977). Perhaps these various populations winter in separate localities, with different DDT usage. In contrast to Alaskan Peregrines, Alaskan Gyr Falcons winter locally and feed on resident prey; they have low pesticide levels in their eggs, negligible shell thinning, and no declines in breeding success or numbers (Cade et al 1971).

SUB-LETHAL EFFECTS ON BREEDING

Shell-thinning

Shell-thinning is evident when recent shells are compared with those which were collected in the past and stored in museums and private collections. Thickness is not always measured directly, but according to the formula: weight/length × breadth (Ratcliffe 1967). This index is highly correlated with shell thickness and enables measurements to be made on museum shells without breaking them. From studies on egg shells, two main trends became apparent, both in Europe and in North America (Ratcliffe 1970, Hickey & Anderson 1968, Anderson & Hickey 1972, 1974):

1. Shell-thinning was more marked in the bird-eating and fish-eating species than in the mammal-eaters (Table 62). More than 15% thinning was found in populations of Peregrines, Merlins, bird-eating accipiters, Ospreys, Bald Eagles and White-tailed Eagles. Less than 10% thinning was found (with some local exceptions) in Golden Eagles, various buteonine hawks, harriers and kestrels. Also, in any one area the bird and fish-eaters were usually first to show shell-thinning and the first in which whole populations were affected.*

2. 1947 was the first year of widespread shell thinning, the year when DDT came into wide use as an agricultural insecticide. For many species, shells were available back to 1870, yet none showed any significant variation in shell index until 1947 or later (see Figure 43 for Sparrowhawk). Shells from agricultural areas changed first and others later, presumably because of the time taken for DDT to disperse. Gunn (1972) claimed that DDT usage in the late nineteen-forties was insufficient to affect raptor egg shells on such a scale. His criticism was largely quelled by the subsequent demonstration of DDE in the shells and membranes of 12 out of 13 Peregrine eggs collected in Britain in 1947–52 inclusive, but not in eggs collected in 1933 and 1936 and analysed as controls (Peakall *et al* 1976). DDE was also found in five Peregrine eggs collected in California in 1948–50, but not in one from 1935 (Peakall 1974).

Using recent eggs, the degree of shell thinning and the amount of DDE in the egg contents were highly correlated with one another, the thinnest shells occurring in the most contaminated eggs. This relationship was found in all raptors for which a reasonable sample of eggs was available, and also in some other birds. On the other hand, no correlations were found between shell-thinning and other organo-chlorine compounds, unless these other compounds were in turn correlated with DDE (Anderson *et al* 1969, Risebrough *et al* 1970,

* The thinnest shells on record were not from raptors, but from Brown Pelicans (*Pelecanus occidentalis*) in California, which were contaminated by the effluent from a DDT factory. Most eggs were so thin-shelled that they collapsed on laying, but the few intact ones showed an average of 34% thinning (Keith *et al* 1970, Lamont *et al* 1970). The extreme showed 95% thinning (Risebrough *et al* 1971).

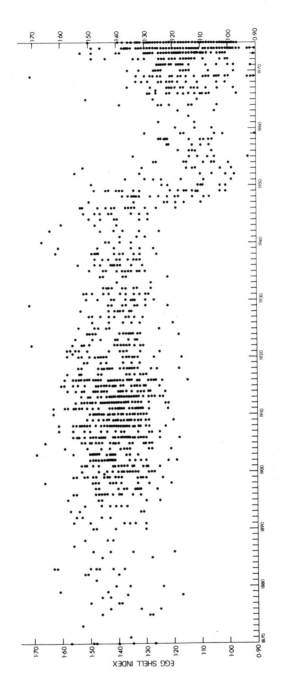

Figure 43. Shell thickness index of British Sparrowhawks, 1870–1975. Shells became thin abruptly from 1947, coincident with the widespread introduction of DDT in agriculture. Each spot represents the mean shell-index of a clutch, and more than 1,000 clutches are represented from all regions of Britain. Shells available in museum and private collections. From I. Newton, unpublished.

The European Kestrel uses an extremely wide range of nest-sites. (Upper) Female at ground nest on ;ney Islands. Photo: A. Gilpin. Male bringing prey to female (lower) in an old crow nest in S. England. Photo: F. 3lackburn.

28   (Upper) Young European Kestrels in tree-cavity nest. Photo: R. T. Smith. Gyr Falcon (lower) at nest in arc
Canada. Over much of the arctic the breeding success of this falcon fluctuates from year to year, according
the availability of the Ptarmigan which are the main prey in the winter and spring. Photo: D. Muir.

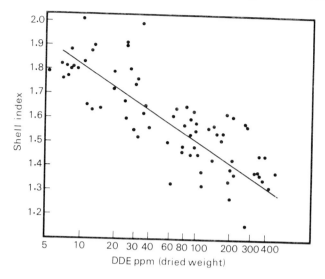

Figure 44. Shell-thinning in relation to egg DDE levels in Peregrines. With DDE on a logarithmic scale, a straight line relationship is obtained. Re-drawn from Peakall et al 1975.

Vermeer & Reynolds 1970, Cooke 1973, Newton & Bogan 1978). Thus compounds of the DDT group were the only ones that were consistently associated with shell thinning in wild populations.* Of course it was not the DDE in the egg that caused the shell-thinning, but that in the mother which the egg level reflected.

Among particular species, the relationship was linear between shell index and log DDE concentration (Figure 44). This was typical of the dose-response effect for many drugs and chemicals and it meant that further change in shell-index became progressively less with increasing DDE concentration. However, some families of birds were more sensitive to a given DDE concentration than were others: with 4–5 ppm DDE in their fresh eggs, raptors showed more than 15 % shell-thinning, whereas pelicans and herons showed about 10%, and songbirds and gamebirds less than 1 %. Thus raptors were particularly vulnerable to DDE both because of their positions high in food chains and because of their high sensitivity. Conversely, gamebirds were much less vulnerable because of their positions low in food chains and because of their low sensitivity.

Within the raptor family, species varied little in their response to DDE. Variations in shell-thinning shown by different species were due

* Regarding supposed effects of dieldrin on shell-thinning, we must discount the conclusions of Lockie et al (1969) on Golden Eagles, since DDE was not examined and of Enderson & Berger (1970) on Prairie Falcons, since in this study DDE could equally plausibly be implicated.

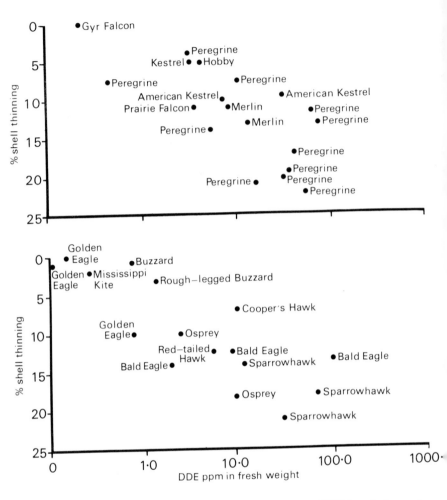

Figure 45. Mean shell-thinning in relation to mean DDE levels in eggs from different raptor populations; falcons above, others below. Using regression methods, the following equations have been obtained for different species to describe the relationship between shell-thinning and DDE levels.

For shell index (S.I.):
Sparrowhawk: S.I. = 1·568–0·186 × Log DDE, r = 0·359, P < 0·001 (Newton & Bogan 1974).
Merlin: S.I. = 1·257–0·224 × Log DDE, r = 0·370, P < 0·01 (Hodson 1975).
   For shell thickness as a percent of pre-DDT level (S.T.):
Prairie Falcon: S.T. = 95·78–18·31 × Log DDE, r = −0·72, P < 0·01 (Enderson & Wrege 1973).
Peregrine: S.T. = 101·68 − 16·65 × Log DDE, r = 0·75, P < 0·01 (Cade et al 1971).

more to different DDE levels, than to a different response to the same DDE level. This was apparent when the mean shell-indices and egg DDE levels of different populations were put on the same graph (Figure 45). The points for all populations lay more or less on a straight line, implying for them all a similar broad relationship between shell index and DDE level. In other words, all these raptor species in six genera reacted similarly to DDE. This was true at least as far as the mean response of the population was concerned, though there was probably some individual variation within populations.

Another important conclusion can be drawn from the shell-thinning data: namely that, whenever a population showed an average of more than 16–18% shell thinning over several years, its numbers declined. This held when comparing different species, different populations of the same species, and the same population at different periods. Within a species, moreover, the fastest declines occurred among populations showing the thinnest shells.

In conclusion, fairly precise relationships exist in raptors between egg DDE levels and shell indices, and between shell indices and population trends. The raptor species studied so far show remarkable consistency in these respects (though not all population declines are due to thin-shelled eggs, see Chapters 13 and 15). As a family, raptors are more sensitive than some other bird-families to DDE.

Egg breakage
Shell-thinning proved to be important because it was associated with egg breakage and with reduced breeding output. Cooke (1975) measured the shell strength of Sparrowhawk and Heron eggs, and found, not surprisingly, that it depended on shell thickness. In these species both the mamillary and the palisade layers of the shell were affected, and all four measures (shell index, shell strength, palisade and mamillary thickness) were highly correlated with DDE concentrations.

In the wild, breakage seemed to follow the same pattern in all species studied (Ratcliffe 1970, Newton 1976). The eggs of a clutch were usually broken one by one over several days, often early in incubation. Each egg was first pricked or dented, then the shell collapsed into several pieces, at which stage the female ate the contents and nibbled or carried away the shell. After this, the only signs, if any, that breakage had occurred, were bits of shell in the nest bottom or on the ground, or dried yolk on remaining eggs. Thus breakage was often recorded as egg disappearance.

In several studies, eggs from broken clutches were found to have thinner shells than eggs from intact clutches (Newton & Bogan 1978). Breakage might thus have resulted solely from the mechanical weakness of the shells, and the subsequent eating of the eggs could have been the normal response of the female for dealing with breakages that occurred naturally. Pieces of shell were sometimes found in the pellets regurgitated by the female near the nest (Newton 1973b, Snyder et al 1973). However, it is not certain whether organo-chlorines also influenced the birds' behaviour, resulting in clumsiness, or in the deliberate stabbing of intact eggs, as seen in contaminated Herons (Milstein *et al* 1970).

Other breeding failures

The breeding output of contaminated populations is further reduced by the failure of many incubated eggs to hatch, and sometimes by non-laying, desertion of clutches, and mortality of small young. The role of organo-chlorine in these kinds of failure is hard to assess, because they all occur naturally (Chapter 8), and have been less studied than shell thinning.

Unhatched eggs are usually fertile, but found to be addled (rotten and watery) or to contain dead embryos. Embryo death occurs at any stage. It might be due either to (a) impaired gas and water exchange through a thin shell (thin shells have fewer pores and lose water more slowly than normal ones); (b) a direct toxic effect of the residues on the embryo, or to (c) poor incubation by an affected adult (Peakall *et al* 1973, Peakall & Peakall 1973). Some contaminated American Kestrels showed poor brooding and parental care, while Cooper's Hawks showed desultory nest building (one bird) and the failure of the female to take food from the male and feed the young (two birds) (Lincer 1972, Snyder *et al* 1973). But again all such behaviour sometimes occurs naturally, so it is hard to assess the role of organo-chlorines.

The possibility of a detrimental influence of behaviour on breeding could be ruled out for some Ospreys in eastern North America. Eggs were exchanged between Connecticut and Maryland nests to find whether the poor breeding of Connecticut Ospreys was due to eggs or parents. Thirty Connecticut eggs incubated by Maryland parents hatched at the usual low Connecticut rate, but 45 Maryland eggs incubated by Connecticut parents hatched at the usual high Maryland rate. So eggs rather than parents were implicated. Average organo-chlorine levels (wet weight) in Connecticut eggs were 8·9 for DDE, 0·61 for dieldrin, and 15·0 for PCB, compared with 2·4, 0·25, and 2·6 in Maryland eggs (Wiemeyer *et al* 1975).

Because the levels of different organo-chlorines in eggs were often correlated with one another, it was hard to find which compound was important in each kind of breeding failure. This problem was investigated using data from 395 Sparrowhawk clutches from different parts of Britain (Newton & Bogan 1978). The levels of organo-chlorines found in unhatched eggs were examined in relation to the success of the remaining eggs in the clutch, and detailed statistical procedures were used to distinguish the roles of different compounds (Table 63). DDE was significantly related to the extent of egg addling between clutches, as well as to shell-thinning and breakage. PCB was related to the extent of addling, but not to shell-thinning, and neither type of compound showed any relationship with initial clutch size (before any breakages), clutch desertions or the amount of nestling mortality. Other compounds were present at only low levels and were unrelated to any aspect of breeding. Hence, in this study both DDE and PCB were implicated in breeding failures, but some kinds of failure occurred independently of any organo-chlorine. In less detailed studies on other species, clutches that were addled, broken or deserted contained more organo-chlorine than did clutches in which one or more eggs hatched. Most studies strongly implicated DDE but did not distinguish between the role of DDE and other compounds (Table 63). An oft-asked question was how much of a particular organo-chlorine an egg could contain and still hatch. This question proved hard to answer precisely, because of individual variation in response and because different compounds could act together to cause failure (see later).

Effects on production of young

Increased egg breakages and addling greatly reduced the number of young which a population could produce. Declines in breeding output were evident in several species, comparing pre-pesticide with recent populations, and also comparing recent populations exposed to different levels of contamination.

To mention a few examples, post-1947 drops in breeding rate were documented for Sparrowhawks, Ospreys and White-tailed Eagles in Europe, and for Cooper's Hawks, Peregrines, Kestrels, Ospreys, Bald Eagles and Red-shouldered Hawks in North America (Koeman *et al* 1972, Newton 1974, Helander 1975, 1977, White & Cade 1977, Henny 1972). Comparisons involving recent populations containing different organo-chlorine levels were made for Peregrines and Sparrowhawks in Britain, and for Peregrines, Prairie Falcons, Ospreys and Bald Eagles in North America (Ratcliffe 1972, Newton & Bogan 1978, Fyfe *et al* 1976, Henny 1977a, Wiemeyer *et al* 1975, Krantz *et al* 1970). In all these instances, populations producing fewest young had the highest organo-chlorine levels in their eggs. Among different Sparrowhawk populations in Britain, the correlation with breeding success was much stronger for DDE than for PCB, and non-existent for dieldrin at the low levels found, thus confirming the trend mentioned above for individual birds.

In addition to the mean differences between populations, differences in breeding success among individuals in the same population have also been related to organo-chlorine levels in eggs. This kind of correlation was found, for example, among Sparrowhawks in Britain, among Prairie Falcons and Merlins in Canada, and among Marsh Harriers in Sweden. In each case, to judge from analyses of addled eggs, the clutches that produced least young had the highest DDE and PCB levels (Newton & Bogan 1974, Fyfe *et al* 1976, Odsjö & Sondell 1977).

Laboratory studies

All the evidence so far given for wild birds is circumstantial, as it is based on correlations. Confirmation that organo-chlorines could produce the effects observed in the wild came from feeding experiments on captive birds. These experiments served better than field studies to define the levels of particular compounds that were needed to produce effects in different species, and to distinguish between the effects of different compounds. They also showed effects that would have been hard to detect in the wild. They usually involved comparing the breeding performance of birds fed on a 'clean' diet with that of other birds fed on diets contaminated to different extents. The American Kestrel, Barn Owl and Screech Owl were the only predatory species used, and other studies were made with more easily kept gamebirds, ducks, doves and finches. Results are summarised in Table 64.

Organo-chlorines were found to have detrimental effects at all stages of breeding, to affect behaviour, hormone and enzyme action, and to alter metabolism by affecting the thyroid (Jefferies 1973). As in wild raptors, the major causes of poor breeding success in most species were shell-thinning, egg breakage and reduced hatchability (Table 64). Mortality of young was also important in some species; it usually occurred soon after hatch, and was associated with the uptake of organo-chlorines into the body during resorption of the yolk sac. At this stage, with more than half the yolk fat already metabolised, organo-chlorine concentrations in the remaining fat had more than doubled (Snyder et al 1973, Newton & Bogan 1977). Retarded gonad development, delayed egg laying and low-weight eggs occurred in a few species. Egg production (clutch size) seemed to be the parameter least affected by organo-chlorines, unless levels in food were so high that they caused obvious ill health or loss of appetite. Fertility, as measured by the number of embryonated eggs, was affected in certain species, but did not contribute greatly to breeding failure.

The various behavioural changes observed took the form of increased aggression (sometimes involving the killing of newly-hatched young), reduced discriminatory behaviour and alertness, poor incubation, and reduced territorial activity (Jefferies 1973). All these features in the wild were likely to lead to decreased survival of young, and to decreased survival and breeding success of adults. Not all captive species showed all effects, however, and markedly different levels of organo-chlorines were needed to produce the same effects in birds of different families. Gallinaceous birds, for example, proved more resistant to shell-thinning than did ducks and doves, and these in turn were more resistant than raptors (Cooke 1973), again confirming the field evidence given earlier.

Results are tabulated according to whether they were obtained under DDT/DDE, cyclodienes (mainly dieldrin) or PCBs. All three groups of compounds reduced the hatchability of eggs; in general cyclodienes were more toxic to embryos than PCBs, and these in turn were more toxic than DDT/DDE. The latter were the main compounds associated with shell-thinning. In the shell gland, DDE inhibited the action of carbonic anhydrase, the enzyme normally necessary to supply the carbonate ions used in shell formation (Peakall 1970). This was the mechanism by which shells were thinned. In contrast, PCBs and dieldrin were exonerated as shell-thinners in almost all experiments. PCB caused deformities and growth depression in chickens, and chromosome aberrations in Ring Doves (Carlson & Duby 1973, Peakall *et al* 1972). There was some evidence that PCB had more effect on breeding output when fed for a long time than for a short time (regardless of concentration), and that PCB, DDT and dieldrin had more effect on particular birds when they had been fed to the parents of these birds as well (Baxter *et al* 1969, Peakall *et al* 1972, Carnio & McQueen 1973).

The presence of one compound in the body influenced the level at which another was stored, and when several were present at once their effects were sometimes additive, so that a bird could suffer from a collective organo-chlorine load, even though the individual compounds were insufficient on their own to cause harm. In their effect on breeding, DDE and PCB together led to more delayed laying and to lower breeding output than did DDE or PCB alone (Risebrough & Anderson 1975). Such interactions made it hard to extrapolate to field situations, for wild birds were usually exposed to a greater range of compounds than were captive ones. When captive birds were taken off a contaminated diet, it sometimes took many months for their tissue organo-chlorines to fall to levels which no longer affected breeding. In one experiment with Black Ducks fed on DDE, females still produced thinner-shelled eggs and fewer young than control birds after two years on clean food (Longcore & Stendell 1977).

Much less is known about the effects of organo-chlorines on the male. DDT was found to inhibit testes growth in young cockerels (Burlington & Lindeman 1950), but in mature birds it had no appreciable effect on sperm production, except at concentrations that were anyway sufficient to cause death (Albert 1962). Similar results on sperm production were obtained for adult Bald Eagles (Locke *et al* 1966). Hence, as far as one can judge from so few studies, it is mainly through their effects on the female that organo-chlorines influence breeding.

DIRECT MORTALITY

In the wild, heavy kills of many kinds of birds were repeatedly associated with local applications of organo-chlorines in seed dressings, or with the release of these chemicals into rivers as industrial effluents. The normal behaviour of raptors makes them particularly vulnerable, for at the prospect of an easy meal, they are attracted to moribund prey, and may eat animals they would not normally catch. Once contaminated, their own hunting ability is impaired, leading them even more to seek out disabled prey. Symptoms of organo-chlorine poisoning include loss of appetite and weight, muscular dystrophy, and poor co-ordination, paralysis and tremoring. To quote from one of many similar reports, in one Dutch province in 1966, corpses of 54 Buzzards were found, and in the same province in 1967 those of 18 Buzzards, five Kestrels, two Marsh Harriers, two Montagu's Harriers and 18 Sparrowhawks (Ensink 1968), all related through chemical analyses to the use of aldrin, dieldrin and heptachlor as seed dressings (Koeman et al 1969). These same chemicals were similarly involved in the deaths of many Sparrowhawks, Kestrels and Hen Harriers in Britain, as well as in those of thousands of seed-eating and other birds (Prestt & Ratcliffe 1972). PCBs were implicated in a large wreck of Guillemots in the Irish Sea in 1969 (Parslow & Jefferies 1973).

Apart from seed dressing incidents, deaths in other areas occurred apparently as a result of food chain concentration, sometimes after body fat had been depleted. DDT was thus implicated in the deaths of individual Bald Eagles, heptachlor in the death of a Merlin, and dieldrin in the deaths of Bald Eagles, Peregrines, Lanner Falcons, Rough-legged Buzzards, Sparrowhawks and others (Reichel et al 1969, Bogan & Mitchell 1973, Prestt et al 1968, Jefferies & Prestt 1966). That all these deaths were due to organo-chlorines was challenged on the grounds that brain levels were sometimes much lower than in birds of other species experimentally poisoned in laboratories; but on the other hand birds may die at lower levels in the wild than in the laboratory where food is freely available. Also, in laboratories, great differences in sensitivity to certain organo-chlorines were evident between species, and between individuals of any one species – hence the use of the LD50 (lethal dose for 50%) as a measure of toxicity. In extensive tests, involving feeding many different pesticides to various bird species, endrin was consistently the most toxic compound, but aldrin and

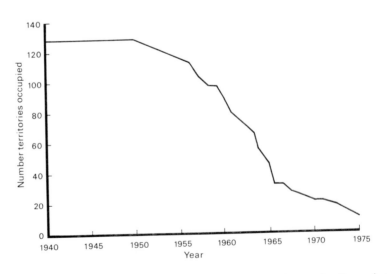

Figure 46. *Downward trend in the occupancy of certain Peregrine territories in Sweden. Re-drawn from Lindberg 1977.*

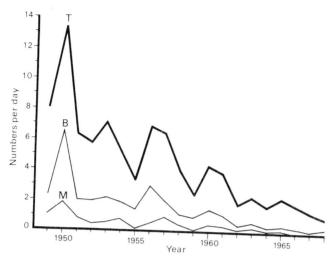

Figure 47. Average numbers of raptors seen per observation day in part of Bavaria, 1949–68. T – Total, B – Buzzard, M – Marsh Harrier. Re-drawn from Bezzel 1969.

dieldrin were also much more toxic than DDE/DDT, and these in turn were more toxic than PCBs (Tucker & Crabtree 1970, Heath et al 1972, Hill et al 1975). However, PCBs often contained highly poisonous impurities, such as chlorinated dibenzofurans and dioxins.

POPULATION DECLINES

Evidence came not only from the casual observations of thousands of birdwatchers, but also from (a) detailed studies and surveys of particular populations (Ratcliffe 1972, White & Cade 1977, Henny 1977, Lindberg 1977); (b) from long-term counts of migrants passing well-known observation points (see Chapter 11; Rosen 1966, Edelstam 1972 and Ulfstrand et al 1974 for Sweden; Hackman & Henny 1971 and Nagy 1977 for eastern North America); and (c) from sample counts of birds seen in particular areas in different years (Ash 1965, Bezzel 1969). Some populations that declined after the introduction of organo-chlorine insecticides, had previously been stable over several decades. Examples of such counts are given in Figures 46 and 47 and the following three main trends emerged.

(1) On both continents, the declines were most marked in the bird-eaters and less marked or non-existent in the mammal-eaters, thus fitting the trends in DDE levels and shell-thinning discussed earlier. The Peregrine became extinct as a breeder over large parts of its range by 1965, eighteen years after the introduction of DDT. So far as is known, the species no longer breeds in the United States east of the Rockies.

(2) In regions where DDT came into wide use at a later date, populations showed correspondingly late declines. The most clear cut were in the Peregrines that bred in Alaska but wintered in Central and South America (White & Cade 1977). Their decline did not become apparent until after 1967, linked with the later widespread use of DDT on their winter range.

(3) Whereas in much of North America the declines were relatively slow and long-term, in parts of eastern North America and western Europe they steepened abruptly in the late nineteen-fifties, and were more in the nature of 'crashes'. The Sparrow-hawk almost disappeared from much of England within two years from 1958, and nearly as rapidly from Holland and Denmark (Anderson & Hickey 1974). This difference was associated with relatively greater cyclodiene usage in Europe than in much of North America, and with greater mortality of full grown birds.

How much were the population declines due to reduced breeding rates and how much to extra mortality of adults? The situation probably differed between areas. In North America, the annual mortality of three declining raptor species estimated from ring recoveries was the same during the DDT era as before, and population decline could be attributed solely to known declines in breeding rates (Henny 1972). The Osprey showed the biggest drop in breeding success and the most rapid decline in population, the Kestrel showed the lowest declines, and the Red-shouldered Hawk was intermediate in both respects (Table 65). Moreover, in several Osprey populations whose reduced breeding rates differed, the rate of population decline was estimated assuming no differences in adult mortality, and these estimated rates agreed closely with those observed (Henny & Ogden 1970). This again implied that population decline was due to poor breeding rather than to increased mortality, and that there was no improved survival among adults to compensate for their lowered production.

In the absence of such detailed studies for Europe, the likely conclusion is that the slow population declines in North America resulted primarily from reduced breeding rates, associated with DDT use, but similar declines in western Europe were accentuated into crashes by extra mortality of adults, associated with the greater cyclodiene use. For some species the numbers killed at seed dressing incidents must have represented a large part of the local population. And the incidents were so widespread around 1960 that the raptors over large areas must have been affected. Moreover, they occurred in spring, after most natural mortality had occurred, so are likely to have had the maximum impact on populations (Chapter 13). I do not wish to imply a firm division between events in Europe and America, for no doubt on both continents intermediate situations occurred.

RECOVERY IN BREEDING RATES AND POPULATIONS

Certain species soon showed improved shell-thickness, breeding or numbers in areas where organo-chlorine usage was reduced appreciably, and where residual populations remained. In Britain, restrictions came into effect progressively from 1962. In the Peregrine between successive national surveys in 1963 and 1970, the numbers of occupied territories increased from 44% to 56%, and those producing young rose from 16% to 21% (Ratcliffe 1972). Most improvement was in northern and western districts, where pesticide use had always been lowest. Since this date, numbers and nest success have continued to improve. The Sparrowhawk had likewise largely recovered in numbers in northwest districts by 1970, and had reappeared over much of the southeast by 1975. However, depressed nesting success was still evident in both species in 1975, and may have slowed recovery. In parts of middle Europe, too, improvements in the numbers of both species were apparent by 1970–75.

The proportion of Golden Eagles in western Scotland that successfully reared young doubled from 31% to 69% following the 1966 ban on dieldrin in sheep dips, while at the same time the average dieldrin residues in eggs dropped significantly from 0·87 ppm to 0·38 ppm (Lockie *et al* 1969). Eagles in eastern Scotland, which ate little sheep carrion showed no change in nest success over this whole period, and negligible levels of dieldrin in their eggs.

In North America, DDT use dropped from a peak in 1959 to nearly zero in 1973. By 1975 Bald Eagles had stopped declining, while Ospreys and Cooper's Hawks had begun to recover in numbers and nest success (Hamerstom *et al* 1975, Henny 1977, Nagy 1977). In California, where declines in nearly all raptors were apparent from 1950, at least three species began to increase after 1967 (Brown 1973).

SUMMARY OF EVIDENCE

As some people have doubted that agricultural pesticides could have had such marked effects on birds of prey, it is appropriate in this last section to summarise the main lines of evidence.

First, the shell-thinning and poor breeding were unprecedented within ornithological history, and likely to be linked with an unprecedented change in the environment.

Second, shell-thinning closely followed the widespread use of DDT, and this in turn was followed by steady population declines in much of North America, or by abrupt declines in parts of Europe, where heavy mortality was associated with the use of cyclodienes in spring cereal dressings. The subsequent improvement in shell thickness, breeding success and populations on both continents followed restrictions in organo-chlorine usage.

Third, the geographical pattern of these changes coincided with the geographical pattern of pesticide use; shell-thinning and population decline were most marked in regions where pesticide use was greatest, and affected resident and migrant populations.

Fourth, analyses of tissues and eggs showed that these compounds were present in all raptors examined in recent years, at highest level in bird-eating and fish-eating species, and at levels comparable to those which produced similar effects in controlled experiments. Organo-chlorines were also found at lower level in all the many prey-species analysed.

Fifth, the amount of residue in individual clutches was closely correlated with the shell-thinning and success of those clutches, as shown by comparing different individuals in a population, different populations of a species, and different species. Raptors were found to be especially vulnerable, partly because of their high exposure near the top of food-chains, but also because of their high physiological sensitivity to organo-chlorines.

The case for the involvement of organo-chlorines in some population declines of raptors thus became extremely compelling. Moreover, each stage in the sequence of events leading to decline has now been well documented. Critics of the case have produced no argument that stands up to examination, nor any satisfactory alternative explanation for the phenomena observed.

CONCLUDING REMARKS

Considering the short time they have been used, the organo-chlorine compounds have had more devastating effects on raptor populations than have any natural factors or any other poisons. The fate of these birds is the proper concern of the ecologist, but its indication of an important environmental problem is of general concern. As discussed in this chapter, it is now clear that persistent organo-chlorine pesticides and PCBs are (a) biologically active at levels far lower than those that are lethal; are (b) affecting populations far removed from areas of application; are (c) able to cause extinctions, at least of certain predatory birds; and are (d) now widely distributed in soils, water and organisms throughout the world.

Recent governmental action on the control of pesticide use has usually taken these facts into account. Again value judgements are involved. DDT has relieved much human suffering and increased crop yields in many parts of the world. However, some of the insects against which it was used are now resistant to it, and there are other pesticides available which are less persistent. The continued use of organo-chlorines could damage fisheries and wildlife resources important to man, as well as promote further extinctions of raptor populations. Regrettably, an understanding of the complex repercussions of organo-chlorine use were for many years deliberately confused and obscured by manufacturers and other vested interests.

The extent to which any country can limit the use of organo-chlorines must obviously depend on local conditions. Thus a case might be made for their temporary use in some tropical regions where human health is at stake, or where development is otherwise impossible, but over the rest of the world their withdrawal seems a wise precaution to protect natural resources. Although the use of DDT in the United States, Canada, the Soviet Union and many European countries has declined impressively, there is no clear evidence that its use has declined on a global scale. Moreover, even in countries in which organo-chlorines have been legally banned, pressure groups are continually pushing for their reintroduction for some purpose or other. They will no doubt remain a major factor affecting raptors for many years to come, and we could well see the extirpation of several more populations. Unlike the insect pests against which many of them are used, there is yet no sign that any wild bird is developing resistance, probably because the effects have been too drastic and rapid in such scarce and slow-breeding animals. If there is an immediate lesson from the DDT saga, it is to avoid the more persistent synthetic pesticides, however satisfactory in other respects they may seem to be.

Bald Eagle

CHAPTER 15

Other pollutants and pesticides

Although the organo-chlorines have formed the main identified threats to fishery and wildlife resources in recent years, other chemicals have been suspected, but less well studied. They include the heavy metals, such as mercury, lead and cadmium. Heavy metals are natural components of any environment, and in trace amounts some are often essential for the good nutrition and health of organisms. But they are so concentrated in certain industrial wastes that they present a hazard to wildlife and to man. So far, only mercury among these metals has been implicated in the population declines of raptors.

Contamination of the environment by mercury first gained attention in the nineteen-fifties, when many deaths and mental disturbances occurred among people who ate shell-fish from the polluted Minamata Bay in Japan. The condition became known as 'Minamata disease'. Soon afterwards, widespread contamination of wild animals became apparent, not only in Japan, but also in Sweden, Holland, Canada and elsewhere (Borg et al 1969, Fimreite et al 1970). In all these countries, the effluent from chlor-alkali plants was a major source, together with the waste from mining, pulp and paper industries. Such waste had been produced since the last century and was probably the main source of mercury in the aquatic environment. In addition, dozens of mercurial compounds were used from the nineteen-twenties as fungicides in seed dressings. The most toxic and most persistent were the so-called 'alkyl-mercury' compounds (including methyl-mercury), and these formed the main source of mercury in the terrestrial environment.

Compared with the other mercurial compounds, the alkyl ones were also important because (in methyl form) they were soluble in water, as well as in fat. They could readily penetrate membranes, and thus get into blood cells and reach the brain. They could pass readily from one animal to another and, like the persistent organo-chlorines, they were found at highest levels in bird-eating and fish-eating species (Figure 48). Their magnification was much less marked in terrestrial food-chains than in aquatic ones, with concentration factors of two to three times compared with hundreds or thousands.

EFFECTS ON BIRDS

Mercury often was found to kill birds outright, especially when used as alkyl-mercury in seed dressings. Levels of mercury in birds found dead in such incidents were similar to levels in birds that were experimentally poisoned in laboratories. Field casualties included mainly granivorous species, but also many others, including the raptors that fed on the corpses (Borg *et al* 1969). In the Dutch province of Zeeland, in one year, 103 raptors and 111 owls were found dead from this cause (Wedts de Swart 1969, Koemen *et al* 1969). The widespread killing of such large numbers presumably contributed to the general decline of European raptors in the nineteen-fifties and 'sixties.

The effects of sub-lethal doses of mercury on the breeding of wild birds proved hard to assess, because in both northern Europe and

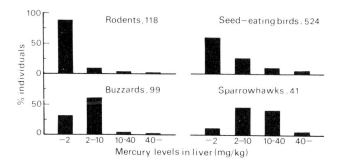

Figure 48. Mercury levels in two prey-predator pairs, based on carcasses analysed in Sweden in 1964. Note the increase between prey and predator. Numbers are specimens examined. Details from Borg et al 1969.

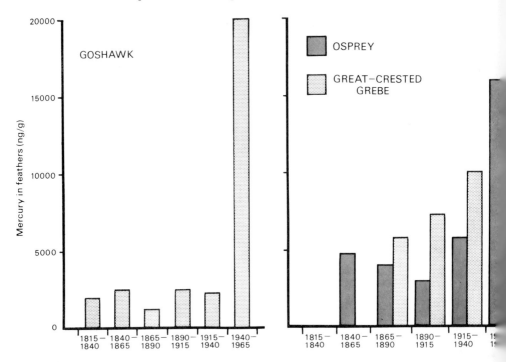

Figure 49. *Average mercury levels in feathers of museum Goshawks, Ospreys and Great Crested Grebes in Sweden at different periods. Redrawn from Jensen et al 1975.*

Canada most birds contaminated with mercury were also contaminated with organo-chlorines at sufficient levels to affect them (Jensen *et al* 1972, Keith & Gruchy 1972). The evidence for the involvement of mercury was chiefly that the levels found in addled eggs of White-tailed Eagles and other raptors often exceeded the amounts known to impair the breeding of other species in captivity (about 2 ppm). In laboratory studies, organic mercury was found to reduce the number of eggs laid (Mallard), reduce the hatchability of eggs (Mallard, Black Duck and Pheasant), increase the mortality of young (Mallard, Black Duck, Pheasant), and effect behaviour (Mallard) (Fimreite *et al* 1970, Heinz 1974). Inorganic mercury at similar concentrations had no detrimental effects on the breeding of Quail (Hill & Shaffner 1976).

The ability of alkyl-mercury to concentrate from prey to predator was shown in an experiment in which four Goshawks were fed on chickens that had themselves been fed on wheat dressed with methyl-mercury (Borg *et al* 1970). After two weeks, the hawks showed losses in appetite and weight, together with incoordination, paralysis and tremoring, and all died within 30–47 days of the start of feeding. On analysis, the mercury levels in their tissues were found to be several times more concentrated than in their food. Brain levels of 30–40 ppm

(Upper) American Kestrels breed only in cavities and are often limited in breeding numbers by shortages of ~~~es; male on left at Flicker hole, and female on right at natural hole. Photos: D. Muir. (Lower) Hobby at nest with ~~ung. A summer migrant to Europe, the Hobby is also unusual among falcons in its narrow choice of nest-sites, ~eeding only in trees, in the old stick nests of crows and other birds. Photo: F. V. Blackburn.

30   Peregrines at cliff nests. (Upper) Female with eggs, in Canada; and (lower) with chicks, in Scotland. Photo D. Muir (upper) and R. T. Smith (lower).

were considered lethal. Similar results were obtained in Canada, using Red-tailed Hawks (Fimreite & Karstad 1971). In other experiments on Pheasants and poultry, the mercury was found to accumulate to high levels in body tissues within a few days, and to take more than six months to disappear after the experiment.

The low numbers and poor breeding success of White-tailed Eagles in northern Europe have been widely publicised. Organo-chlorine and alkyl-mercury compounds were present at high levels in addled eggs and in dead adults, and shell-thinning was widespread (Jensen *et al* 1972, Helander 1977). In Finland, both types of compounds were used less than in Sweden, and the Baltic was probably the main source of contamination. Finnish breeding pairs were said to have dropped from about 35 in the late nineteen-forties to 12 in 1964. No nesting was recorded in 1966, and eight birds were found dead. The five analysed all had mercury at sufficient levels to have killed birds in laboratories (Henriksson *et al* 1966). Among Sparrowhawks in Germany, differences in breeding success between two areas were related to differences in the mercury levels in eggs, but as organo-chlorines were also present, the results were again hard to assess (Bednarek *et al* 1975).

LONG-TERM CHANGES IN ENVIRONMENTAL LEVELS

Using sensitive techniques, mercury can be shown to be present practically everywhere, mostly in minute amounts. In order to confirm contamination from human activity, it is necessary to demonstrate a rise of mercury content above this natural level. The general levels in various bird species can be estimated from the mercury content of their feathers, which depends on the amount circulating in the blood during feather formation. Normally the mercury concentration in feathers is about 7–8 times that in fresh muscle. By use of dated museum skins, it proved possible to estimate the levels of mercury present in various birds from the nineteenth century up to the present time (Berg *et al* 1966).

In Swedish Goshawks, the initial low content of mercury in feathers prevailed until about 1940, after which it rose rapidly to about ten times the original mean (Figure 49). In the nineteen-twenties, inorganic mercury compounds were used as seed dressings, followed in the nineteen-thirties by alkoxy-alkyl and phenyl-mercury compounds. But in 1940, alkyl-mercury compounds were introduced into Sweden, and from about 1950 dominated the market. The simultaneous appearance of an increased mercury concentration in birds' feathers helped to confirm that seed-dressing agents of the alkyl-mercury type were an important source of terrestrial pollution. Such compounds were replaced by the less persistant alkoxy-alkyl-mercury from 1966, and a dramatic decrease in the mercury level of terrestrial birds was obvious from the next spring, first in herbivorous species, then in carnivorous ones, such as the Goshawk. Over the next few years, the amounts in the

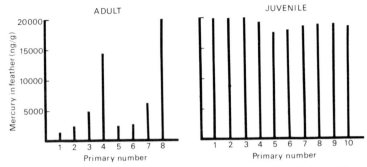

Figure 50. Mercury levels in the primary feathers of an adult and juvenile Osprey in Sweden. The adult grew most of its feathers in African winter quarters, while the juvenile grew all its feathers in Swedish breeding quarters. The levels reflect the lower contamination of the African environment. From Jensen et al 1972.

feathers of various birds fell to about the same levels as prevailed in the last century (Westermark et al 1975, Odsjö & Sondell 1977).

The progressive pollution of the aquatic environment was documented, using dated museum feathers from fish-eating Ospreys and Great Crested Grebes *Podiceps cristatus* (Figure 49). Whereas the feathers of Goshawks showed a sudden rise in mercury content following the introduction of alkyl-mercury in agriculture, those of the fish-eaters showed a more gradual rise, starting near the end of the last century. Mercury contamination of the Swedish water environment evidently followed the general increase in industrial activity from the nineteenth into the twentieth century. In whatever form the mercury was released into water, it was generally present in feathers as the alkyl form (methyl-mercury), having been converted in the bodies of aquatic organisms (Jensen et al 1972). Comparison of seabirds with earlier specimens from museums also reflected the great pollution of the Baltic in recent decades. Fitting with this, White-tailed Eagles from the south of Sweden showed increased mercury levels in feathers, but those from the north did not (Berg et al 1966).

The fact that most northern birds are migratory provided an opportunity to compare contamination levels in Sweden with those elsewhere (Johnels et al 1968). A nestling Osprey hatched in Sweden had a large amount of mercury in all the wing primaries, but an adult returning to Sweden from Africa showed a different result (Figure 50). Some of its primaries had about the same amount as the nestlings, while others in the same set had much less. The latter had been grown in winter quarters, suggesting much lower levels of mercury in the African environment. Two feathers (numbers 3 and 7) had mercury levels intermediate between feathers grown in Sweden and feathers grown towards the end of the stay in Africa. They were produced 2–3 months after the bird's departure from Sweden and suggested a half-life of methyl-mercury for incorporation into Osprey feathers of about this duration.

Because of improved control over pollution and pesticide use, heavy metals now pose less widespread problems than they did in the nineteen-sixties. In addition, we now know that animals are better able to cope with heavy metals at sub-lethal level in their bodies than they are with organo-chlorines. This is presumably because these metals occur naturally in the environment, so that animals already possess ways of dealing with at least small amounts. In birds, the growing feathers are clearly an important route for the excretion of mercury.

OTHER PESTICIDES

As we have seen, two aspects of pollutants and pesticides are important: their direct toxicity and their persistence and ability to concentrate in food chains. Regarding direct toxicity, extensive tests on non-raptorial birds have confirmed that the organo-chlorine, organo-phosphorus and organo-metallic compounds contain the most toxic pesticides. Some of them are among the most poisonous substances known (Table 59). The carbamates and carboxylates are generally much less toxic to birds and mammals (Tucker & Crabtree 1970, Hill et al 1975). Whatever the chemical, large birds usually need bigger doses to kill them than do small birds, but birds from different families also differ greatly in their sensitivity irrespective of body size.

The organo-phosphorus compounds are much less persistent than the organo-chlorines, but because of their high toxicity, they have often caused mass mortality of birds in areas of usage. In one two-month period in Holland in 1960, 27,000 birds were found dead and dying around newly sown fields, associated with the use of parathion, and the total number killed was probably nearer 200,000 including eight species of raptors (Mörzer-Bruyns 1963). In parts of Africa, large numbers of raptors are said to die during Quelea control operations in which parathion is sprayed from aircraft, but I have seen no published figures. It is not only the compounds sold as insecticides that are non-selective, for 50 Snail Kites were killed following an application of molluscicide in Surinam (Vermeer et al 1974).

Another way in which pesticides might reduce raptor numbers is through destroying prey-species. No good data are available on this aspect, but it would be surprising if the carrying capacity of many agricultural areas for raptors was now as great as in pre-pesticide days. Long use of insecticides tends to simplify ecosystems, which also makes for instability, so that particular animal populations fluctuate more markedly than previously.

THALLIUM AND OTHER RODENTICIDES IN ISRAEL

One of the most thorough, yet unintentional, extirpations of a large and diverse raptor population resulted from the attempts of the Plant

Protection Department in Israel to control the rodents which occasionally damaged crops (Mendelssohn 1972). Such rodents were eaten by the many species of raptors that wintered in Israel, and also by the smaller numbers that bred there. Among the wintering species, the main rodent-eaters (and their estimated numbers per 100 km^2) included the Kestrel (120), the eagles *Aquila clanga* and *A. pomarina* (20), Buzzard (100), Black Kite (40) and Pallid Harrier (40).

Beginning with a vole outbreak in 1949/50, wheat grains coated with thallium sulphate were spread over wide areas, each grain containing the lethal dose for one vole. Thallium is slow acting, and even voles which had eaten many grains survived for up to two days, carrying enough poison in their bodies to kill a much larger animal. These semi-paralysed and slow-moving voles soon attracted predators of all kinds, including the larger vultures. From the fifth day after baiting, moribund and dead raptors were found in the fields, and subsequent analyses confirmed the presence of thallium in their tissues. The presumed course of events, from grain through rodent to raptor, was easily duplicated in the laboratory.

Such schemes against voles were carried out repeatedly, so that raptors all over the country had many poisoned prey available to them several times a year. After each large scale operation, raptor carcasses containing thallium were found and, after a few years, big reductions in populations became apparent. During 1949–60, most operations were carried out in winter, with a few in summer as 'preventative measures'. The wintering raptor population had almost disappeared by 1955/56, whereas the breeders took about another five years to reach their recent low levels (Table 66). The Ravens and Herons which ate rodents also declined. The only common breeding raptor which maintained its numbers was the Snake Eagle and this was the only species that did not eat rodents. However, no change was noted in the migrant populations which passed rapidly through the country twice each year. All this occurred before the general declines of birds-of-prey in Europe that were associated with organo-chlorine use, but in Israel organo-chlorines may have contributed in later years, and slowed recovery. Other human influence could be ruled out, as there was no animosity towards raptors in Israel, and no-one had a vested interest in destroying them.

The poisoned grains were also eaten by seed-eating birds, whose marked decline may have contributed to the disappearance of the once-common Sparrowhawks and bird-eating falcons. Some seed-eating birds suffered especially big declines, and the Rook *Corvus frugilegus* disappeared completely from an estimated wintering population of 100,000. Meanwhile, some insectivorous birds increased and spread enormously, an event which Mendelssohn attributed to reduced predation from raptors.

In 1965 thallium sulphate was replaced by fluoracetamide, which is more toxic, but at usual doses does not give rise to secondary poisoning. This was followed by slight improvements in the numbers of certain raptors, first among breeding Kestrels and later among winter-ing Buzzards and Black Kites. But some species received a further set-back at this time as a result of a scheme to reduce Jackals *Canis aureus*, in which tens of thousands of chicks injected with fluoracetamide were broadcast over vast areas. Also in later years, grain poisoned with endrin was repeatedly used against sparrows, larks and other seed-eating birds. One extensive campaign against Skylarks *Alauda arvensis* in autumn 1970 coincided with the raptor migration through the country. The influence was considerable, and dead or moribund raptors were found up to 200 km south of target areas. The use of endrin for this purpose was later forbidden.

By the time the wintering raptors had recovered further, another attempt was made in the winter of 1975/76 to get rid of voles from 8 km² of alfalfa in northern Israel (Mendelssohn & Paz 1977). Voles preferred alfalfa as a food, and would not eat the poison baits. So the farmers sprayed the organophosphorus compound, azodrin, from the air to cover the crop. Many birds-of-prey had been attracted to the area by the abundance of voles, and in three months 116 raptors and 29 owls were found dead from secondary poisoning. Another 67 birds that were still alive when found were rehabilitated and released again. Casualties were mainly Black Kites, Buzzards and Kestrels, but they also included eagles.

Considering the variety of species involved, the pest controllers of Israel have probably had more effect on the raptors of their country than those of any other western area. Executed without regard to side-effects, these events re-emphasise how completely and effectively a particular agricultural policy can obliterate whole raptor populations within a few years. As in other countries, the fast-breeding pest-species have remained a problem, while some other animals, apparently previously held in check by the predators, have increased and become pests themselves.

CONCLUDING REMARKS

The events of recent years, in which one chemical after another has proved detrimental to fish and wildlife, must naturally make anyone wonder how many other substances are being released into the envi-ronment and with effects which may not become apparent for years to come. Some of the existing problems were discovered almost by accident, usually by people with no concern with industry or pest-control. A major need, therefore, is for knowledge of new pesticides and pollutants on which the spotlight has not yet fallen. The biologist cannot make toxicological tests or routine analyses for chemicals he does not know about. The number of poisonous chemicals entering the environment is so great that a priority is for more effective ways of predicting those likely to give most trouble. Chemical structure can provide only a rough guide to the potential effects of a pollutant, because minor changes in structure which take place in the animal's body can cause great differences in effects.

So far in this book, I have described five different ways in which particular raptor populations were poisoned to near extinction in the years since 1950: (1) deliberate destruction through use of poisoned meat baits, as for several species in parts of Europe; (2) incidental destruction through raptors taking poisoned meat baits intended for other predators, such as wolves or foxes in parts of Europe; (3) secondary poisoning through raptors taking poisoned prey, such as rodents in Israel; (4) secondary poisoning through raptors taking prey that had themselves been poisoned incidentally during attempts to control some other pest, as when seed-eating birds took grain dressed with insecticide in various parts of Europe and North America; (5) secondary poisoning resulting from food-chain contamination with persistent organo-chlorine or mercury compounds, which increased mortality, reduced breeding success, or both, in large parts of the 'developed' world. In any one region, more than one factor has sometimes been involved, and particular species have declined because of different factors in different regions. When the above events are added to the effects of habitat destruction, shooting and other direct persecution, it is often hard to specify for a given population which factor was most important in causing decline. In particular areas, populations have generally increased somewhat in the periods between successive declines, but increasingly since 1945 populations over large areas have been held by chemical means well below the level that the environment would otherwise support. The biggest problems have arisen mainly from the use of compounds which are non-selective, persistent and cumulative, but also to some extent from unnecessary, careless or exaggerated use. It would be unrealistic to expect that these problems will not recur again and again.

SUMMARY

Mercury has been implicated in the population declines and breeding failures of raptors in Sweden and elsewhere. The field evidence is hard to assess because birds with high mercury levels usually contained high organo-chlorine levels as well; however, the mercury levels involved were similar to those that caused death or breeding failure in captive birds. Alkyl-mercury used in seed dressings caused much direct mortality among raptors and other birds in Sweden and elsewhere in the years around 1960. Of other pesticides, the organophosphorus compound, parathion, has also caused much mortality among birds of prey. Following several years of intense poisoning of rodents and other crop pests in Israel, the entire raptor population was almost eliminated through secondary poisoning, but has recovered somewhat since.

CHAPTER 16

Conservation management

To counter widespread population declines, many attempts have been made in recent years to increase raptor numbers, either by management of the birds themselves or of their habitat and food sources. Management principles that have been applied for decades to game animals can be applied equally to the conservation of birds-of-prey. Conserving raptors is more difficult, however, because of the greater land areas needed to sustain populations, because of the more specialised nesting requirements of raptors, and because of conflicts with other human activities and resulting unsympathetic attitudes. For raptors in general, three main factors have been identified as causing declines (or limiting numbers), namely, restriction and degradation of habitat, persecution by man and contamination by toxic chemicals. The sections below consider the particular forms of management that have been applied against each of these factors, and unless stated otherwise, I shall assume that the objective of management is to increase a breeding population or halt its decline. The role of breeding from captive birds is left for the next chapter.

HABITAT MANAGEMENT

On a world scale, habitat destruction has already accounted for

Illustration: Lammergeier

bigger reductions in raptor and other wildlife populations than has any other factor; and with the continuing growth in human population and development, it is still the most serious threat in the long term. Irrespective of any other depressing influence, habitat sets the ultimate limit on the size and distribution of any wild population. Habitat destruction takes two forms, the reduction of a former widespread habitat to tiny fragments (e.g. wetlands in many regions) or the degradation of a former habitat by land use practices which lead to reductions in prey (e.g. cultivation of natural grassland). In the first instance the raptor population is restricted in distribution but, within the remaining habitat fragments, it may live at no less a density than before; in the second instance, the population shows no restriction in distribution, but lives at much lower density than before. In practice, most raptor populations are affected simultaneously by both forms of habitat destruction. Two measures have been taken to counter these threats: (1) find the remaining areas of good habitat, and preserve as many as possible; and (2) increase the carrying capacity of certain areas so that they will support more raptors than previously.

Nest-sites

The carrying capacity of any environment for raptors is usually set by nest-sites or food supplies, and whichever is in shortest supply can limit the number of breeding pairs. For some species, a shortage of nest-sites has been rectified by adding sites artificially. Thus densities of Kestrels have been increased by adding nest-boxes, those of Prairie Falcons by making ledges and nest-holes in earth cliffs, and those of Ospreys and Bald Eagles by providing artificial nests or nest platforms (Chapter 3). Peregrines, Hobbies, Merlins and others have used baskets or other artificial tree-nests on occasion, so their numbers could perhaps also be increased in some areas if enough of such sites were provided (e.g. Fyfe 1978, Saurola 1978). For large raptors it is sometimes necessary only to saw off the top of a conifer tree to provide a base on which a bird can build.

The provision of safe nest-sites has also improved the breeding success of some populations, either by enabling certain pairs to move from roadsides to places safer from human disturbance, or by reducing losses to predators or bad weather (Postupalsky & Stackpole 1974). The rare Snail Kites of Florida often build in substandard sites, in emergent marsh vegetation too weak to support their nests in wind; but in recent years precarious nests have been transferred to nearby special baskets mounted on short poles embedded in the marsh. The birds accepted the change and bred more successfully (Kern 1978). This sort of procedure is worthwhile management only if there is reason to believe that improved nest success might lead to an increase in subsequent breeding numbers, but not if other factors are limiting the population. Some people object that platforms and other man-made sites spoil natural environments, but this is a fault in nest-site design, not in the concept of management by nest-site provision. White-tailed Eagles in Sweden have used man-made but natural-looking nests in trees where the birds could not build for themselves (Helander 1977, Saurola 1978).

The preservation of existing nesting places is particularly important for species with special needs. The large nest-trees of certain eagle species have increasingly been protected against lumber activities, an obviously beneficial step in view of the scarcity of such trees and the time they take to replace themselves. In parts of North America the traditional nest-cliffs of Peregrines have received special protection, which includes a ban on any modification of the surrounding area that might render the cliff unattractive, including increased human presence. Along the Alaska pipeline, areas of undisturbed habitat two miles in diameter were left round each Peregrine cliff, areas of one mile diameter round each Bald Eagle and Osprey nest, and of 0·5 miles around each Rough-legged Buzzard nest. One of the pumping stations was relocated to protect a Peregrine cliff, and part of the pipeline highway was shifted at considerable expense to avoid other nesting areas (Olendorff & Kochert 1977). In North America, a consideration of birds-of-prey has become usual in land use planning and development.

Food supplies

Raising the carrying capacity of an area for breeding raptors through increase in food supply is more difficult than managing nest-sites, because it usually entails changing the land use so as to promote an increase in prey. Often the best that can be achieved is to preserve existing areas of good habitat, or prevent their further degradation. In North America and Africa, the larger national parks provide some excellent raptor habitat, capable of maintaining large populations; but in more heavily peopled countries, most areas that can be preserved in this way are too small to support many birds. This is especially true of the large species that require huge areas.

An attempt to control the development of wilderness, with raptors in mind, has been made around the Snake River in northern Idaho (Olendorff & Kochert 1977). The area of interest lies within an extensive tract of government-owned land, on which there is continual pressure for agricultural development. Yet it holds what is probably the densest raptor population in North America, with more than 200 breeding pairs of five species (including 117 pairs of Prairie Falcons) in 60 km of river canyon (Chapter 2). All these birds depend on the ground squirrels and jackrabbits that dwell in abundance in the surrounding sage and grass covered tablelands. To save the raptors, it is clearly necessary to protect not only the nesting cliffs, but also a sufficient area of foraging habitat nearby. Research has included fitting birds with radio transmitters to find how far from the canyon they range to feed, and in this way the area necessary to maintain the population can be more precisely gauged. Initial boundaries extended only three kilometres from the canyon, but it was soon found that Prairie Falcons ranged up to 27 km for their food. At the time of writing it has still not been decided how much land will be allocated to this unique breeding concentration.

Increasingly in recent years, the need has arisen for rapid means of habitat assessment, so that the good areas can be found, and if possible preserved. Areas are usually selected because of the numbers and variety of species which breed there (as in the case above), or because they hold some particularly rare species. Other nature reserves may be chosen on different criteria, such as 'naturalness' or 'representativeness', but areas that are good for raptors tend also to be good for other animals. This is to be expected, for the predators would not be there without a good prey base to maintain them. For raptors, areas often vary in quality, according to human land use. Among open habitats, natural grasslands hold more prey than do similar areas used for ranching, and ranchlands in turn hold more prey than do cultivated lands. Annual ploughing greatly reduces small mammal populations by eliminating their food and cover, and thus renders croplands of limited value to mammal-feeding predators. Anyone who doubts these facts should visit the prairies, and compare the richness of predator and prey populations in regions under natural grass with the poverty of their equivalents in regions under wheat. In forest habitats, natural areas of varied structure and tree composition generally support more wildlife than do the managed, uniform stands of conifers or eucalypts (Newton & Moss 1977). All this results from the tendency of modern land-use practices to simplify habitats, and to channel large parts of the annual production into crop plants or domestic stock, leaving little for wildlife. It is as though each step towards more intense land use has a further cost in wildlife terms. These points are important in assessing habitats, and also in predicting the effects of further development.

Such phenomena are by no means restricted to the temperate regions. In several tropical and subtropical areas of South America, Reichholf (1974) attempted to relate the numbers of raptors seen to various features of the local environment. The extent of human land management showed the best correlation. As human influence on the land intensified, the number of species and individuals seen declined — except for the scavengers, which increased (Table 67). Likewise in southern Africa, Cade (1969) noted from roadside counts the great abundance of raptors in natural and semi-natural areas, compared to their scarcity in heavily farmed areas.* Some recent studies have actually documented the decline or disappearance of raptor populations coincident with an intensification in land use. For instance, in Rhodesia, Black Eagles declined over several years from 26 to 14 pairs in an area of heavily used tribal land, but maintained their numbers in a

* The contrast between natural and cultivated areas is sometimes confounded by variations in soil fertility, the most productive areas being taken for agriculture and the poorest left in a more natural state. This tends to reduce the differences that would occur if both types of area were treated similarly, and in extreme cases the farmland may be richer in wildlife than nearby natural areas. This can happen especially in arid climates where the cultivated land is artificially watered, and thus provides a striking contrast with the neighbouring desert.

neighbouring protected area (Gargett 1978).

Even quite small changes in forestry or agricultural procedure have had big effects on raptor food supplies. Vultures and other carrion-feeding birds that survived the change from wild to domestic animals have recently experienced a big reduction of food in many areas, owing to improvements in veterinary medicine, which have reduced the mortality of cattle and sheep on open range. In many countries the problem is accentuated by the existence of laws requiring the immediate burial of carcasses, in the interests of disease control. Only in remote mountain areas, where stock is less accessible, has the supply of carrion been maintained.

The Old World Lammergeier is particularly instructive in this respect, as its present distribution is correlated more with stock-keeping methods than with anything else. The species is often commensal with primitive pastoral cultures, such as those in Ethiopia or Tibet, which keep large populations of domestic animals in harsh mountain conditions with poor veterinary services. In these places the Lammergeier thrives in large numbers, benefitting from the abundance of carrion. But in other parts of its natural range where modern stock-keeping and modern sanitation prevail, the Lammergeier is extinct, rare or fast declining (Brown 1977).

In parts of Spain and South Africa, attempts have been made to counter the general shortage of carrion by setting up 'vulture restaurants', at which carcasses or slaughter offal are continually supplied (Iribarren 1977). This helps a minority of birds, and also provides opportunities for people to watch them, but it does not nearly redress the huge food loss on the rangelands. The system works, however, so the opportunities are open to develop it on a larger scale. A similar feeding programme was started for California Condors to offset the shortage of food in the nesting areas (Wilbur *et al* 1974). The birds soon began to eat the extra food (road-killed deer) and within three years they had produced more young than in the previous three.

Several objections were voiced against the last programme: (1) the enormous cost of maintaining the feeding; (2) the difficulty of ensuring that Condors rather than other animals ate the food; (3) that constant feeding at one locality would attract bears and other carnivores which might harm Condors; (4) that Condors might cease to search for natural food, become sedentary and dependent on the carcasses provided; (5) that they would lose some of the caution necessary to survive; and (6) that supplemental feeding is unnatural and aesthetically undesirable. In the event, most of the objections proved groundless, apart from the expense. Those who objected on aesthetic grounds were perhaps unaware that Condor feeding habits had been unnatural since the mid-19th century when the native game herds were replaced by cattle and sheep. Moreover, for the species to survive in the long-term, such feeding may become essential, and if carcasses are placed sporadically over the range instead of at regular feeding stations, the managed situation could closely approach the natural.

Not all human activities on the land are detrimental to birds-of-prey, and sometimes as certain species decline, others increase. Over much of the northern world, the initial destruction of tree cover to make way for cultivation would have favoured open-country raptors at the expense of forest ones, and in parts of the western United States the White-tailed Kite has spread in recent years, associated with this continuing process (Eisenmann 1971). Several other species have spread across the Great Plains in America as a result of tree planting, which has provided nest-sites (Chapter 3). In parts of Britain, Peregrines probably owe their high densities to the popular interest in pigeon racing, which ensures a continual supply of prey in regions scarce in wild food. The scavenging raptors that abound in some tropical and subtropical towns could not be maintained without the garbage and other human waste. But like similar populations in Europe in past centuries, these can be expected to disappear as urban hygiene improves.

Human management of water resources also has varying effects on raptors. Wetland areas are destroyed not only by specific drainage operations, but also in arid areas by a general lowering of water tables, due to over-use of water in towns and croplands. In Arizona, Black Hawks and Grey Hawks *B. nitidus* formerly bred in abundance along several major rivers, but are now confined to a few stretches where surface water still flows (Snyder & Snyder 1975). On the other hand, the construction of dams and reservoirs has facilitated the spread of Ospreys and Bald Eagles in other parts of the United States. In Oregon in 1976, about 47% of 318 known Osprey nests were associated with reservoirs (Henny *et al* 1978).

Overhead wires

The proliferation of power poles and wires in recent decades has had two main effects on raptors. On one hand, it has provided perching and nesting places in terrain lacking natural sites, and has thus allowed certain species to extend their breeding range (Chapter 3). On the other hand, it has killed large numbers by electrocution or collision. The Cape Vulture suffers a large mortality from electrocution in parts of South Africa and the same applies to the Golden Eagle in parts of the United States (Chapter 12). Known losses of several hundreds of birds per year in such slow-breeding species can soon cut deeply into populations, though the losses involve mainly immatures.

Using trained eagles and slow motion film, Nelson & Nelson (1977) were able to find exactly how the electrocutions occurred, and to suggest ways in which the poles could be altered to prevent it. For filming they used mock poles and wires, specially built for the purpose by the Idaho Power Company. Future poles could be built to a new design, while existing ones could be modified either by changing the positions of the wires, or by adding a special perch for eagles above the wires. It was not necessary to alter every pole, for the eagles used only a minority that were on hillocks or other raised situations. On the basis of dead birds found on the ground, the authors estimated that 95% of electrocutions could be avoided by changing only 2% of the poles. Not only did certain power companies alter their poles, but they also attached shaded nesting platforms and thus extended the breeding range of eagles and other raptors through some western deserts. The platforms were soon used by Golden Eagles, Red-tailed Hawks, Bald Eagles and Ospreys. On the other hand, some poles were adjusted to discourage perching, as was deemed desirable near to gamebird lekking areas.

In East Germany, about 30 iron platforms were built for Ospreys on the tops of power poles and almost all were used in each successive year. In this case the platforms were put up not only for conservation purposes, but also to forestall the trouble to the power company caused by birds trying to nest on poles without platforms (Sietke, in Saurola 1978).

PROTECTION AGAINST HUMAN INTERFERENCE

Coincident with growing public appreciation of birds-of-prey, bounty schemes have increasingly given way to protective legislation. At the time I write this, 14 European countries afford full protection to all birds-of-prey, 16 afford partial protection (certain species, certain regions or certain seasons), while one country (Malta) gives no protection (Conder 1977). The species which receive least protection over the continent as a whole include the Goshawk and Sparrowhawk, Marsh Harrier, Buzzard and Rough-legged Buzzard, mainly as a result of pressure from game interests. Canada, the United States, Japan and the Soviet Union protect all birds of prey. In this last country, it was the collation of research results on the effects of raptors on gamebird and waterfowl populations that led to the replacement of bounty schemes by protection laws. This was a big step affecting, as it did, one-sixth of the world's surface (Galushin 1977).

The protection of some species and not others proved ineffective in some countries, because hunters did not distinguish the different species and shot them all indiscriminately. Full legislation has met with varying success in different countries, as attitudes towards it have ranged from respect to scorn. In Britain, for example, the law has been almost wholly ignored by many landowners and their gamekeepers, perhaps because they know it is hard to enforce and that fines are derisory. But legal protection is probably important in projecting the official view, and in changing attitudes in the long-term, especially when accompanied by education. It has also provided a basis for more concerted protection, such as the guarding of individual nests.

Nest guarding has played an integral part in the management of Ospreys and Kites in Britain, of White-tailed Eagles in Germany, and of Peregrines in several parts of the world. The opportunities for human interference are thereby reduced, and the chances of successful nesting are improved. In Sweden in 1967–74, no less than 500 people were involved in voluntary watches of the remaining Peregrine nests (Lindberg 1977). For all these species, rewards have sometimes been paid to landowners, foresters and gamekeepers, following a successful fledging.

Limiting access in the breeding season to areas containing nests has met with success on federal land in North America (Olendorff & Kochert 1977). Moreover, around the nests of Bald Eagles and other rare species it has become usual to establish buffer zones, in which forestry and other activities are prohibited. Sometimes two zones have been established, an inner zone free at all times from modification or human activity, and an outer zone in which activities were restricted to the non-breeding season. Similar measures have been taken to protect White-tailed Eagle nests in Sweden (Helander 1977). This does not of course protect the feeding areas, for which other measures are needed. Limits to buffer zones are often set somewhat arbitrarily, and research is needed on how much disturbance different species will stand.

So rapid has been the growth in birdwatching in recent years, that even the well-meaning people concerned – by sheer numbers – threaten the breeding of rare raptors in certain areas. In this context, too, bird populations should be regarded like any other self sustaining natural resource – one which is valuable for recreation and which can stand a certain amount of use, but when overtaxed will be destroyed. Fortunately, most interested people accept the need to control the disturbance they create.

MANAGEMENT AGAINST TOXIC CHEMICALS

The only long-term solution to this problem is to reduce the use of the chemicals involved, so that their concentration in the environment falls, and not only in northern countries but in the world as a whole. In northern countries this has been achieved by substituting other chemicals which are less toxic or less persistent than the offending ones. Experience has shown that, after a ban on usage, the levels of organochlorines in raptors can take several years to decline. Meanwhile, various measures have been taken to counter the effects of organochlorines, until environmental levels fall sufficiently to enable the birds to survive on their own. The decline of many contaminated stocks is due not to increased mortality but to depressed breeding success. Hence, decline can be halted if the production of young can be raised.

In Peregrines, the method of 'double-clutching' is sometimes used. The first batch of eggs is removed from a nest in the wild, and the female usually lays a second clutch, which is left for her to raise. The first clutch is then incubated artificially, and the resulting young can either be hand-raised for later release, put under wild foster parents with depleted broods of their own, or cross-fostered under another species (Chapter 17). By this method, Monneret (1977) doubled the production of four Peregrine pairs in France in one year, while Fyfe *et al* (1978) achieved similar success in 1974 with the three known pairs in Alberta. In the last area, the procedure was repeated in succeeding years, and extra young were added from captive birds. The territorial population increased to six pairs in 1976, and to seven in 1977. Several of these breeders were seen to be ringed and the one ring that was read through a telescope was on a bird produced by captive parents; so it seemed that the management was having some effect. Many eggs from contaminated birds are already cracked or pitted when obtained from the wild and would surely be broken if left in the nest. But by glueing the cracked eggs, or sanding down the shells of those eggs that will not lose water, and by partially covering pitted and dried eggs with paraffin wax, some such eggs can be encouraged to hatch in incubators (Burnham, in Cade 1977).

With Ospreys another method has been used, namely the transfer of eggs from a stable, healthy population to a contaminated, decreasing one. The pairs were allowed to hatch and raise the adopted young themselves. The donor population could stand the loss of eggs without declining, and the recipient one benefitted from the addition of young. In one experiment in 1968–70, Spitzer transported 45 eggs or nestlings from Chesapeake Bay to failing pairs in eastern Long Island. The unaltered hatching success of the transported eggs (44%, 20 out of 45) and the high fledging success of the fostered young (85%, 45 out of 53) showed that parental care of the Long Island Ospreys was adequate, that enough food was available, and that human disturbance was not a major problem. The reciprocal transfer of eggs from Long Island Sound to Chesapeake Bay nests produced the expected opposite results, and the whole experiment helped to confirm that breeding failure in eastern Long Island resulted from chemical contamination. A few years later, in 1972–73, seven of the fostered Ospreys were found at nests within 50 km of where they had fledged (in Cade 1974). Smaller scale experiments have been done on White-tailed Eagles and on Bald Eagles, and the latter entailed shifting young 3,000 km from Minnesota to Maine (Helander 1977, Hamerstrom 1977).

All these methods depend on the fact that raptors readily accept strange eggs and nestlings, and will even accommodate switches from eggs to young, or from small young to large ones (Fyfe *et al* 1978). This is presumably because in normal life raptors have not needed to evolve the ability to distinguish their own young until after fledging, and in this respect differ from such colonial birds as gulls and penguins whose young intermix from an early age. Such methods also depend on the fact that most nestling raptors return to breed in the same general area where they were raised (Chapter 10). The effect of adding extra young is therefore felt on the population concerned, and is not wholly dissipated by wide dispersal. Such procedures may thus prolong the life of particular populations, even though these young might in turn become contaminated and fail in their breeding. But it is easier to maintain an existing population, with its breeding and migratory traditions, than to start a new one after the old has disappeared (Chapter 17).

A third method being tried in Sweden is the large scale feeding of White-tailed Eagles in winter with pesticide-reduced slaughter offal and carcasses (Helander 1978). The hopes are to improve the survival of young birds through their first winter (normally a stage of heavy mortality), to keep more young birds within the regions concerned and prevent them moving to other areas where they are more likely to be shot, and to reduce the dependence of young and adult birds on natural prey, which is more heavily contaminated. Since 1970 when the experiment started, the number of feeding stations has increased steadily from 62 to 96, the tons of food provided from 19 to 82, and the number of immature eagles seen per feeding station from 0·8 to 1·7. If the experiment works fully, the pesticide loads in the birds should in time decline, and their breeding and survival should improve, but results to date are inconclusive.

In addition to these direct measures, many populations of susceptible species have been monitored in recent years to keep track of trends in numbers and breeding success, and also of pesticide levels in eggs. This is so with Peregrines over much of the northern hemisphere, with Bald Eagles in America, White-tailed Eagles in Europe, and other species at more restricted localities. Results provide 'the conservation case' against continued organo-chlorine use, a way of assessing the effects of restrictions, and also indicate which populations are under greatest threat. Time and again, raptors have proved their value as indicators of trends in pollutant levels in the environment.

More research is needed in tropical areas where most DDT is now being applied, and where practically nothing is known of trends in raptor populations. It is said that DDT remains in the environment a shorter time in tropical than in temperate areas but this needs checking. It should not be hard to identify threatened populations for immediate attention. For example, the Aplomado Falcon *F. femoralis* of Central America would seem to be especially vulnerable, because its entire population is resident in a cotton growing region, where DDT is widely used.

INCREASING THE BREEDING RATE OF RARE EAGLES

In several of the eagle species which hatch two young, but raise only one because of sibling aggression, the second young can often be saved by human intervention (Chapter 7). The chief proponent of this method is Meyburg (1975, 1977), who has experimented mainly with Lesser Spotted Eagles in Slovakia. At each attempt, the second young was removed soon after hatch and placed under a common foster species, such as Buzzard or Black Kite. About a week before fledging, it was returned to the original nest, at which stage it was accepted by its sibling (which was no longer aggressive) and by its parents. The

Feeding sequence of Peregrines in NE Scotland. Male arrives with food and calls female from nest; male approaches female, transferring food from feet to bill; male passes food to female, bill to bill; male leaves and female feeds young. Photos: H. Papke.

32 Breeding Peregrines in captivity; Cornell University project. In the USA east of the Rockies, Peregrines we
exterminated completely by the use of DDT and other persistent pesticides, and the aim of the project is to rai
birds for re-introduction to the region now that DDT is no longer used. Photos: The Peregrine Fund.

parents had no obvious difficulties in providing enough food, and in each case the two young fledged successfully. If no foster parents were available, the extra young was hand-reared, but was exchanged every ten days or so with its sibling in the nest to discourage either of them from imprinting on man. Both methods were applied repeatedly with success. Under natural conditions, the Lesser Spotted Eagle has not been known to raise more than one young.

The same method has been tried successfully on the Black Eagle, Booted Eagle and African Hawk Eagle, which normally would seldom raise two chicks (Gargett 1970, Osborne 1966, Snelling 1975). The rare western subspecies of the Imperial Eagle in Spain hatches up to four young, and raises up to three, but here again production can be increased if the last-hatched chick of a large brood is placed in a nest containing a single young of the same age or an addled clutch (Meyburg 1975, 1977). One chick was even placed in a nest that had been robbed a week earlier, and was reared successfully. With care, the production of the whole population could probably be increased by 50% per year.

Existing work has thus proved that the method is feasible, but in no endangered population has it yet been applied on any scale. The amount of work involved is small compared with other forms of management, especially where the nests are already known, as is often the case. As yet, it is not known whether twosomes survive as well as single young after leaving the nest, or whether they have the same chance of entering the breeding population. The method is of course worthwhile only where it can prevent a decline or lead to an increase in breeding pairs, so it is important to make sure beforehand that enough habitat and food are available.

REINTRODUCTIONS

Several attempts have been made to reintroduce species to areas where they were eliminated by human action, but where the habitat is still suitable. This is an obviously sensible move, for the long-term security of any species depends partly on its maintaining a wide distribution. At present, schemes are underway for White-tailed Eagles in Scotland, Bald Eagles and Peregrines in eastern North America, and Lammergeiers in Switzerland (Love et al 1978, Cade 1976, Geroudet 1977). The first three schemes employed the falconry technique of 'hacking', whereby large young were put out and fed until they became self-sufficient. The young were obtained from nests in other regions or, in the case of the Peregrine, from captive stock (Chapter 17). All these schemes have been successful thus far, but at the time of writing, none has reached the stage at which the young have nested themselves.

Reintroductions are by no means easy to achieve, and several previous attempts failed, either because too few birds were involved, or because conditions were still unsuitable. In an earlier attempt to reintroduce the White-tailed Eagle to Fair Isle off Scotland, only four young were released and, although they became self-sufficient, two died on the island and the others soon moved away and were not seen again (Dennis 1968–69). Likewise, attempts were made from 1970 to reintroduce the Griffon Vulture to the Massif Central in France, but of four young birds released, three were subsequently found shot. In the nineteen-sixties, the reinforcement of the much reduced Dutch Goshawk population in Gelderland Province with 40 German birds was also without effect, despite several breeding attempts. In this case, the failure was almost certainly due to pesticide poisoning, because more than 20 birds were found dead (Bijleveld 1974). This emphasises the need to ensure that the habitat is fit for occupation before the releases are made. Successful re-introductions of other bird species, such as the Capercaillie *Tetrao urogallus* to Scotland, involved quite large numbers of birds and more than one attempt in different localities before a population became established (Ritchie 1920).

Some people oppose reintroductions on the grounds that birds might recolonise naturally. Often in these days, however, the next nearest populations are far distant from the area of concern, and have also reached low levels themselves, so that the chance of natural recolonisation is almost nil. The unexpected and widespread nature of recent population declines underlines the value of ensuring that species are at all times as widely distributed within their natural range as possible.

REHABILITATION

Several centres have been set up in recent years to which injured raptors can be taken, treated and, if feasible, returned to the wild. At some such centres, the birds are 'hacked back' by falconry technique, and provided with food until they become self-sufficient; but since the birds are seldom seen again there is no way of assessing success. Treatment and hacking back have most potential for rare species, in which each bird saved could make a difference to the population. The same procedures on individuals of common species serve mainly to increase expertise, and have given rise to big advances in the veterinary medicine and anaesthesia of raptors, as well as to increased proficiency in surgical techniques (Cooper 1972, 1975, 1977, Redig 1978). This knowledge is becoming increasingly important in captive populations, and also in wild ones, in which little is known of the role of disease and pathogens (Chapter 12). In addition, Mendelssohn (1972) achieved considerable success in treating large numbers of raptors suffering from pesticide poisoning; he described the stages of poisoning and defined those from which recovery was possible, so that effort was thereafter no longer wasted on birds that were too far gone.

CONCLUDING REMARKS

So far, management has aimed chiefly at preventing the loss of particular endangered populations; but in the long term more difficult problems will have to be faced. For example, consider the fragmentation of habitats. Scientists are increasingly being asked what is the minimal area necessary to support a self-sustaining population of a given species? Is it necessary for each such population to retain contact with its neighbours through exchange of individuals, and if so, how far apart can such populations be? Questions such as these could be answered by research, but this is not true of other questions. For example, how much raptor habitat should be kept? How many eagles do we want? Where should they be? Decisions like these entail difficult value judgements that have to take account of other pressing needs on land, as well as the general public opinion. They have not been faced before because wild animals (other than game) have not normally been considered in land development. And as in game management, a cost is involved, in that it entails putting aside areas that might be used more intensively for some other purposes, or at least curbing the further development of such areas. As it happens, much could be achieved in raptor conservation at little extra cost, merely by greater control on certain activities. Thus a lot of otherwise suitable habitat is at present rendered unfit for occupation by raptors because of excess human presence around nest-sites and other crucial places, because of unnecessary shooting, or because of the use of persistent pesticides. It is chiefly in curbing these activities that immediate progress could be made. At the same time there are some genuine conflicts that must be resolved, including some unacceptably high predation on domestic stock.

The factors that affect bird-of-prey populations are complex and various, and no species can be properly managed without some knowledge of its needs, and its limiting factors. Early attempts to maintain populations were based solely on protection from human interference, and achieved a limited success. Later attempts have used more versatile means, and have the potential for much greater success. Here lies a conflict within the conservation movement itself. Many feel that endangered birds should not be subjected to the disturbance caused by research. But the policy of absolute protection is likely to work only if human interference is the factor that is holding numbers down. Protection can do little or nothing for populations that are limited by habitat or by pesticide use. Protection policies have held sway in the conservation of the California Condor for decades, but the species has continued to decline. Perhaps it is now too late to gain the information necessary to preserve the bird in the wild, but without more research and some alternative form of management (as above) the species seems doomed.

The hundreds of raptors that have benefited in the last decade from conservation management seem puny indeed compared with the huge numbers killed in the same period by hunting and pesticides. By concentrating on the most threatened populations, however, the achievements of management have been considerable, and should increase as the various schemes described above reach fruition. But in the long run, you may feel, none of these efforts will be worthwhile without more control of human population growth, of development for its own sake, of pollution and pesticide use, and of the gamekeepers and hunters who continue to wreak so much slaughter on the remaining European raptors.

SUMMARY

Measures to counter the effects of habitat destruction have entailed surveys to find the best areas of remaining habitat for preservation, and the active management of other areas to increase raptor numbers. The latter has often been achieved by providing artificial nest-sites where a shortage of natural sites was limiting breeding density. The provision of food to populations limited by low prey supplies has been tried only on a small scale. To do it on a large scale would entail changing land use practices. Against persecution by man, protective legislation has had varying effects in different countries, but has proved generally hard to enforce. However, it has provided a basis for nest guarding and other protection, and for reducing disturbance at nest-sites. In populations declining from organo-chlorine contamination, attempts have been made to increase the production of young by shifting eggs and young from stable populations to declining ones, by inducing birds to lay repeat clutches so as to increase production, and by putting the eggs and young of captive birds into nests of wild bids. Some early attempts to reintroduce species to their former range failed, because too few birds were involved or because the habitat was unsuitable at the time; but some current schemes have more chance of success. Conservation management has so far been aimed chiefly at endangered populations to increase their sizes, but in the long term, objectives will need careful definition to take account of other land use, and will probably involve some difficult value judgements.

CHAPTER 17

Breeding from captive birds

Although raptors are fairly easy to keep in captivity, until recently few people tried seriously to breed them. The role of captive breeding in raptor conservation has only recently been realised, and since 1970 has become the main hope in some regions of saving the Peregrine. To this end, breeding programmes have been established in North America and Europe, both by government and university departments, and by private individuals. The biggest projects are at Cornell in New York (under Tom Cade) and at Edmonton in Canada (under Richard Fyfe). The stated aims are (a) to ensure the continued survival of endangered stocks, at least in captivity, and (b) to produce birds for release into the wild, either to re-stock vacated range or to augment depleted populations. The method also has potential as a source of birds for falconry, hopefully reducing the pressures on wild stocks, and for behavioural and other research (Nelson 1972). It is a field in which biologists, falconers and aviculturalists have worked together for a common cause.

The success of breeding projects must be judged in stages, the first being the establishment of self-perpetuating captive stocks and the

Illustration: American Kestrels

277

production of young in appreciable numbers; the second the successful establishment of these young in the wild; and the third, the successful breeding of these young and the development of a wild population. All the projects on diurnal birds-of-prey known to me are still (in 1977) too young to have reached the third stage, but an older Swedish project on Eagle Owls has resulted in released birds breeding successfully in the wild (Broo 1977).

PRODUCTION OF YOUNG

So far, more than 30 species of diurnal raptors, from small falcons to large vultures, have produced young in captivity. In the list below, an asterisk indicates that the young produced have in turn bred successfully in enclosures. Species include, among the falcons: American Kestrel*, European Kestrel*, Lesser Kestrel, Red-footed Falcon, Merlin, Red-headed Falcon, Eleonora's Falcon, Lanner Falcon*, Lagger Falcon, Saker Falcon, Prairie Falcon*, Peregrine* and Gyr Falcon*; and among the accipitrine hawks: Common Buzzard, Red-tailed Hawk, Rough-legged Buzzard, Long-legged Buzzard, Ferruginous Hawk, Goshawk, Cooper's Hawk, Sparrowhawk, Sharp-shinned Hawk, Harris' Hawk, Golden Eagle, Steppe Eagle, Black Kite, Red Kite, White-tailed Eagle, Bald Eagle, White-backed Vulture, Griffon Vulture, Egyptian Vulture, Lammergeier, and Andean Condor. Several others have produced fertile eggs and the list of successes is growing year by year (Mendelssohn & Marder 1970, Kenward (ed) 1971–74, 1977, Bird & Lague 1974, Maestrelli & Wiemeyer 1975, Cade & Weaver 1976, Fyfe 1976, Cade et al 1977, Psenner 1977, Fiedler 1977).

In some species the number reared is impressive. In 1973–77 in North America, more than 300 Peregrines were produced, more than 200 Prairie Falcons, and smaller numbers of other species. American Kestrels have been bred so successfully that they have become the standard laboratory raptor for pesticide and other research (Porter & Wiemeyer 1969, 1970); a private breeder obtained 61 young from two pairs in eight years (Koehler 1968). These various records show that captive breeding of a range of species is possible, and that at least for some species it can become a routine business, yielding large numbers of offspring, which can themselves be bred from. From initial results, Cade (1973) estimated that, from 30 Peregrine pairs, 200 young could be produced in a year, more than the entire pre-DDT annual production of the United States population east of the Mississippi River.

Species differ in the ease with which they can be bred in captivity, and in the conditions they require to do so. It has proved best to obtain Peregrines from the wild as feathered nestlings (Table 68), but for other large falcons this was less important, and for American Kestrels, wild caught birds bred just as well as hand-reared ones (Cade & Temple 1977). Whatever way the birds were housed when young, they bred best as pairs in individual chambers, each chamber containing an acceptable nest-site and a suitable arrangement of perches and open space for courtship and copulation (Hurrell 1973, Cade & Temple 1977). When breeding south of the normal range, it sometimes proved helpful to use artificial light to lengthen the day and bring the birds into breeding condition at an appropriate date (Chapter 6, Weaver & Cade 1974, Nelson 1972, Nelson & Campbell 1973).

Sometimes birds may not accept one another (for like humans, not all individuals get along together), or may not copulate effectively so as to produce fertile eggs. Not all breeders possess enough birds to allow a choice of mate, and the problem of getting fertile eggs has been overcome to some extent by the use of artificial insemination. Indeed this has proved the only way to get fertile eggs from some females, especially those that are sexually imprinted on humans (see later), but it is an art that not everyone can achieve. In the so-called 'co-operative technique', semen from an imprinted copulating male is transferred by hand to the oviduct of an imprinted soliciting female. In this way Goshawks, Red-tailed Hawks, Golden Eagles and Prairie Falcons have produced young (Berry 1972, Temple 1972, Grier 1973). The 'forced massage technique' developed on poultry has also been used extensively on Peregrines to circumvent the problem of incompatible partners (Cade & Temple 1977). In an experimental study of American Kestrels, the fertility of eggs was 15% higher by AI than by normal matings, with no difference in subsequent hatching success; twice-weekly inseminations were enough to achieve fertile clutches (Bird et al 1977). In similar research on large falcons, a fertile egg required a minimum insemination time before laying of about 60 hours, and the maximum period from insemination to the last fertile egg was seven days (Boyd 1978).

As in the wild, some birds will lay more eggs than usual if each egg is removed as laid, or they will lay a second clutch if their first is taken (Chapter 7). This provides an obvious means of producing more young (Mendelssohn & Marder 1970). More than 20 eggs have been obtained from individual American Kestrels in a season by removing the eggs as laid, and 16–20 eggs from individual Peregrines by removing the clutches as laid (Porter 1975, Cade & Temple 1977). There is, however, a seasonal decline in clutch size and in the fertility and hatchability of eggs which makes it unprofitable to continue removing eggs beyond a certain point. A common procedure is to incubate the first eggs artificially, and allow the adults (if they are reliable) to incubate later ones. Young hatched in incubators are then either hand-reared (preferably in groups to reduce the risk of imprinting on man), or given to other pairs that have infertile eggs or depleted broods. Generally, such pairs accept strange young and raise them with no difficulty. At Cornell, young falcons are hand-fed until there is an opportunity to put them with adults, usually at 2–3 weeks. The young are introduced when last clutches are removed. With enough food, pairs can raise larger broods than normal, and more than one brood in a year, provided that the first brood is removed before the second is added. If the second batch of young is at an earlier stage of development than the first, the adults adjust their behaviour accordingly.

Hatching success in incubators is usually less than in the nest, but total production is generally greater by 'double clutching' than by relying on one clutch alone. Some breeders use incubators from the start, others only after leaving the eggs for some days under the mother (Fyfe 1976). This last procedure sometimes greatly increases the subsequent hatchability of the eggs in incubators, but a compromise is necessary if another clutch is wanted, for the longer a female has incubated, the less the likelihood that she will re-lay. Raptor eggs have also been incubated successfully under bantams. Rearing usually presents few problems if the young can be fed from the entire carcasses of small birds or mammals, which may be ground up beforehand; if the young are fed on meat alone, this has to be supplemented with bone meal and vitamins, if rickets or other problems are to be avoided (Fyfe 1976, McElroy 1974). Domestic fowl, gamebirds and rodents fed on uncontaminated cereal grains may provide a diet freer of pesticides than may birds collected for this purpose from the wild. By use of these various techniques, it needs only a few proven pairs to produce large numbers of young, the particular procedure adopted depending on the individual birds and on the facilities available.

Despite the initial successes, some worries remain. To begin with, the original captive stocks were often small, and there is the risk of in-breeding. Second, not all individuals will breed successfully in captivity (only 50% of Peregrine pairs have bred), and by selecting those that do, the stock may be changed slowly in a way that makes it less fitted for life in the wild. Both problems might be overcome to some extent by taking fresh birds from time to time or by extracting semen from wild males to inseminate captive females. Bird *et al* (1977) have gone some way towards assessing the quality of semen and finding ways of storing it. It would probably be wise to confine releases to birds with only a brief ancestry in captivity.

PREPARATION FOR RELEASE

The problems are how to establish inexperienced birds in the wild with the maximum likelihood that they will survive on their own, return to the desired area to breed, and nest with their own kind in appropriate and safe places. Among birds in general, fixations to social companions, to type of nest-site, to habitat and locality are greatly influenced by experience during certain critical periods in the individual's early life. If raptors are taken from the nest young enough, and reared without the company of conspecifics, they will imprint on humans, first accepting them as parents, then as social companions, and eventually as sex partners (Lorenz 1935). In effect, this lasts for the life of the bird, though by various means it has occasionally been possible to get imprinted birds to copulate with their own species (Nelson 1977).

The age at which sexual imprinting occurs in young raptors is not precisely known, but the last half of the nestling period is probably important. In normal life, imprinting would ensure that individuals paired with members of their own species rather than another. Imprinting on man may be avoided if birds of the same species are raised in groups; in later life they will then behave sexually towards their own kind. It is assumed that raptors raised singly by other raptor species will likewise imprint on their foster parents, whereas those raised in groups will not, but no serious study has been made. Broods of young Prairie Falcons produced in captivity have been raised to independence by three *Buteo* species in the wild; and one such young taken back into captivity at fledging later mated with another Prairie Falcon and produced fertile eggs (Fyfe 1976). To rear young Peregrines, hand-puppets with appropriate facial markings have been used during feeding to reduce the risk of imprinting, but again the method has not been properly assessed (Campbell 1977). Birds imprinted on man can be used, artificially inseminated, for breeding in captivity, or as foster parents, but it would probably be pointless to release them into the wild.

The evidence for imprinting to nest-site in raptors is based on the existence of local traditions in the types of sites used, and the fact that traditions have changed (Chapter 5). It is assumed that adult raptors favour whatever kind of site they themselves were raised in. Young ducks that are hatched in baskets or boxes more readily accept these artificial structures in later life than do young hatched in natural sites, and in this way nest-site preferences have been deliberately changed. Such management can be used on raptors in one of two ways: either to encourage the birds to nest in the same kind of sites as the local wild population; or to change traditions, so that the birds come to nest in different kinds of sites, if this is deemed desirable. Captive-hatched Peregrines have been artificially raised on city buildings in Canada in the hope that they will return there to nest in future years (Fyfe 1977). This is not a new habit for Peregrines, but it is hoped to establish it on a large scale in areas where it did not occur naturally. The reasoning is that, for a species in such great demand by falconers and others, cities provide safer nesting places than do the well-known traditional nesting cliffs, as well as a good food supply, in the form of feral pigeons, which contain less pesticide than their rural counterparts.

Another much discussed possibility is the release of young Peregrines from tree-nests in parts of their range where Peregrines are not known to have used trees. If tree-nesting took hold, it would in theory make the birds harder for men to find (and hence safer) than if they remained on traditional cliffs, and might enable them to spread over large areas at present unavailable. This kind of management tends to be more controversial, because it entails changing not only the traditions of the birds themselves, but also what the interested public have come to expect of their birds. So far as I know, the method has only been tried experimentally, with Prairie Falcons in Canada, and with no definite result, as tree-nesting is not known to have occurred subsequently (Fyfe 1978).

The evidence for imprinting to area, in raptors as in other birds, is that individuals normally return to breed near where they were born, though some species shift much further on average than do others (Chapter 10). Knowledge of the natal area must presumably be gained in the post-fledging period, because before this stage the young of many species cannot properly see their environment, and after it, they move away. Some young Ospreys returned to breed in the area in which they fledged, after being transplanted as eggs or small chicks from another area (Chapter 16). A more extensive series of experiments on a songbird, the Collared Flycatcher *Ficedula albicollis*, gave similar results, in that birds taken from one area and reared in another returned only to the area where they had been reared and not to their original home (Lohrl 1959). Experiments on Short-tailed Shearwaters *Puffinus tenuirostris* and other birds have shown that the sensitive period, when the individual imprints on its locality, is restricted to a few days (Serventy 1967).

Even at a later stage of life, it may be possible to accustom raptors to particular areas. If trained birds are flown regularly in the same area, they develop strong attachments to it, and even defend it against intruders (Cade 1974, McElroy 1974). They quickly learn the locations of their mews, their bath tub or any other significant object in their environment, and will return there from several kilometres (Cade 1974). From these facts, one might predict that, with proper treatment, a raptor pair could become established in an area and come to accept an appropriate structure for nesting. When more is known about these various learning and associative processes, they can be put to use in management, but if they are ignored, any effort to restore birds to vacated range will probably fail. The whole field is in need of more research.

METHODS OF RELEASE INTO THE WILD

So far, three main techniques have been used to introduce captive birds to the wild. The first method is to put young (or eggs) from captive birds into the nests of wild ones, whose own eggs fail to hatch. The young are then raised and fly in the normal way. This method entails least effort and interference from the observer, and minimal contact between birds and man, but can be used only in areas where residual populations remain. In this way the production of a declining Peregrine population in Alaska was more than doubled in 1975 (Cade 1975). The second method is similar, but entails use of another species as foster parents. The advantages are the same, except for the possible danger that the young might imprint on the wrong species. As a trial, the method is being used with Prairie Falcons as foster parents to re-introduce Peregrines into an Idaho area where Peregrines used to breed, and also into a Colorado area, where they still breed in reduced numbers (Cade 1974, 1977). The third method of releasing captive birds entails the traditional falconry practice of 'hacking back', whereby young are put out and fed regularly on a safe natural or artificial nest. Once on the wing, the young continue to return for food until they have learned to hunt for themselves. At this stage the falconer would normally take them back into captivity, and begin training them, but in release programmes, the young are left to their own devices. This method requires more effort than previous ones, but over a fairly short period. It has been used to establish White-tailed and Bald Eagles, Peregrines and Prairie Falcons in the wild, and in many other species as a means of releasing unwanted hand-reared young (Dennis 1968–69, Love et al 1979, Cade 1974–77).

The Cornell Peregrine releases in the eastern United States have been made at several localities and from natural cliff-sites or from man-made towers. The procedure involved 4-week old nestlings that until then had been reared by adults of their own species, yet could tear up food for themselves. Each batch of young was placed in an enclosed artificial eyrie, guarded by a caretaker, and fed remotely through a chute so that they did not associate food with man. After two weeks, when the birds' flight feathers developed, the wire front of the enclosure was opened and the young were allowed to fly out when they were ready, usually in a day or two. The young were fitted with radio-transmitters to keep track of them in the post-fledging period. They continued to return faithfully for the food provided in the artificial eyrie, but gradually moved further afield, exploring their environment. After two weeks they began suddenly to chase and hunt other birds, and two to three weeks later they had become proficient hunters, and could kill enough wild prey to sustain themselves. At this stage they began to lose their attachment to the release site, and gradually ceased to collect the food provided. They stayed away for longer and longer each day, or skipped some days entirely, and then left the area altogether about six weeks after fledging. In all these respects they showed no obvious differences in behaviour from wild birds. In their first winter, they dispersed fairly widely (judged from sightings and re-trappings), but they did not migrate. This was interesting because, although they were bred from northern birds (race *tundrius*) which normally winter in South America, they behaved much like the original birds of the area, which re-affirms the importance of local conditions in influencing migratory habits (Chapter 11). Some lone birds returned the following spring to the release sites, or to nearby areas but, at the time of writing, none has bred. One marked male which returned as a 2-year old even hunted and provided food for the current year's young.

About three-fourths of all young survived the period from nest departure to independence. In the absence of parents to protect them, some of the young released from natural cliff-sites were taken by Great Horned Owls. Other young were shot or electrocuted. All these cases provided information useful to the siting of future releases. A total of 13 birds flew successfully in 1975, 18 in 1976 and 33 in 1977. In the Canadian scheme further north, birds were put out in Edmonton and Ottawa in 1976 and 1977, and one was later recovered in South America, the normal wintering area for such northern birds.

Methods which depend on the release of young naive birds are inevitably wasteful, because many such young will surely perish before reaching breeding age (in many raptor species more than 60% of young die in their first year, Chapter 12). It might be less wasteful to keep the young birds until they are ready to breed, allow them (using falconry techniques) to develop hunting skills, and at the same time condition them to the area where they will be set free. They could first be hacked back in the normal way, to ensure appropriate conditioning to brood mates and nest site, before being taken back into captivity. Kenward (1974) has provided figures showing that the first-year mortality is less among birds kept by experienced falconers than it is in the wild, providing that mortality during the period of acquisition and transportation is kept low. The disadvantage is the enormous prolonged effort required, and the need to rely on a small number of falconers. The birds may also become habituated to man, and may thus be more likely to be shot after release, though in general, released birds soon go wild again. Hence birds released in this way may well survive better than birds released as juveniles, but less well than wild adults.

A questionnaire survey in Britain revealed that 80–93% of trained falcons and 50–67% of trained hawks were eventually lost or released (Kenward 1974). There are several records of breeding Goshawks which had jesses on, showing that at least in this species, falconers' birds have survived and bred in the wild. Goshawks are gradually spreading in Britain and if, as is believed, the original stock came from falconers' escapes and releases, this has been the cheapest, easiest and most successful reintroduction programme to date. There is also at least one British record of a jessed Peregrine breeding.

GENETICS

Much discussion has centred on which genetic stock should be used for reintroductions. When the plan is merely to supplement an existing stock, it is best done with birds of the same provenance. In this way, the stock is not diluted by birds which might be less well adapted to the region and the original gene pool is preserved. When the plan is to re-introduce a species into vacated range, ideally this should be done with a genetic stock as similar as possible to that which originally inhabited the area. This will usually be from a nearby area, as with the White-tailed Eagles brought from Norway to Scotland. When the original stock has gone completely from a very wide area, as with the Peregrines in eastern North America, two options have been considered. The first is to put in entirely one stock, preferably one which is close in size, appearance and behaviour to the original. The other is to put in a mixture of different stocks, in order to give a maximum of genetic variation, and then let natural selection take its course. Biologically, the latter is perhaps the best course, because it gives more raw material on which selection can operate, and thus increases the chance of a well-adapted stock evolving quickly. For it to work, however, the different stocks must be compatible, and must interbreed freely to produce fertile offspring (as happens between Peregrines from different regions). If the different stocks were not compatible, then a great deal of effort would have been wasted.

HYBRIDISATION

A quite different problem arises from the production of interspecies hybrids and their escape to the wild. As a side development of captive breeding, hybrids have already been produced between several congeneric raptors (father named first): Red-tailed Hawk × Common Buzzard, Red Kite × Black Kite, Kestrel × Lesser Kestrel, Peregrine × Saker Falcon; and using artificial insemination, Peregrine × Gyr Falcon, Prairie Falcon × Shahin (a Middle East Peregrine) (H. Mendelssohn, Morris 1972, Cade & Weaver 1976, Boyd & Boyd 1975). These followed some earlier cases in zoos (Gray 1958). Unlike hybrids between some other birds, however, those from falcons are fertile, and have produced viable eggs or young when back-crossed with a parent species or with a third species (Boyd 1978). Thus, a Shahin × Prairie Falcon named 'Sharie' was crossed with a female Merlin to produce a 'Shahmerie'. With the technique of artificial insemination available, the possibilities are endless. Some hybridisation is perhaps justifiable on scientific grounds, for study of taxonomy or behaviour, but many people are likely to breed hybrids for fun or in an attempt to produce a new kind of hunting hawk. Concerned voices have already been raised that escaped hybrids might interbreed with wild birds, and adulterate the natural stock. If hybrids did pair with wild birds, the effects would presumably depend on the relative numbers involved; a few hybrids could have no appreciable impact on a large wild population subject to the full force of natural selection, but they could well alter the gene pool of a small population, reduced to a handful of pairs. That wild animals can be changed by large numbers of escaped domesticated ones is shown by the virtual extinction of Rock Doves *Columba livia* in their original form from much of their British range; but it is perhaps premature to be concerned about raptors in this way. One way round the problem would be a legal requirement for all hybrids flown for falconry to be sterilised or imprinted on man; this would prevent their breeding with wild birds, but it might not stop escaped hybrids from holding valuable nest-sites.

There are at least three records of mixed pairs in the wild. An escaped Red-tailed Hawk mated with a Common Buzzard and pro-

duced fertile eggs in Scotland, while naturally formed Red Kite ×
Black Kite pairs were found in Germany and Sweden, at least one of
which fledged young (Murray 1970, Sylvén 1977). The three wild
species were all rare in the areas concerned, and individuals may have
had difficulty in finding mates of their own species.

THE ROLE OF CAPTIVE BREEDING IN CONSERVATION

Although relatively new for raptors, breeding from captive birds is
almost certainly here to stay as a conservation technique, but at this
stage it is hard to predict its role. It will probably lead to the
re-establishment of Peregrines in eastern North America, and perhaps
in other areas, while for other species it could become even more
important. For example, in the Mauritius Kestrel *F. punctatus* in 1973,
only five birds were known in the wild, and of these only one was a
male, and no young were fledged. One pair had earlier been taken into
captivity, and produced eggs, so captive breeding could in this species
play a crucial role (Temple, in Cade 1973). To date, however, breeding
has been unsuccessful because of a number of difficulties, and in the
meantime the wild population has increased a little. The method may
also help the declining California Condor, especially considering the
ease with which the related Andean Condor has been bred. The first
successes were with zoo birds, but all four pairs of Andean Condors
taken in recent years by the United States Fish and Wildlife Service
have now produced young. No doubt there are other species in similar
plight for which breeding in captivity may become important in the
near future, at least as a last resort measure. For it is in the last stages of
a population decline that captive breeding is likely to prove most
valuable.

The main drawback is the cost and effort involved. By 1980, it was
predicted, one large breeding unit could produce up to 250 young
Peregrines per year for restocking purposes. Taking this as the maximum annual production, and using mortality and breeding statistics
appropriate to a stable Peregrine population, 1,250 young Peregrines
would need to be released over a 5-year period in order to establish 122
breeding age birds in a total wild population of 432 birds (Cade &
Temple 1977). Possibly survival and breeding would be better in an
expanding than in a stable population, but these figures give some idea
of what is needed in a reintroduction programme that is to succeed in a
short time. It is calculations like these, and the costs involved, that
show the value of maintaining existing wild stocks wherever possible,
rather than allowing such stocks to become extinct and having to start
from scratch. They also emphasise what slim chances of success some
of the other programmes have, with fewer, slower maturing birds. At
best it will take very many years for a self-sustaining population of
eagles to form, during which time the project is the more vulnerable,
because the birds are so few.

Throughout its short life, the practice of breeding raptors in captivity has formed a focus for controversy. While some feel that this is the only way to restore certain endangered species, others are violently opposed not only to the breeding, but also to the keeping of any birds-of-prey. They would rather see a rare species left to its fate in the wild than see any of the remaining individuals in captivity. The same people often argue strongly against the continuance of falconry, chiefly because they think that many modern populations cannot stand the consequent loss of young. My own view is that captive breeding offers the best chance of restoring certain endangered remnant populations, providing of course that the environment into which they will be released is suitable at the time.

SUMMARY

At the time of writing, birds of more than 30 raptor species have been bred in captivity, some in numbers large enough for release programmes or for extensive laboratory work. Breeding for reintroduction to the wild is a three-step process, involving the production of young in captivity, their successful release into the wild and the subsequent establishment of a natural breeding population. The biggest current scheme involves the reintroduction of the Peregrine to eastern North America, though other schemes are concerned with adding captive-produced young to remnant wild populations, in which breeding is poor because of pesticide contamination. More information is needed on the process of imprinting to parents, to nest-sites and to home area, partly to avoid making mistakes and partly to improve the effectiveness of reintroductions. The role of captive breeding in conservation should become clearer with time, but it offers promise for several endangered species.

Young white-tailed Eagle

CHAPTER 18

Conclusions

It will be apparent from previous chapters that some aspects of population regulation in raptors have been studied intensively, others rather little, and some scarcely at all. As a result, some steps in the arguments developed in this book rest on a firm basis of facts, others on suggestive but insufficient evidence, and yet others on untested ideas. In this final chapter, I shall summarise those findings that seem best established, and discuss their relevance to studies on other birds and to conservation.

Some raptors nest as solitary pairs spaced fairly evenly through the habitat, while others nest in loose groups or dense colonies. As in other birds, solitary-nesting is associated with fairly evenly-dispersed food supplies, and communal nesting is associated with patchy and sporadic food supplies. Sometimes, however, species which are usually solitary are found nesting in concentrations where the only available nest-sites are clumped together in a wide foraging area. This occurs, for example, where isolated woods or cliff escarpments are found in otherwise open terrain.

In any environment, a ceiling on breeding density is set by food or nest-sites, whichever is in shortest supply. If nest-sites are freely available, the pairs of solitary species space themselves more widely where prey is scarce than where prey is plentiful. In some landscapes, a shortage of nest-sites restricts the numbers of breeding pairs to a level

Illustration: Gyr Falcon

288

lower than the available food would support, and in some areas pair numbers have been increased by providing nest-sites artificially. Raptors which have a fairly stable (often varied) food supply show fairly stable breeding densities, and some populations have shown remarkable constancy over long periods. In contrast, raptors which depend on a restricted number of cyclic prey species fluctuate greatly in numbers from year to year, paralleling the changes in their prey. A given species may show fluctuating densities in one region, and stable densities in another, depending on the food supply. In addition, sudden or long-term changes in the numbers of breeding raptors have often accompanied sudden or long-term changes in food supplies. Thus, much of the variation in the breeding density of any given raptor species can be attributed to variations in the carrying capacity of the environment, either from one district to another or, in cyclic species, from one year to another. At least in solitary species, the regulation is brought about by spacing behaviour, which limits breeding density in relation to resources, and in some cases gives rise to a 'surplus' population of non-breeding adults.

In general, wintering densities seem even more closely related to food than do summer ones. This is partly because the birds are no longer confined to the vicinity of nests, and are more free to move around and to exploit prey in areas lacking nest-sites. Often, raptor densities change abruptly from month to month and from place to place, in strict accordance with fluctuations in the numbers of prey. Some Palaearctic species wintering in Africa continually move from one area of temporary prey abundance to another.

In recent times, many populations have been well below the level that the environment would support, chiefly because of human activities. Large areas of otherwise suitable habitat have been emptied of raptors by deliberate killing, by excessive human presence (as in recreation activities), and by use of organo-chlorine pesticides and other toxic chemicals. Persecution has been especially prevalent in Europe, but its effect on populations has varied, mainly according to the size and breeding rate of the species concerned. Some large, slow-breeding eagles and vultures were eliminated altogether from large parts of their former range, and at the time of writing they are still absent, despite the continued presence of apparently suitable habitat. These and other raptors have fluctuated in numbers over the years, declining and contracting in range with each major campaign against them, and increasing and spreading again between times. Over much of the northern world from 1947, DDT and other persistent organo-chlorine compounds caused severe declines in certain populations. The bird-eating and fish-eating raptors were most affected, and the Peregrine disappeared as a breeder from one-third of North America. Declines occurred because of reduced breeding rates, associated with shell-thinning, so that not enough young were produced to offset the usual losses of adults, or because of a combination of reduced breeding rates and increased mortality of full-grown birds. Mass mortality was especially prevalent in some European countries around 1960, when dieldrin and other organo-chlorine and alkyl-mercury compounds were used as seed dressings.

The environmental carrying capacity for raptors has declined over the years, through the destruction and degradation of habitats. This has come from the increasing conversion of natural lands to cultivated ones, and from the intensification of cultivation procedures on existing farmland which has further reduced the prey populations on which raptors depend. A few raptors have benefited from human activities, however, notably the urban scavengers.

As in other birds, life-histories in raptors are linked with body-size, large species tending to live longer, begin breeding at a later age, and produce fewer young at each attempt than small species. All this in turn has certain effects on population dynamics: large species typically have a substantial non-breeding sector (because of deferred maturity), a slow turnover of breeding birds (because of low mortality), and a long recovery period after a decline (because of slow breeding); in contrast, small species have a small non-breeding sector, a rapid turnover of breeding birds, and a short recovery period after a decline. Within each species, much of the natural variation in breeding rate can be linked with variation in food supply, which influences the numbers of pairs that lay, their clutch and brood sizes and other aspects of reproduction. Predators of eggs or young are locally important in lowering breeding rates, especially where nest-sites are easy to reach.

My overall conclusion is that, in the absence of human intervention, almost every aspect of the natural population ecology of a given raptor species can be explained in terms of food: its dispersion over the countryside, its density in different areas and in different years; the extent of its numerical fluctuations; its breeding seasons and breeding rates; and its seasonal movements and dispersals. In some of these relationships, social behaviour is important as an intermediary between the birds and their resources, for example in the adjustment of breeding density to food-supplies, in influencing which individuals obtain nesting territories and which do not, and which individuals stay in the breeding area for the non-breeding season and which move out. For obvious reasons, predators are less important to raptors than to other birds, and their main effect seems to be indirect, in influencing the kinds of places where raptors put their nests, and hence in some localities restricting their breeding numbers. Similarly, although many diseases and pathogens have been identified in raptors, there is yet no evidence that they have influenced raptor populations adversely.

Our present understanding of what determines raptor numbers has come more from comparisons between populations than from long-term studies of particular species, of the kind done on other birds (Lack 1966). One effect of this has been to highlight the influence of habitat quality on densities in a way not brought out in other studies. For birds in general, three questions have repeatedly been asked about territorial behaviour and its role in population regulation. The first concerns the survival value to the individual. In raptors, territorialism in the breeding season clearly provides a nest-site and a place to mate with minimal interference from other individuals, and in some solitary species it also provides a feeding area from which other birds are excluded or at least restricted (and thus where food is presumably more readily available). Whether solitary or colonial, a bird cannot breed without a nesting territory. There is thus no doubt about the value of a territory to the individual. The second question is whether territorial behaviour restricts breeding density, or merely spaces out individuals already limited in some other way. In solitary raptors, territorial behaviour clearly restricts breeding density, as is shown by the proven existence of 'surplus' adults, which breed only when a gap is made available through the death or removal of a territory owner. Moreover, from the instances in the literature, such surplus birds are present in natural low density populations, as well as in high density ones. The third question is whether territorial spacing is related to the food resources of the environment. In raptors, the correlation between nest spacing and food supply in areas where nest-sites are not limiting strongly implies that this is so, as does the change in spacing observed in populations subject to change in food supply. Since individuals hold larger territories in areas where food is scarce, this automatically sets breeding density at a lower level than where food is plentiful. Some ornithologists have long tended to confuse the consequences of territoriality for the individual with its consequences for the population, supposing that it served primarily to regulate population density. On present understanding of natural selection, the most rational view is that territoriality serves to ensure the individual certain needs, including a restriction on other individuals in the vicinity. And with all birds behaving territorially for their own sakes, a limit to total population density is achieved incidentally.

The influence of other dispersion systems on population density is less clear than in the case of mutually exclusive feeding territories. Where pairs occupy such territories, it is easy to see how the number of pairs may be limited by the number of territories an area can provide. But where ranges overlap widely, so that several pairs use any one piece of ground, the means by which numbers are restricted is less obvious, for with no territorial defence of the hunting area, it would seem easy for extra birds to move in. However, all the birds in an area are utilising and disturbing prey, so with continued immigration, there will inevitably come a point when there are more birds present than the area will support, so that the weakest or least efficient hunters must leave or die. This point will be reached sooner in poor food areas than in good ones, so that even in the absence of territorial behaviour, it is easy to imagine how numbers might be regulated in relation to resources. It is presumably through such competition on feeding areas that the numbers of colonial raptors are limited in areas where nest-sites are freely available. (Among vultures at a carcass, for example, the effects of the social hierarchy on who feeds first are readily apparent.) However, the behaviour involved in regulating the size of colonies and the spacing between them has not been studied.

Amongst ornithologists, there is also confusion over the role of birth and death rates in setting population levels. Some raptors are evidently limited in numbers by the carrying capacity of the habitat, which can be independent of their population dynamics. It is theoretically possible for two populations of a species to have identical birth and death rates, but to live at quite different densities if they occupy areas of different carrying capacity. Conversely, it is possible for other populations to show the same density, yet widely divergent rates of turnover. In populations normally limited by resources, birth and death rates influence density independently only if there is a drop in the births/deaths ratio, so that the population declines below the level that the resources would support. If the unfavourable ratio persists, and is not countered by immigration, the population will decline to extinction, as happened often in the northern hemisphere during the main period of organo-chlorine use.

An understanding of population regulation in raptors is of practical importance chiefly in conservation. Since the birds have nearly always become rare because man has killed them or destroyed their haunts, preservation is achieved either by stopping the killing or by setting aside areas of natural habitat as reserves. The more recent problem of toxic chemicals can be solved in the long-term only by reducing usage, but since several such compounds act mainly on breeding, endangered populations can sometimes be sustained by increasing the breeding rate, using artificial means. Though simple in principle, the whole-hearted conservation of raptors is often thwarted because it conflicts with other forms of land-use or with powerful vested interest. In Europe it has so far proved impossible to persuade hunters and gamekeepers to stop killing raptors, and over much of the world many species live at such low natural densities that exceptionally large areas are needed to conserve them effectively. My own view is that reserves alone will not be enough to protect raptors in the long-term, and that sizeable populations will persist only if the birds are encouraged to recover and maintain themselves in land used primarily for other purposes. The need for at least a limited destruction (or removal) of raptors in places of serious conflict with human interest is likely to continue, and the methods of achieving this are sufficiently well known to need no further elaboration. Compared to pest-species generally, raptors have proved especially easy to 'control', largely because their low breeding rates make them slow to recover from reductions.

The conservation problems of recent years have repeatedly shown that it is not enough to let nature alone, nor merely to protect against direct human interference. Research may often be needed to define the problem and some form of active management to rectify it. Effective conservation usually entails some assessment of breeding stocks, of numbers, distribution and nest success. Only in this way can the most endangered stocks be identified and funds used to best effect. With any depressed population, the first problem is to find what has caused the decline, or is limiting the potential for increase. If it is then thought desirable, remedial measures can be tried, and the response of the population monitored. This last step is important, for only if numbers respond appropriately has the limiting factor been identified correctly. Free from the depressing influence of one factor, the population may then increase until it comes up against another limit, which must in turn be alleviated if numbers are to increase further. Curiously, the main opponents of active management in recent years have not been those with vested interests against raptors, but some emotive bird protectionists, to whom any form of intervention with a 'natural' population is anathema.

The study of population dynamics is also relevant to falconry, for as in other forms of cropping, it is important to limit the numbers removed, so that the remainder can quickly make good the losses. Comparison of breeding rates in different populations will often indicate which can best stand the loss, and how many birds might be taken each year without causing long-term decline. The harvest is best concentrated on the youngest birds, for these form the most expendable segment of any population, and by removing them from the nest before much natural mortality has occurred, there is the maximum opportunity for their removal to be compensated by improved survival in the remaining birds.

Most of the ideas on population expressed in this book are based on comparisons of one form or another, on correlations and associations, and thus fall into the category of 'circumstantial evidence'. It is important in future years that ideas be tested by further work, and particularly by experiment, for only in this way can they be properly checked and superceded if necessary. Some experiments, like those on the provision of nest-sites or the removal of territory owners, can be simple, but they need to be based on a sound ecological knowledge of the species concerned, for otherwise they may be badly designed or misinterpreted. Some ideas based on observation seem at present to be impossible to test by experiment, but perhaps in time they will yield to ingenuity. If this book leads anyone to check, and accept or discard, the ideas it contains, then it will have served its purpose.

African Hawk-eagle

Bibliography

ABBOTT, C. G. 1911. The home-life of the Osprey. London: Witherby.

ALBERT, T. F. 1962. The effect of DDT on the sperm production of the domestic fowl. Auk 79: 104–7.

ALI, S. & RIPLEY S. D. 1968. Handbook of the birds of India and Pakistan, Vol. 1. London: Oxford University Press.

AMADON, D. 1959. The significance of sexual differences in size among birds. Proc. Amer. Phil. Soc. 103: 531–6.

AMADON, D. 1975. Why are female birds of prey larger than males? Raptor Research 9: 1–11.

AMES, P. L. 1966. DDT residues in the eggs of the Osprey in the north-eastern United States and their relation to nesting success. J. appl. Ecol. 3 (suppl.): 87–97.

AMES, P. L. & MERSEREAU, G. S. 1964. Some factors in the decline of the Osprey in Connecticut. Auk 81: 173–85.

ANDERSON, D. W. & HICKEY, J. J. 1972. Eggshell changes in certain North American birds. Proc. Int. Orn. Congr. 15: 514–40.

ANDERSON, D. W. & HICKEY, J. J. 1974. Eggshell changes in raptors from the Baltic region. Oikos 25: 395–401.

ANDERSON, D. W., HICKEY, J. J., RISEBROUGH, R. W., HUGHES, D. F. & CHRISTENSEN, R. E. 1969. Significance of chlorinated hydrocarbon residues to breeding pelicans and cormorants. Canad. Field-Nat. 83: 91–112.

ANGELL, T. 1969. A study of the Ferruginous Hawk: adult and brood behaviour. Living Bird 8: 225–41.

ANTHONY, A. J. 1976. The Lappet-faced Vultures of the Gonarezhou. Bokmakierie 28: 54–6.

ASH, J. S. 1960. Birds of prey numbers on a Hampshire game preserve during 1952–9. Brit. Birds 53: 285–300.

ASH, J. S. 1965. A reduction in numbers of birds of prey in France. Bird Study 12: 17–26.

ASHFORD, W. J. 1928–9. Peregrine Falcon nesting on the ground in Hampshire. Br. Birds 22: 190.

AZEVEDO, J. A., HUNT, E. G. & WOODS, L. A. 1965. Physiological effects of DDT on pheasants. Calif. Fish Game 51: 276–93.

BAKER, J. R. 1938. The evolution of breeding seasons. Pp. 161–77 in 'Evolution', ed. G. R. de Beer. London: Oxford University Press.

BALFOUR, E. 1955. Kestrels nesting on the ground in Orkney. Bird Notes 26: 245–53.

BALFOUR, E. 1957. Observations on the breeding biology of the Hen Harrier in Orkney. Bird Notes 27: 177–83, 216–24.

BALFOUR, E. 1962. The nest and eggs of the Hen Harrier in Orkney. Bird Notes 30: 69–73, 145–52.

BALFOUR, E. 1970. Iris colour in the Hen Harrier. Bird Study 17: 47.

BALFOUR, E. & CADBURY, J. 1974. A population study of the Hen Harrier, Circus cyaneus, in Orkney. Pp. 122–8 in 'The natural environment of Orkney', ed. R. Goodier. Nature Conservancy Council, Edinburgh.

BALFOUR, E. & CADBURY, J. C. 1979. Polygyny, spacing and sex ratio among Hen Harriers Circus cyaneus (L.) in Orkney, Scotland. Ornis Scand. (in press).

BALGOOYEN, T. G. 1976. Behaviour and ecology of the American Kestrel (Falco sparverius L.). Univ. Calif. Publ. Zool. 103: 1–85.

BANNERMAN, D. A. 1956. The Birds of the British Isles, Vol. 5. Edinburgh & London: Oliver & Boyd.

BANNERMAN, D. A. 1968. History of the birds of the Cape Verde Islands. Edinburgh & London: Oliver & Boyd.

BANZHAF, W. 1937. Der Seeadler. Dohrniana 16.

BATES, G. G. 1976. Breeding of sub-adult Golden Eagle. Bird Study 23: 284.

BAXTER, W. L., LINDER, R. L. & DAHLGREN, R. B. 1969. Dieldrin effects in two generations of penned hen pheasants. J. Wildl. Manage. 33: 96–102.

BEAMAN, M., FISHER, P., ROUND, P., HEREWARD, A., HEUBECK, M. & PORTER, R. F. 1979. The migration of raptors and storks through northeast Turkey. Sandgrouse 1, in press.

BEAMAN, M. & GALEA, C. 1974. The visible migration of raptors over the Maltese Islands. Ibis 116: 419–31.

BEAMAN, M., PORTER, R. F. & SØGÅRD, S. 1979. A decade of observations on the migration of raptors and storks through the Bosphorus and Sea of Marmara region. Sandgrouse 1, in press.

BEDNAREK, W., HAUSDORF, W., JÖRISSEN, U., SCHULTE, E. & WEGENER, H. 1975. Über die Ausiwirkungen der chemischen Umwelthelastung auf Greifvögel in zwei Probeflächen Westfalens. J. Orn. 116: 181–94.

BEEBE, F. L. 1960. The marine Peregrines of the northwest Pacific coast. Condor 62: 145–89.

BEECHAM, J. J. & KOCHERT, M. N. 1975. Breeding biology of the Golden Eagle in southwestern Idaho. Wilson Bull. 87: 506–13.

BELISLE, A. A., REICHEL, W. L., LOCKE, L. N., LAMONT, T. G., MULHERN, B. N., PROUTY, R. M., DeWOLF, R. B. & CROMARTIE, E. 1972. Residues of organochlorine pesticides, polychlorinated biphenyls and mercury, and autopsy data for Bald Eagles, 1969 and 1970. Pestic. Mon. J. 6: 133–8.

BELON, P. 1555. L'histoire de la nature des oyseaux, avec leurs descriptions et naifs portraicts. Paris.

BELOPOLSKIJ, L. O. 1971. (Migration of Sparrowhawk on Courland Spit.) Notaki ornitologiczne 12: 1–12.

BENGSTON, S–A. 1967. Observations of the reproductive success in 26 nests of the Marsh Harrier Circus aeruginosus in Skåne Province, Sweden. Ool. Rec.41: 23–8.

BENSON, C. W. 1951. A roosting site of the eastern Red-footed Falcon F. amurensis. Ibis 93: 467–8.

BENSON, C. W., BROOKE, R. K., DOWSETT, R. J. & IRWIN, M. P. S. 1971. The birds of Zambia. London: Collins.

BENSON, G. B. G. 1958–9. Some notes on Marsh Harriers. Bird Notes 28: 407–11.

BENT, A. C. 1938. Life histories of North American birds of prey. US Nat. Mus. Bull. 170.

BERG, W., JOHNELS, A., SJÖSTRAND, B. & WESTERMARK, T. 1966. Mercury content in feathers of Swedish birds from the past 100 years. Oikos 17: 71–83.

BERNIS, F. 1975. (Migration of Falconiformes and Ciconia spp. through the Straits of Gibralter. Part 2. Descriptive analysis of summer-autumn 1972.) Ardeola 21: 489–580.

BERRY, R. B. 1972. Reproduction by artificial insemination in captive American Goshawks. J. Wildl. Manage. 36: 1283–8.

BEUSEKOM, C. F. van. 1972. Ecological isolation with respect to food between Sparrowhawk and Goshawk. Ardea 60: 72–96.

BEZZEL, E. 1969. Ergebnisse quantitativier Greifvogelbeobachtungen in Oberbayern. Ornithol. Mitt. 21: 29–36.

BIJLEVELD, M. 1974. Birds of prey in Europe. London: Macmillan.

BIRD, D. M. & LAGUE, P. C. 1974. Successful captive breeding of American Rough-legged Hawks. Raptor Research 8: 77.

BIRD, D. M., LAGUE, P. C. & BUCKLAND, R. B. 1977. Artificial insemination and semen production of the American Kestrel. Proc. ICBP World Conf. on birds of prey, Vienna, 1975: 347–8.

BITMAN, J., CECIL, H. C., HARRIS, S. J. & FRIES, G. F. 1969. DDT induces a decrease in eggshell calcium. Nature, Lond. 224: 44–6.

BLACK, H. 1978. Feeding ecology and behaviour of the Bat Hawk Machaerhamphus alcinus. Proc. Symp. Afr. Pred. Birds: 105. Pretoria: Northern Transvaal Ornithological Society.

BLUS, L. J., GISH, C. D., BELISLE, A. A. & PROUTY, R. M. 1972. Logarithmic relationship of DDE residues to eggshell thinning. Nature, Lond. 235: 376–7.

BLUS, L. J., NEELEY, B. S., BELISLE, A. A. & PROUTY, R. M. 1974. Organochlorine residues in Brown Pelican eggs: relation to reproductive success. Environ. Pollut. 7: 81–91.

BOEKER, E. L. & RAY, T. D. 1972. Golden Eagle population studies in the southwest. Condor 73: 463–7.

BOGAN, J. A. & MITCHELL, J. 1973. Continuing dangers to Peregrines from dieldrin. Br. Birds 66: 437–9.

BOGAN, J. & NEWTON, I. 1977. Redistribution of DDE in Sparrowhawks during starvation. Bull. environ. Contam. & Toxicol. 18: 317–21.

BOOTH, B. D. M. 1961. Breeding of the Sooty Falcon in the Libyan desert. Ibis 103A: 129–30.

BORG, K., ERNE, K., HANKO, E. & WANNTORP, H. 1970. Experimental secondary methyl mercury poisoning in the Goshawk (Accipiter g. gentilis L.). Environ. Pollut. 1: 91–104.

BORG, K., WANNTORP, H., ERNE, K. & HANKO, E. 1969. Alkyl mercury poisoning in terrestrial Swedish wildlife. Viltrevy 6: 301–79.

BOUDOINT, Y., BROSSET, A., BUREAU, L., GUICHARD, G. & MAYAUD, N. 1953. Étude de la biologie du Circaëte Jean le Blanc. Alauda 21: 86–127.

BOYD, L. & BOYD, N. 1975. Hybrid falcon. Hawk Chalk 14: 53–4.

BOYD, L. 1978. Hybridisation of falcons by artificial insemination. Raptor Research.

BRAAKSMA, S., KNIPPENBERG, W. H. Th. & LANGENHOFF, V. 1959. Enige broedvogels in Noord-Brabant. Limosa 32: 206–12.

BRECKENRIDGE, W. J. 1935. An ecological study of some Minnesota Marsh Hawks. Condor 37: 268–76.

BRIGGS, D. M. & HARRIS, J. R. 1973. Polychlorinated biphenyls influence on hatchability. Poultry Sci. 52: 1291.

BRITTON, W. M., & HUSTON, T. M. 1973. Influence of polychlorinated biphenyls in the laying hen. Poult. Sci. 52: 1620.

BROEKHUYSEN, M. & G. 1974. Black-shouldered Kite (Elanus caeruleus) builds a nest on a telephone pole. Bokmakierie 26: 36.

BROEKHUYSEN, G. J. & SIEGFRIED, W. R. 1970. Age and moult in the Steppe Buzzard in southern Africa. Ostrich Suppl. 8: 223–37.

BROLEY, C. L. 1947. Migration and nesting of Florida Bald Eagles. Wilson Bull. 59: 3–20.

BROO, B. 1977. Project Eagle Owl in southwest Sweden. Proc. ICBP World Conf. on birds of prey, Vienna, 1975: 338–43.

BROOKE, R. K. 1965. Roosting of the Black-shouldered Kite Elanus caeruleus (Desfontaines). Ostrich 46: 43.

BROOKE, R. K., GROBLER, J. H. & IRWIN, M. P. S. 1972. A study of the migratory eagles Aquila nipalensis & A. pomarina (Aves: Accipitridae) in Southern Africa, with comparative notes on other large raptors. Occ. Pap. natn. Mus. Rhod., 1972 B5(2): 61–114.

BROOKS, A. 1927. Breeding of immature hawks. Condor 29: 245–6.

BROUN, M. 1949. Hawks aloft: the story of Hawk Mountain. New York: Dodd, Mead.

BROWN, L. H. 1952/53. On the biology of the large birds of prey of Embu District, Kenya Colony. Ibis 94: 577–620; 95: 74–114.

BROWN, L. H. 1955. Eagles. London: Michael Joseph.

BROWN, L. H. 1955a. Supplementary notes on the biology of the large birds of prey of Embu District, Kenya Colony. Ibis 97: 38–64.

BROWN, L. H. 1960. The African Fish Eagle Haliaëtus vocifer, especially in the Kavirondo Gulf. Ibis 102: 285–97.

BROWN, L. H. 1966. Observations on some Kenya eagles. Ibis 108: 531–72.

BROWN, L. H. 1970. African birds of prey. London: Collins.

BROWN, L. H. 1970a. Some factors affecting breeding in eagles. Ostrich, suppl. 8: 157–67.

BROWN, L. H. 1972. Natural longevity in wild Crowned Eagles Stephanoaetus coronatus. Ibis 114: 263–5.

BROWN, L. 1976. Birds of prey: their biology and ecology. London: Hamlyn.

BROWN, L. 1976a. British birds of prey. London: Collins.

BROWN, L. 1976b. Eagles of the world. London: David & Charles.

BROWN, L. H. 1977. The status, population structure, and breeding dates of the African Lammergeier *Gypaetus barbatus meridionalis*. Raptor Research 11: 49–80.

BROWN, L. H. & AMADON, D. 1968. Eagles, hawks and falcons of the world. London: Country Life Books.

BROWN L. H. & CADE, T. J. 1972. Age classes and population dynamics of the Bateleur and African Fish Eagle. Ostrich 43: 1–16.

BROWN, L. H., GARGETT, V. & STEYN, P. 1977. Breeding success in some African eagles related to theories about sibling aggression and its effects. Ostrich 48: 65–71.

BROWN, L. H. & HOPCRAFT, J. B. D. 1973. Population structure and dynamics in the African Fish Eagle *Haliaeetus vocifer* (Daudin) at Lake Naivasha, Kenya. E. Afr. Wildl. J. 11: 255–69.

BROWN, L. H. & WATSON, A. 1964. The Golden Eagle in relation to its food supply. Ibis 106: 78–100.

BROWN, P. M., BUNYAN, P. J. & STANLEY, P. I. 1977. The investigation and pattern of occurrence of animal poisoning resulting from the misuse of agricultural chemicals. J. Forens. Sci. Soc., 17: 211–21.

BROWN, V. K., RICHARDSON, A., ROBINSON, J. & STEVENSON, D. E. 1965. The effects of aldrin and dieldrin on birds. Fd. Cosmet. Toxicol. 3: 675.

BROWN, W. R. 1973. Winter population trends in the Marsh, Cooper's and Sharp-shinned Hawks. Amer. Birds 27: 6–7.

BROWNING, M. R. 1974. Comments on the winter distribution of the Swainson's Hawk (*Buteo swainsoni*) in North America. Amer. Birds 28: 865–7.

BURLINGTON, H. & LINDEMAN, V. F. 1950. Effect of DDT on testes and secondary sex characters of White Leghorn cockerels. Proc. Soc. exp. Biol., N.Y. 74: 48–51.

BURNS, F. L. 1911. A monograph of the Broad-winged Hawk (*Buteo platypterus*). Wilson Bull. 23: 139–320.

BUXTON, A. 1947–8. Bird notes from Horsey. Trans. Norfolk & Norwich Nat. Soc. 1947: 302–5; 1948: 355–60.

CADE, T. J. 1955. Experiments on winter territoriality of the American Kestrel, *Falco sparverius*. Wilson Bull. 67: 5–17.

CADE, T. J. 1960. Ecology of the Peregrine and Gyrfalcon populations in Alaska. Univ. Calif. Publ. Zool. 63: 151–290.

CADE, T. J. 1969. The status of the Peregrine and other falconiforms in Africa. Pp. 289–322 in 'Peregrine Falcon populations: their biology and decline', ed. J. J. Hickey. Madison, Milwaukee and London: Univ. Wisconsin Press.

CADE. T. J. 1973–7. The Peregrine Fund, Cornell Newsletter Nos. 1–5.

CADE. T. J. 1974. Plans for managing the survival of the Peregrine Falcon. Raptor Research Foundation, Raptor Research Report 3: 89–104.

CADE, T. J., LINCER, J. L., WHITE, C. M., ROSENEAU, D. G. & SWARTZ, L. G. 1971. DDE residues and eggshell changes in Alaskan falcons and hawks. Science 172: 955–7.

CADE, T. J. & TEMPLE, S. A. 1977. The Cornell University falcon programme. Proc. ICBP World Conf. on birds of prey, Vienna, 1975: 353–69.

CADE, T. J. & WEAVER, J. D. 1976. Gyrfalcon and Peregrine hybrids produced by artificial insemination. The Journal NAFA 15: 42–7.

CADE, T. J., WHITE, C. M. & HAUGH, J. R. 1968. Peregrines and pesticides in Alaska. Condor 70: 170–8.

CALL, D. J. & HARRELL, B. E. 1974. Effects of dieldrin and PCB's upon the production and morphology of Japanese quail eggs. Bull. Environ. Contam. Toxicol. 11: 70–7.

CAMERON, A. C. 1974. Nesting of the Letter-winged Kite in western Queensland. Sunbird 5: 89–95.

CAMPBELL, J. A. & NELSON, R. W. 1975. Captive breeding and behaviour of Richardson's Merlins, 1974. The Journal NAFA, 14: 24–31.

CAMPBELL, J. 1977. Hand rearing and other alternatives. Pp. 55–8 in 'Papers on the veterinary medicine and domestic breeding of diurnal birds of prey', ed. J. E. Cooper & R. E. Kenward. Oxford: British Falconers' Club.

CAMBELL, R. W., PAUL, H. A., RODWAY, H. S. & CARTER, H. R. 1977. Tree-nesting

Peregrine Falcons in British Columbia. Condor 79: 500–1.

CARLSON, R. W. & DUBY, R. T. 1973. Embryotoxic effects of three PCBs in the chicken. Bull. environ. Contam. & Toxicol. 9: 261–6.

CARNIO, J. S. & McQUEEN, D. J. 1973. Adverse effects of 15 ppm p, p'-DDT on three generations of Japanese Quail. Can. J. Zool. 51: 1307–12.

CARRICK, R., KEITH, K. & GWYNN, A. M. 1960. Fact and fiction on the breeding of the Wandering Albatross. Nature 188: 112–14.

CASH, C. G. 1914. History of the Loch an Eilein Ospreys. Scott. Nat. 25: 149–58.

CAVÉ, A. J. 1968. The breeding of the Kestrel, *Falco tinnunculus* L., in the reclaimed area Oostelijk Flevoland. Netherlands J. Zool. 18: 313–407.

CHEYLAN, G. 1973. Notes sur la compétition entre l'Aigle royal *Aquila chrysaëtos* et l'Aigle de Bonelli *Hieraaetus fasciatus*. Alauda 41: 303–12.

CLAPHAM, C. S. 1964. The birds of the Dahlac Archipelago. Ibis 106: 376–88.

CHRISTENSEN, S., LOU, MÜLLER & WOHLMUTH, H. 1979. Spring migration of raptors in southern Israel and Sinai. Sandgrouse 1, in press.

COCHRAN, W. W. 1972. A few days of the fall migration of a Sharp-shinned Hawk. Hawk Chalk 11: 39–44.

COCHRAN, W. W. 1975. Following a migrating Peregrine from Wisconsin to Mexico. Hawk Chalk 14: 28–37.

CODY, M. L. 1966. A general theory of clutch size. Evolution 20: 174–84.

CODY, M. L. 1971. Ecological aspects of reproduction. Pp. 461–512 in 'Avian Biology', Vol. 1, ed. D. S. Farner. London & New York: Academic Press.

COLE, L. C. 1954. The population consequences of life history phenomena. Quart. Rev. Biol. 29: 103–137.

COLLING, A. W. & BROWN, E. B. 1946. The breeding of Marsh and Montagu's Harriers in North Wales in 1945. Br. Birds 39: 233–43.

CONDER, P. 1977. Legal status of birds of prey and owls in Europe. Proc. ICBP World Conf. on birds of prey, Vienna, 1975: 189–93.

CONRAD, B. 1977. Die Giftbelastung der Vogelwelt Deutschlands. Vogelkundliche Bibliothek 5: 1–68.

COOKE, A. S. 1973. Shell thinning in avian eggs by environmental pollutants. Environ. Pollut. 4: 85–152.

COOKE, A. S. 1975. Pesticides and eggshell formation. Symp. zool. Soc. Lond. 35: 339–61.

COON, N. C., LOCKE, L. N., CROMARTIE, E. & REICHEL, W. L. 1969. Causes of Bald Eagle mortality 1960–1965. J. Wildl. Dis. 6: 72–6.

COOPER, J. E. 1972. Veterinary aspects of captive birds of prey. Hawk Trust, Newent. Glos.

COOPER, J. E. 1973. Post-mortem findings in East African birds of prey. J. Wildl. Dis. 9: 368–75.

COOPER, J. E. 1975. First aid and veterinary treatment of wild birds. J. Small Anim. Pract. 16: 579–91.

COOPER, J. E. 1976. Clinical conditions of East African birds of prey. Trop. Anim. Hlth. Prod. 8: 203–11.

COOPER, J. E. 1977. Anaesthetics and other drugs. Pp. 16–21 in 'Papers on the veterinary medicine and domestic breeding of diurnal birds of prey', ed. J. E. Cooper & R. E. Kenward. Oxford: British Falconers' Club.

COULSON, J. C. 1966. The influence of the pair bond and age on the breeding biology of the Kittiwake Gull *Rissa tridactyla*. J. anim. Ecol. 35: 269–79.

CRAIGHEAD, J. J. & CRAIGHEAD, F. C. 1956. Hawks, owls and wildlife. Pennsylvania: Stackpole Co.

CROMARTIE, E., REICHEL, W. L., LOCKE, L. N., BELISLE, A. A., KAISER, T. E., LAMONT, T. G., MULHERN, B. M., PROUTY, R. M. & SWINEFORD, D. M. 1975. Residues of organochlorine pesticides and polychlorinated biphenyls and autopsy data for Bald Eagles, 1971–72. Pestic. Mon. J. 9: 11–14.

CROOK, J. H. 1965. The adaptive significance of avian social organisations. Symp. Zool. Soc. Lond. 14: 181–218.

CUNEO, F. 1968. Notes on breeding the King Vulture *Sarcoramphus papa* at Naples Zoo. Int. Zoo Yb. 8: 156–7.

DAHLGREN, R. B. & LINDER, R. L. 1970. Eggshell thickness in pheasants given dieldrin. J. Wildl. Manage. 34: 226–8.

DAHLGREN, R. B. & LINDER, R. L. 1971. Effects of polychlorinated biphenyls on pheasant reproduction, behaviour and survival. J. Wildl. Manage. 35: 315–19.

DALLING, J. 1975. Lanners in central Salisbury: the first four years. Honeyguide 84: 23–6.

DARE, P. 1961. Ecological observations on a breeding population of the Common Buzzard *Buteo buteo*. Ph.D. thesis, Exeter University.

DAVIES, P. W. & DAVIS, P. E. 1973. The ecology and conservation of the Red Kite in Wales. Br. Birds 66: 183–224, 241–70.

DAVIS, T. A. W. & SAUNDERS, D. R. 1965. Buzzards on Skomer Island. Nature in Wales 9: 116–24.

DE BONT, A. 1952. Brèves communications. Le Gerfaut 42: 255.

DEMENTIEV, G. P. & GLADKOV, N. A. 1954. (Birds of the Soviet Union), Vol. 1. Moscow: State Publishing House.

DENNIS, R. H., 1968, 1969. Sea Eagles. Fair Isle Observatory Report 21: 17–21: 22: 23–9.

DENT, G. 1939. A case of bigamy in Montagu's Harrier. Br. Birds 33: 51–2.

DEPPE, H. J. 1972. Einige Verhaltensbeobachtungen in einem Doppelhorst von Seeadler (*Haliaeetus albicilla*) und Wanderfalke (*Falco peregrinus*) in Mecklenburg. J. Orn. 113: 440–4.

DeWITT, J. B. 1955. Effects of chlorinated hydrocarbon insecticides upon Quail and Pheasants. J. agric. Fd. Chem. 3: 672–6.

DeWITT, J. B. 1956. Chronic toxicity to Quail and Pheasants of some chlorinated insecticides. J. agric. Fd. Chem. 4: 863–6.

DICKSON, R. C. 1974. Hen Harriers' hunting behaviour in south-west Scotland. Br. Birds 67: 511–13.

DIXON, J. B. 1937. The Golden Eagle in San Diego County, California. Condor 39: 49–56.

DOBINSON, H. M. & RICHARDS, A. J. 1964. The effects of the severe winter of 1962/63 on birds in Britain. Br. Birds 57: 373–434.

DOBZHANSKY, T. 1950. Evolution in the tropics. Amer. Sci. 38: 209–21.

DROST, R. & SCHÜZ, E. 1940. Von den Folgen des harten Winters 1939–40 für die Vogelwelt. Der Vogelzug 11: 161–91.

DUNACHIE, J. F. & FLETCHER, W. W. 1966. Effects of some insecticides on the hatching rate of hens' eggs. Nature, London. 212: 1062–3.

DUNKLE, S. W. 1977. Swainson's Hawks on the Laramie Plains, Wyoming. Auk 94: 65–71.

DUNNE, P. J. & CLARK, W. S. 1977. Fall hawk movement at Cape May Point, N. J. – 1976. New Jersey Audubon 3: 114–24.

DUNSTAN, T. C. 1968. Breeding success of Osprey in Minnesota from 1963 to 1968. Loon 40: 109–12.

EARHART, C. M. & JOHNSON, N. K. 1970. Size dimorphism and food habits of North American owls. Condor 72: 251–64.

EDELSTAM, C. 1972. The visible migration of birds at Ottenby, Sweden. Vår Fågelvärld, Suppl. 7.

EISENMANN, E. 1971. Range expansion and population increase in North and Middle America of the White-tailed Kite (*Elanus leucurus*). Amer. Birds 25: 529–36.

ELGOOD, J. H., FRY, C. H. & DOWSETT, R. J. 1973. African migrants in Nigeria. Ibis 115: 1–45, 375–411.

ELLIS, D. H., SMITH, D. G. & MURPHY, J. R. 1969. Studies on raptor mortality in western Utah. Great Basin Nat. 29: 165–7.

ELTON, C. S. 1942. Voles, mice and lemmings. Oxford: University Press.

ELTRINGHAM, S. K. 1975. Territory size and distribution in the African Fish Eagle. J. Zool., Lond. 175: 1–13.

ENDERSON, J. H. 1960. A population study of the Sparrow Hawk in east-central Illinois. Wilson Bull. 72: 222–31.

ENDERSON, J. H. 1964. A study of the Prairie Falcon in the central Rocky Mountain region. Auk 81: 332–52.

ENDERSON, J. H. 1965. A breeding and migration survey of the Peregrine Falcon. Wilson Bull. 77: 327–39.

ENDERSON, J. H. 1969. Peregrine and Prairie Falcon life tables based on band-recovery data. Pp. 505–08 in 'Peregrine Falcon populations: their biology and decline', ed. J. J. Hickey. Madison, Milwaukee and London: Univ. Wisconsin Press.

ENDERSON, J. H. & BERGER, D. D. 1968. Chlorinated hydrocarbon residues in Peregrines and their prey species from northern Canada. Condor 70: 149–53.

ENDERSON, J. H. & BERGER, D. D. 1970. Pesticides: eggshell thinning and lowered production of young in Prairie Falcons. BioScience 20: 355–6.

ENDERSON, J. H., TEMPLE, S. A. & SWARTZ, L. G. 1973. Time lapse photographic records of nesting Peregrine Falcons. Living Bird 11: 113–28.

ENDERSON, J. H. & WREGE, P. H. 1973. DDE residues and\eggshell thickness in Prairie Falcons. J. Wildl. Manage. 37: 476–8.

ENGLAND, M. D. 1976. Montagu's Harrier retrieving displaced nestling. Br. Birds 69: 499.

ENSINK, H. J. A. 1968. Verslag betreffende vergiftiging, speciaal van vogels, in de provincie Drente in het voorjaar van 1967. Rapport van het Comité tot instandhouding van de Drentse fauna, Assen.

EVANS, P. R. & LATHBURY, G. W. 1973. Raptor migration across the Straits of Gibraltar. Ibis 115: 572–85.

EVERETT, M. J. 1971. The Golden Eagle survey in Scotland. Br. Birds 64: 49–56.

FANNIN, A. & WEBB, D. 1975. Notes on the breeding of the Crowned Eagle. Honeyguide 82: 36.

FEARE, C. J., TEMPLE, S. A. & PROCTER, J. 1974. The status, distribution and diet of the Seychelles Kestrel, *Falco araea*. Ibis 116: 548–51.

FENNELL, C. M. 1954. Notes on the nesting of the Kestrel in Japan. Condor 56: 106–7.

FENTZLOFF, C. 1975. Erfolgreiche Zucht und Adoption von Seeadlern (*Haliaeetus albicilla*). Deutscher Falkenorden 1975: 28–40.

FENTZLOFF, C. 1976. Seeadlerschutz 1975 erfolgreich. Die Welt der Tiere 1: 5–7.

FERGUSON-LEES, I. J. 1951. The Peregrine population of Britain. Bird Notes 24: 200–5, 309–314.

FIEDLER, W. 1977. On the captive breeding of the White-tailed Sea Eagle (*Haliaeetus albicilla*) and Griffon Vulture (*Gyps fulvus*) in Schönbrunn Zoo. Proc. ICBP World Conf. on birds of prey, Vienna, 1975: 372–5.

FIMREITE, N. FYFE, R. W., KEITH, J. A. 1970. Mercury contamination of Canadian prairie seed eaters and their avian predators. Canad. Field-Nat. 84: 269–76.

FIMREITE, N. & KARSTAD, L. 1971. Effects of dietary methyl mercury on Red-tailed Hawks. J. Wildl. Manage. 35: 293–300.

FISHER, H. I. 1975. The relationship between deferred breeding and mortality in the Laysan Albatross. Auk 92: 433–41.

FISHER, R. A. 1930. The genetical theory of natural selection. Oxford: University Press.

FITCH, H. S. 1974. Observations on the food and nests of the Broad-winged Hawk (*Buteo platypterus*) in northeastern Kansas. Condor 76: 331–60.

FITCH, H. S., SWENSON, F. & TILLOTSON, D. F. 1946. Behaviour and food habits of the Red-tailed Hawk. Condor 48: 205–37.

FITZNER, R. E. 1978. The ecology and behaviour of the Swainson's Hawk (*Buteo swainsoni*) in southeastern Washington. Ph.D. thesis, Washington State University, Pullman.

FITZNER, R. E., BERRY, D., BOYD, L. L. & RIECK, C. A. 1977. Nesting of Ferruginous Hawks (*Buteo regalis*) in Washington 1974–75. Condor 79: 245–9.

FIUCZYNSKI, D. 1978. Zur Populationsökologie des Baumfalken (*Falco subbuteo* L.,

1758). Zool. Jb. Syst. Bd. 105: 193–257.

FIUCZYNSKI, D. & WENDLAND, V. 1968. Zur Populationsdynamik des Schwarzen Milans (*Milvus migrans*) in Berlin. Beobachtungen 1952–1967. J. Orn. 109: 462–71.

FLOWER, S. S. 1923. List of birds of prey 1898–1923, with notes on their longevity. Cairo: Govt. Press.

FLOWER, S. S. 1938. Further notes on the duration of life of animals – IV. Birds. Proc. Zool. Soc., Lond. 108A: 195–235.

FOWLER, J. M. & COPE, J. B. 1964. Notes on the Harpy Eagle in British Guiana. Auk 81: 257–73.

FRERE, H. T. 1886. Changes of plumage in the Kestrel. Zoologist 1886: 180.

FYFE, R. 1969. The Peregrine Falcon in northern Canada. Pp. 101–14 in 'Peregrine Falcon populations: their biology and decline', ed. J. J. Hickey. Madison, Milwaukee & London: Univ. Wisconsin Press.

FYFE, R. W. 1976. Rationale and success of the Canadian Wildlife Service Peregrine breeding project. Canad. Field-Nat. 90: 308–19.

FYFE, R. W. & ARMBRUSTER, H. I. 1977. Raptor research and management in Canada. Proc. ICBP World Conf. on birds of prey, Vienna, 1975: 282–93.

FYFE, R. W., ARMBRUSTER, H., BANASCH, U. & BEAVER, L. J. 1978. Fostering and cross-fostering of birds of prey. Pp. 62–71 in 'Birds of prey management techniques', ed. T. A. Geer. Oxford: British Falconers' Club.

FYFE, R. W., CAMPBELL, J., HAYSOM, B. & HODSON, K. 1969. Regional population declines and organochlorine insecticides in Canadian Prairie Falcons. Canad. Field-Nat. 83: 191–200.

FYFE, R. W., RISEBROUGH, R. W. & WALKER, W. 1976. Pollutant effects on the reproduction of the Prairie Falcons and Merlins of the Canadian prairies. Canad. Field-Nat. 90: 346–355.

FYFE, R. W., TEMPLE, S. A. & CADE, T. J. 1976. The 1975 North American Peregrine Falcon survey. Canad. Field-Nat. 90: 228–73.

GAILEY, J. & BOLWIG, N. 1973. Observations on the behaviour of the Andean Condor (*Vultur gryphus*). Condor 75: 60–8.

GALUSHIN, V. M. 1971. A huge urban population of birds of prey in Delhi, India (preliminary note). Ibis 113: 522.

GALUSHIN, V. M. 1974. Synchronous fluctuations in populations of some raptors and their prey. Ibis 116: 127–34.

GALUSHIN, V. M. 1977. Recent changes in the actual and legislative status of birds of prey in the USSR. Proc. ICBP World Conf. on birds of prey, Vienna, 1975: 152–59.

GARBER, D. P. & KOPLIN, J. R. 1972. Prolonged and bisexual incubation by California Ospreys. Condor 74: 201–2.

GARGETT, V. 1970. Black Eagle experiment II. Bokmakierie 22: 32–5.

GARGETT, V. 1970a. Black Eagle Survey, Rhodes Matopos National Park. A population study 1964–69. Ostrich Suppl. 8: 397–414.

GARGETT, V. 1971. Some observations on Black Eagles in the Matopos, Rhodesia. Ostrich Suppl. 9: 91–124.

GARGETT, V. 1975. The spacing of Black Eagles in the Matopos, Rhodesia. Ostrich 46: 1–44.

GARGETT, V. 1977. A 13-year population study of the Black Eagles in the Matopos, Rhodesia, 1964–1976. Ostrich 48: 17–27.

GARGETT, V. 1978. Black Eagles in protected and unprotected habitats. Proc. Symp. Afr. Pred. Birds: 96–102. Pretoria: Northern Transvaal Ornithological Society.

GATES, J. M. 1972. Red-tailed Hawk populations and ecology in east-central Wisconsin. Wilson Bull. 84: 421–33.

GENELLY, R. E. & RUDD, R. L. 1956. Effects of DDT, toxaphene and dieldrin on Pheasant reproduction. Auk 73: 529–39.

GEORGE, J. L. & FREAR, D. E. H. 1966. Pesticides in the Antarctic. J. appl. Ecol., Suppl. 3: 155–67.

GEROUDET, P. 1977. The reintroduction of the Bearded Vulture in the Alps. Proc. ICBP World Conf. on birds of prey, Vienna, 1975: 392–7.

GILBERT, H. A. 1951. Display of Common Buzzard. Br. Birds 44: 411–12.

GLOVER, B. 1952. Movements of birds in South Australia. S. Austr. Ornithol. 20: 82–91.

GLUE, D. E. 1971. Ringing recovery circumstances of small birds of prey. Bird Study 18: 137–46.

GLUTZ von BLOTZHEIM, U. N., BAUER, K. & BEZZEL, E. 1971. Handbuch der Vögel Mitteleuropas, Vol. 4. Frankfurt am Main: Akademische Verlagsgesselschaft.

GOLDBERG, E. D. 1975. Synthetic organohalides in the sea. Proc. Roy. Soc. Lond. B. 189: 277–289.

GOLODUSCHKO, B. S. 1961. Quoted in Meyburg 1970.

GOODMAN, D. 1974. Natural selection and a cost ceiling on reproductive effort. Amer. Nat. 108: 247–68.

GORDON, S. 1955. The Golden Eagle, King of Birds. London: Collins.

GRAVES, J. B., BONNER, F. L., McKNIGHT, W. F., WATTS, A. B. & EPPS, E. A. 1969. Residues in eggs, preening glands, liver and muscle from feeding dieldrin-contaminated rice bran to hens and its effect on egg production, egg hatch, and chick survival. Bull. Environ. Contam. & Toxicol. 4: 375–83.

GRAY, A. P. 1958. Bird hybrids: a check list with bibliography. Buckinghamshire: Farnham Royal.

GREEN, R. 1976. Breeding behaviour of Ospreys Pandion haliaetus in Scotland. Ibis 118: 475–90.

GREENWOOD, A. 1977. The role of disease in the ecology of British raptors. Bird Study 24: 259–65.

GREENWOOD, P. J., HARVEY, P. H. & PERRINS, C. M. 1978. Inbreeding and dispersal in the Great Tit. Nature 271: 52–4.

GREGG, M. 1961. Alaskan Bald Eagles. Alaskan Sportsman 1961: 27–8.

GRIER, J. W. 1969. Bald Eagle behaviour and productivity responses to climbing to nests. J. Wildl. Manage. 33: 961–6.

GRIER, J. W. 1973. Techniques and results of artificial insemination with Golden Eagles. Raptor Research 7: 1–12.

GRIFFIN, C. R. 1976. A preliminary comparison of Texas and Arizona Harris' Hawk (Parabuteo unicinctus) populations. Raptor Research 10: 50–4.

GROMME, O. J. 1935. The Goshawk (Astur atricapillus atricapillus) nesting in Wisconsin. Auk 52: 15–20.

GROSS, A. O. 1947. Cyclic invasions of the Snowy Owl and the migration of 1945–46. Auk 64: 584–601.

GUNN, D. L. 1972. Dilemmas in conservation for applied biologists. Ann. appl. Biol 72: 105–27.

HACKMAN, C. D. & HENNY, C. J. 1971. Hawk migration over White Marsh, Maryland. Chesapeake Sci. 12: 137–41.

HAEGELE, M. A. & HUDSON, R. H. 1973. DDE effects on reproduction of ring doves. Environ. Pollut. 4: 53–7.

HAGAR, D. C. 1957. Nesting populations of Red-tailed Hawks and Horned Owls in central New York State. Wilson Bull 69: 263–72.

HAGAR, J. A. 1969. History of the Massachusetts Peregrine Falcon population, 1935–57. Pp. 123–32 in 'Peregrine Falcon populations: their biology and decline', ed. J. J. Hickey, Madison, Milwaukee & London: Univ. Wisconsin Press.

HAGEN, Y. 1942. Totalgewichts – Studien bei norwegischen Vogelarten. Arch. f. Naturgesch. N. F. 11: 1–132.

HAGEN, Y. 1969. Norwegian studies on the reproduction of birds of prey and owls in relation to micro-rodent population fluctuations. Fauna 22: 73–126.

HALL, E. M. 1947. Concentrated nesting of Marsh Hawks. Condor 49: 211–12.

HALL, G. H. 1955. Great moments in action; the story of the Sun Life falcons. Montreal: Mercury Press. (Also reprinted 1970, Canad. Field Nat. 84: 211–30).

HAMERSTROM, F. 1969. A harrier population study. Pp. 367–85 in 'Peregrine Falcon populations: their biology and decline', ed. J. J. Hickey. Madison, Milwaukee & London: Univ. Wisconsin Press.

HAMERSTROM, F. 1977. Introducing captive-reared raptors into the wild. Proc ICBP World Conf. on birds of prey, Vienna, 1975: 348–53.

HAMERSTROM, F., HAMERSTROM, F. N. & HART, J. 1973. Nest boxes: an effective management tool for Kestrels. J. Wildl. Manage. 37: 400–3.

HAMERSTROM, F., RAY, T., WHITE, C. M. & BRAUN, C. E. 1975. Conservation committee report on status of eagles. Wilson Bull 87: 140–3.

HAMERTON, A. E. 1942. Report on the deaths occurring in the Society's gardens during the year 1941. Proc. Zool. Soc. Lond. 112: 120–37.

HAMMOND, M. C. & HENRY, C. J. 1949. Success of Marsh Hawk nests in North Dakota. Auk 66: 271–4.

HANSON, R. P. 1969. The possible role of infectious agents in the extinctions of species. Pp 439–44 in 'Peregrine Falcon populations: their biology and decline', ed. J. J. Hickey. Madison, Milwaukee & London: Univ. Wisconsin Press.

HANSON, W. C., 1971. The 1966–67 Snowy Owl incursion in southeastern Washington and the Pacific Northwest. Condor 73: 114–16.

HARWIN, R. & J. 1972. Raptor territories. Ostrich 43: 73–6.

HAUGH, J. R. 1975. Derby Hill. Proc. N. Amer. Hawk Mig. Conf., Syracuse, 1974: 11–5.

HAUKIOJA, E. & HAUKIOJA, M. 1970. Mortality rates of Finnish and Swedish Goshawks (*Accipiter gentilis*). Finnish Game Research 31: 13–20.

HAVERSCHMIDT, F. 1953. Observations on the Marsh Harrier with particular reference to clutch-size and nesting success. Br. Birds 46: 258–9.

HAVERSCHMIDT, F. 1970. Notes on the Snail Kite in Surinam. Auk 87: 580–4.

HEATH, R. G., SPANN, J. W., & KREITZER, J. F. 1969. Marked DDE impairment of Mallard reproduction in controlled studies. Nature, Lond. 224: 47–8.

HEATH, R. G., SPANN, J. W., KREITZER, J. F. & VANCE, C. 1972. Effects of polychlorinated biphenyls on birds. Proc. Int. Orn. Congr. 15: 475–85.

HECHT, W. R. 1951. Nesting of the Marsh Hawk at Delta, Manitoba. Wilson Bull 63: 167–76.

HEINZ, G. 1974. Effects of low dietary levels of methyl mercury on Mallard reproduction. Bull. Environ. Contam. & Toxicol 11: 386–91.

HEINTZELMAN, D. S. & NAGY, A. C. 1968. Clutch sizes, hatchability rates, and sex ratios of Sparrow Hawks in eastern Pennsylvania. Wilson Bull 80: 306–11.

HELANDER, B. 1975. The White-tailed Sea-eagle in Sweden. Svenska Naturskyddsföreningen, unpublished report for 1974.

HELANDER, B. 1977. The White-tailed Eagle in Sweden. Proc. ICBP World Conf. on birds of prey, Vienna, 1975: 319–29.

HELANDER, B. 1978. Feeding White-tailed Sea Eagles in Sweden. Pp. 47–56 in 'Bird of prey management techniques', ed. T. A. Geer. Oxford: British Falconers' Club.

HENNY, C. J. 1972. An analysis of the population dynamics of selected avian species. Bureau of Sport Fisheries & Wildlife, Res. Rep. 1. Washington, D.C.: Govt. Printer.

HENNY, C. J. 1977. Birds of prey, DDT, & Tussock Moths in Pacific Northwest. Trans. N. Amer. Wild. & Nat. Res. Conf. 42: 397–411.

HENNY, C. J. 1977a. Research, management and status of the Osprey in North America. Proc. ICBP World Conf. on birds of prey, Vienna, 1975: 199–222.

HENNY, C. J., COLLINS, J. A. & DEIBERT, W. J. 1978. Osprey distribution, abundance and status in western North America: II. The Oregon population. The Murrelet 59: 14–25.

HENNY, C. J. & OGDEN, J. C. 1970. Estimated status of Osprey populations in the United States. J. Wildl. Manage. 34: 214–17.

HENNY, C. J., SMITH, M. M. & STOTTS, V. D. 1974. The 1973 distribution and abundance of breeding Ospreys in the Chesapeake Bay. Chesapeake Sci. 15: 125–33.

HENNY, C. J., SCHMID, F. C., MARTIN, E. L. & HOOD, L. L. 1973. Territorial behaviour, pesticides, and the population ecology of Red-shouldered Hawks in central Maryland, 1943–1971. Ecology 54: 545–54.

HENNY, C. J. & WIGHT, H. M. 1969. An endangered Osprey population: estimates of mortality and production. Auk 86: 188–98.

HENNY, C. J. & WIGHT, H. M. 1972. Red-tailed and Cooper's Hawks: their population ecology and environmental pollution. Pp. 229–50 in 'Population ecology of migratory birds', Symposium Volume, Patuxent Wildlife Research Center.

HENNY, C. J. & VAN VELZEN, W. T. 1972. Migration patterns and wintering localities of

American Ospreys. J. Wildl. Manage. 36: 1133–41.

HENRIKSSON, K., KARPPANEN, E. & HELMINEN, M. 1966. High residue of mercury in Finnish White-tailed Eagles. Ornis Fennica 43: 38–45.

HERBERT, R. A. & HERBERT, K. G. S. 1965. Behaviour of Peregrine Falcons in the New York City Region. Auk 82: 62–94.

HERBERT, R. A. & HERBERT, K. G. S. 1969. The extirpation of the Hudson River Peregrine Falcon population. Pp. 133–54 in 'Peregrine Falcon Populations', ed. J. J. Hickey. Madison, Milwaukee & London: Univ. Wisconsin Press.

HICKEY, J. J. 1942. Eastern populations of the Duck Hawk. Auk. 59: 176–204.

HICKEY, J. J. 1952. Survival studies of banded birds. U. S. Dept. Int. Spec. Sci. Rep. Wildlife 15.

HICKEY, J. J. (Ed) 1969. Peregrine Falcon populations, their biology and decline. Madison, Milwaukee & London: Univ. Wisconsin Press.

HICKEY, J. J. & ANDERSON, D. W. 1968. Chlorinated hydrocarbons and eggshell changes in raptorial and fish-eating birds. Science, N.Y. 162: 271–3.

HICKEY, J. J. & ANDERSON, D. W. 1969. The Peregrine Falcon: life history and population literature. Pp 3–42 in 'Peregrine Falcon populations, their biology and decline,' ed. J. J. Hickey. Madison, Milwaukee & London: Univ. Wisconsin Press.

HILDÉN, O. 1966. (Occurrence, habits and food of the Marsh Harrier (Circus aeruginosus) in Finland). Suomen Riista 18: 82–93. (Finnish, with English summary).

HILDÉN, O & KALINAINEN, P. 1966. Über Vorkommen und Biologie der Rohrweihe, Circus aeruginosus (L.), in Finland. Orn. Fenn. 43: 85–124.

HILL, E. A. 1946. Placing of a leaf: brooding of the Broad-winged Hawk. Aud. Mag. 48: 137–39.

HILL, E. F., HEATH, R. G., SPANN, J. W. & WILLIAMS, J. D. 1974. Polychlorinated biphenyl toxicity to Japanese Quail as related to degree of chlorination. Poult. Sci. 53: 597–604.

HILL, E. F., HEATH, R. G., SPANN, J. W. & WILLIAMS, J. D. 1975. Lethal dietary toxicities of environmental pollutants to birds. Special Scientific Report – Wildlife No. 191. Washington: Fish & Wildlife Service.

HILL, E. F. & SHAFFNER, C. J. 1976. Sexual maturation and productivity of Japanese Quail fed graded concentrations of mercuric chloride. Poult. Sci. 55: 1449–59.

HIRONS, G. 1976. A population study of the Tawny Owl and its main prey species in woodland. D. Phil. thesis, Oxford University.

HODSON, K. 1975. Some aspects of the nesting ecology of Richardson's Merlin (Falco columbarius richardsonii) on the Canadian prairies. M.Sc. thesis, Univ. of Brit. Columbia, Vancouver.

HÖGLUND, N. 1964. Der Habicht Accipiter gentilis Linné in Fennoscandia. Viltrevy 2: 195–270.

HOLDSWORTH, M. 1971. Breeding biology of Buzzards at Sedbergh during 1937–67. Br. Birds 64: 412–20.

HOLSTEIN, V. 1942. Duehøgen Astur gentilis dubius (Sparrman). Copenhagen: Forlag.

HOLSTEIN, V. 1944. Hvepsevaagen Pernis apivorus apivorus (L.). Copenhagen: Forlag.

HOLSTEIN, V. 1950. Spurvehøgen Accipiter nisus nisus (L.). Copenhagen: Forlag.

HOLSTEIN, V. 1956. Musvaagen, Buteo buteo buteo (L.). Copenhagen: Forlag.

HÖLZINGER, J., MICKLEY, M. & SCHILHANSL, K. 1973. Beobachtungen on überwinternden Rotmilanen (Milvus milvus) im Donaumoos bei Ulm. Anz. orn. Ges. Bayern 12: 106–13.

HORN, H. S. 1968. The adaptive significance of colonial nesting in the Brewer's Blackbird (Euphagus cyanocephalus). Ecology 49: 682–94.

HORNADAY, W. T. 1913. Our vanishing wildlife: its extermination and preservation. New York Zoological Society.

HORVATH, L. 1955. Red-footed Falcons in Ohat-woods, near Hortobágy. Acta Zool. Acad. Sci. Hung. 1: 245–88.

HOSKING, E. 1943. Some observations on the Marsh Harrier. Br. Birds 37: 2–9.

HOUSTON, C. S. 1975. Close proximity of Red-tailed Hawk and Great Horned Owl nests. Auk 92: 612–14.

HOUSTON, D. C. 1974. The role of Griffon Vultures Gyps africanus and Gyps ruppellii as scavengers. J. Zool., Lond. 172: 35–46.

HOUSTON, D. C. 1975. Ecological isolation of African scavenging birds. Ardea 55: 55–64.

HOUSTON, D. C. 1975a. The moult of the White-backed and Rüppell's Griffon Vultures *Gyps africanus* & *G. rueppellii*. Ibis 118: 474–88.

HOUSTON, D. C. 1976. Breeding of the White-backed and Rüppell's Griffon Vultures, *Gyps africanus* and *G. rueppellii* Ibis 118: 14–40.

HOUSTON, D. C. 1978. The effects of food quality on breeding strategy in Griffon Vultures (*Gyps spp.*), J. Zool. Lond. 186: 175–84.

HOWARD, R. P. & WOLFE, M. L. 1976. Range improvement practices and Ferruginous Hawks. Journal of Range Manage. 29: 33–7.

HOWELL, A. H. 1932. Florida Bird Life. New York: Coward – McCann.

HUHTALA, K. & SULKAVA, S. 1976. (Breeding biology of the Goshawk). Suomen Luonto 6: 299–300.

HULCE, H. 1886. Eagles breeding in captivity. Forest & Stream 27: 327.

HUNT, W. G., ROGERS, R. R. & SLOWE, D. J. 1975. Migratory and foraging behaviour of Peregrine Falcons on the Texas coast. Canad. Field-Nat. 89: 111–23.

HURRELL, L. H. 1973. On breeding the Sparrowhawk in captivity. Pp. 29–44 in 'A hawk for the bush' by J. Mavrogordato. London: Spearman.

IRIBARREN, J. J. 1977. The present status of birds of prey in Navarra (Spain). Proc. ICBP World Conf. on birds of prey, Vienna, 1975: 381–87.

JACKSON, J. A. 1975. Regurgitative feeding of young Black Vultures in December. Auk 92: 802–3.

JAMES, D. & DAVIS, K. B. 1965. The effect of sublethal amounts of DDT on the discrimination ability of the Bobwhite, *Colinus virginianus* Linnaeus. Am. Zool. 5: 229.

JARVIS, M. J. F. & CRICHTON, J. 1978. Notes on Longcrested Eagles in Rhodesia. Proc. Symp. Afr. Pred. Birds: 17–24. Pretoria: Northern Transvaal Ornithological Society.

JEFFERIES, D. J. 1967. The delay in ovulation produced by pp' – DDT and its possible significance in the field. Ibis 109: 266–72.

JEFFERIES, D. J. 1971. Some sublethal effects of pp' – DDT and its metabolite pp' – DDE on breeding passerine birds. Meded. Fakult. Landbouwwet–enschappen Gent 36: 34–42.

JEFFERIES, D. J. 1973. The effects of organochlorine insecticides and their metabolites on breeding birds. J. Reprod. Fert., Suppl. 19: 337–52.

JEFFERIES, D. J. & PRESTT, I. 1966. Post-mortems of Peregrines and Lanners with particular reference to organochlorine residues. Br. Birds 59: 49–64.

JENKINS, M. A. 1978. Gyrfalcon nesting behaviour from hatching to fledging. Auk 95: 122–8.

JENNINGS, A. R. 1961. An analysis of 1,000 deaths in wild birds. Bird Study 8: 25–31.

JENSEN, S. 1966. Report on a new chemical hazard. New Scient. 32: 612.

JENSEN, S., JOHNELS, A. G., OLSSON, M. & WESTERMARK, T. 1972. The avifauna of Sweden as indicators of environmental contamination with mercury and chlorinated hydrocarbons. Proc. Int. Orn. Congr. 15: 455–65.

JOENSEN, A. 1968. (An investigation on the breeding population of the Buzzard (*Buteo buteo*) on the Island Als in 1962 & 1963.) Dansk Ornith. Foren. Tidsskr. 62: 17–31. (Danish, with English summary).

JOHANNESSON, H. 1975. Activities of breeding Marsh Harriers *Circus aeruginosus*. Vår Fågelvärld 34: 197–206.

JOHNELS, A. G., OLSSON, M. & WESTERMARK, T. 1968. *Esox lucius* and some other organisms as indicators of mercury contamination in Swedish lakes and rivers. Bull. Off. int. Epiz. 69: 1439–52.

JOHNSEN, G. 1929. Rovdyr – og rovfuglstatistikken i Norge. Bergens Museum Arboke 2: 5–118.

JOHNSON, S. J. 1973. Post-fledging activity of the Red-tailed Hawk. Raptor Res. 7: 43–8.

JOHNSON, S. J. 1975. Productivity of the Red-tailed Hawk in southwestern Montana. Auk 92: 732–36.

JONES, F. M. 1946. Duck hawks in eastern Virginia. Auk 63: 592.
JONES, P. J. & WARD, P. 1976. The level of reserve protein as the proximate factor controlling the timing of breeding and clutch-size in the Red-billed Quelea *Quelea quelea*. Ibis 118: 547–74.
JONES, S. G. & BREN, W. M. 1978. Observations on the wintering behaviour of Victorian Peregrine Falcons. Australian Bird Watcher 1978: 198–203.
JOURDAIN, F. C. R. 1928. Bigamy in the Sparrowhawk. Br. Birds 21: 200–2.
JUILLARD, M. 1977. Observations sur l'hivernage et les dorloirs du Milan royal *Milvus milvus* (L.) dans le nord-ouest de la Suisse. Nos Oiseaux 34: 41–57.

KALMBACH, E. R. 1939. American vultures and the toxin of *Clostridium botulinum*. J. Amer. Vet. Med. Ass. 94: 187–91.
KEICHER, K. 1969. Beobachtungen an Schlafplätzen des Wanderfalken (*Falco peregrinus*) auf der Schwäbischen Alb. Anz. orn. Ges. Bayern 8: 545–55.
KEITH, J. A. & GRUCHY, I. M. 1972. Residue levels of chemical pollutants in North American birdlife. Proc. Int. Orn. Congr. 15: 437–54.
KEITH, J. O. 1966. Insecticide contaminations in wetland habitats and their effects on fish-eating birds. J. appl. Ecol., Suppl. 3: 71–85.
KEITH, J. O., WOOD, L. A. & HUNT, E. G. 1970. Reproductive failure in Brown Pelicans on the Pacific coast. Trans. N. Amer. Wildl. Nat. Res. Conf. 35: 56–64.
KEITH, L. B. 1963. Wildlife's ten-year cycle. Madison: Univ. Wisconsin Press.
KEITH, L. B. 1974. Some features of population dynamics of mammals. Proc. Int. Congr. Game Biol 11: 17–58. Stockholm: National Swedish Environment Protection Board.
KEMP, A. C. 1972. The use of man-made structures for nesting sites by Lanner Falcons. Ostrich 43: 65–6.
KEMP, A. C. & M. I. 1975. Observations on the White-backed Vulture *Gyps africanus* in the Kruger National Park, with notes on other avian scavengers. Koedoe 18: 51–68.
KEMP, A. C. & M. I. 1977. The status of raptorial birds in the Transvaal Province of South Africa. Proc. ICBP World Conf. on birds of prey. Vienna, 1975: 28–34.
KEMP. A. C & M. I. 1978. *Bucorvus* and *Sagittarius*: two modes of terrestrial predation. Proc. Symp. Afr. Pred. Birds: 13–6. Pretoria: Northern Transvaal Ornithological Society.
KENDAL, M. D., WARD, P. & BACCHUS, S. 1973. A protein reserve in the pectoralis flight muscle of *Quelea quelea*. ibis. 115: 600–1.
KENWARD, R. E. (ed.) 1970–74. Captive breeding of diurnal birds of prey. Oxford: British Falconers' Club & Hawk Trust.
KENWARD, R. E. 1974. Mortality and fate of trained birds of prey. J. Wildl. Manage. 38: 751–6.
KENWARD, R. E. 1977. Captive breeding – a contribution by falconers to the preservation of Falconiforms. Proc. ICBP World Conf. on Birds of Prey, Vienna, 1975: 378–81.
KENWARD, R. E. 1977a. Predation on released Pheasants (*Phasianus colchicus*) by Goshawks in central Sweden. Viltrevy 10: 79–112.
KENYON, K. N. 1947. Breeding populations of the Osprey in Lower California. Condor 49: 152–8.
KERN, J. A. 1968. Snail Hawks of Okeechobee. Audubon 80: 10–17.
KEYMER, I. F. 1972. Diseases of birds of prey. Vet. Rec. 1972: 579–94.
KLEINSTAUBER, K. 1969. The status of cliff nesting Peregrines in the German Democratic Republic. Pp 209–16 in 'Peregrine Falcon populations: their biology and decline', ed J. J. Hickey. Madison, Milwaukee & London: Univ. Wisconsin Press.
KOEHLER, A. 1968. Über die Fortpflanzung einiger Greifvogelarten in Gefangenschaft. Der Falkner 18: 28–33.
KOEHLER, A. 1970. Red-headed Merlins breed in captivity. Captive breeding of diurnal birds of prey, 1: 16–19.
KOEMAN, J. H., van BEUSEKOM, C. F. & de GOEIJ. J. J. M. 1972. Eggshell and population changes in the Sparrow-hawk (*Accipiter nisus*). TNO nieuws 1972: 542–50.
KOEMAN, J. H., VINK, J. A. J. & De GOEIJ, J. J. M. 1969. Causes of mortality in birds of prey and owls in the Netherlands in the winter of 1968–1969. Ardea 57: 67–76.
KOFORD, C. B. 1953. The California Condor. Nat. Audubon Soc. Res. Rep. 4: 154.

KOPLIN, J. R. 1973. Differential habitat use by sexes of American Kestrels wintering in northern California. Raptor Research 7: 39–42.

KRAAN, van der, C. & STREIN, van, N. J. 1969. Polygamie bij de Blauwe Kiekendief (*Circus cyaneus*). Limosa 42: 34–5.

KRAMER, K. 1973. Habicht und Sperber. Die Neue Brehm-Bücherei. Wittenberg Lutherstadt: Ziemsen Verlag.

KRANTZ, W. C., MULHERN, B. M., BAGLEY, G. E., SPRUNT, A., LIGAS, F. J. & ROBERTSON, W. B. 1970. Organochlorine and heavy metal residues in Bald Eagle eggs. Pestic. Monit. J. 4: 136–40.

KREITZER, J. F. & HEINZ, G. H. 1974. The effect of sublethal dosages of five pesticides and a polychlorinated biphenyl on the avoidance response of *Coturnix* quail chicks. Environ. Pollut. 6: 21–9.

KUMARI, E. 1974. Past and present of the Peregrine Falcon in Estonia. Pp. 230–53 in 'Estonian wetlands and their life'. Tallinn: Valgus.

KURTH, D. 1970. Der Turmfalke (*Falco tinnunculus*) in Münchener Stadtgebiet. Anz. orn. Ges. Bayern 9: 2–12.

KUYT, E. 1962. A record of a tree-nesting Gyrfalcon. Condor 64: 508–10.

KWON, K-C, & WON, P-O. 1975. Breeding biology of the Chinese Sparrowhawk *Accipiter soloensis*. Misc. Rep. Yamashina's Inst. Ornithol. 7: 501–22.

LACK, D. 1954. The natural regulation of animal numbers. Oxford: University Press.

LACK, D. 1966. Population studies of birds. Oxford: University Press.

LACK, D. 1968. Ecological adaptations for breeding in birds. London: Methuen.

LAING, N. C. B. 1965. Notes on the nesting of the Augur Buzzard. Bokmakierie 17: 45.

LAMONT, T. G., BAGLEY, G. E. & REICHEL, W. L. 1970. Residues of op′ – DDD & op′ – DDT in Brown Pelican eggs and Mallard ducks. Bull. environ. Contam. & Toxicol. 5: 231–6.

LASIEWSKI, R. C. & DAWSON, W. R. 1967. A re-examination of the relation between standard metabolic rate and body weight in birds. Condor 69: 13–23.

LASZLO, S. 1941. The habits and plumages of Montagu's Harrier. Aquila 46–51: 247–68.

LEDGER, J. & MUNDY, P. 1975. Research on the Cape Vulture, 1974 progress report. Bokmakierie 27: 2–7.

LEHNER, P. N. & EGBERT, A. 1969. Dieldrin & eggshell thickness in ducks. Nature, Lond. 224: 1218–9.

LEOPOLD, A. S. & WOLFE, T. O. 1970. Food habits of nesting Wedge-tailed Eagles, *Aquila audax*, in south-eastern Australia. CSIRO Wildl. Res. 15: 1–17.

LESHEM, J. 1976. The biology of the Bonelli's Eagle (*Hieraaetus fasciatus fasciatus*) in Israel. M.Sc. thesis, Tel-Aviv University.

LILLIE, R. J., CECIL, H. C., BITMAN, J. & FRIES, G. F. 1974. Differences in response of caged White Leghorn layers to various polychlorinated biphenyls (PCBs) in the diet. Poult. Sci. 53: 726–32.

LINCER, J. L. 1972. The effects of organochlorines on the American Kestrel (*Falco sparverius* Linn.) Ph.D. thesis, Cornell University, Ithaca, New York.

LINCER, J. L. 1975. DDE-induced eggshell-thinning in the American Kestrel: a comparison of the field situation and laboratory results. J. appl. Ecol. 12: 781–93.

LINCER, J. L., CADE, T. J. & DEVINE, J. M. 1970. Organochlorine residues in Alaskan Peregrine Falcons (*Falco peregrinus* Tunstall), Rough-legged Hawks (*Buteo lagopus* Pontoppidan) and their prey. Canad. Field – Nat. 84: 255–63.

LINCER, J. L. & PEAKALL, D. B. 1973. PCB pharmacodynamics in the Ring Dove and early gas chromatographic peak diminution. Environ. Pollut. 4: 59–68.

LINDBERG, P. 1975. (The Peregrine Falcon (*Falco peregrinus*) in Sweden). Stockholm: Svenska Naturskyddsföreningen.

LINDBERG, P. 1977. The Peregrine Falcon in Sweden. Proc. ICBP World Conf. on birds of prey, Vienna, 1975: 329–38.

LINDSTEDT, S. L. & CALDER, W. A. 1976. Body size and longevity in birds. Condor 78: 91–145.

LINT, K. C. 1960. Notes on breeding Andean Condors at San Diego Zoo. Int. Zoo Yb. 2: 82–3.

LIVERSIDGE, R. 1962. The breeding biology of the Little Sparrowhawk (*Accipiter minullus*). Ibis 104: 399–406.

LOCKE, L. N., CHURA, N. J. & STEWART, P. A. 1966. Spermatogenesis in Bald Eagles experimentally fed a diet containing DDT. Condor 68: 497–502.

LOCKIE, J. D. 1955. The breeding habits and food of Short-eared Owls after a vole plague. Bird Study 2: 53–69.

LOCKIE, J. D. 1964. The breeding density of the Golden Eagle and Fox in relation to food-supply in Wester Ross, Scotland. Scot. Nat. 71: 67–77.

LOCKIE, J. D., RATCLIFFE, D. A. & BALHARRY, R. 1969. Breeding success and dieldrin contamination of Golden Eagles in West Scotland. J. appl. Ecol. 6: 381–9.

LOFTS, B. & MURTON, R. K. 1968. Photoperiodic and physiological adaptations regulating avian breeding cycles and their ecological significance. J. Zool., Lond. 155: 327–94.

LÖHRL, H. 1959. Zur Frage des Zeitpunktes einer Prägung auf die Heimatregion beim Halsbandschnäpper (*Ficedula albicollis*). J. Orn. 100: 132–40.

LOKEMOEN, J. T. & DUEBBERT, A. F. 1976. Ferruginous Hawk nesting ecology and raptor populations in northern South Dakota. Condor 78: 464–70.

LONGCORE, J. R., SAMSON, F. B., KREITZER, J. F. & SPAANS, J. W. 1971. DDE thins eggshells and lowers reproductive success of captive Black Ducks. Bull. environ. Contam. & Toxicol. 6: 485–90.

LONGCORE, J. R. & STENDEL, R. C. 1977. Shell thinning and reproductive impairment in Black Ducks after cessation of DDE dosage. Arch. Environ. Contam. Toxicol. 6: 293–304.

LORENZ, K. 1935. Der Kumpan in der Umwelt des Vogels (1). J. Orn. 83: 137–213; 289–413.

LOVE, J., BALL, M. & NEWTON, I. 1978. Sea Eagles in Britain and Norway. Br. Birds 71: 475–81.

LUCANUS, F. von. 1937. Deutschlands Vogelwelte. Berlin: Parey.

LUCAS, J. (Ed.) 1970. Longevity of birds of prey and owls in captivity. Int. Zoo Yb. 10: 36–7.

LUNDEVALL, C. F. & ROSENBERG, E. 1955. Some aspects of the behaviour and breeding biology of the Pallid Harrier (*Circus macrourus*). Proc. Int. Orn. Congr. 11: 599–603.

LUTTICH, S. N., KEITH, L. B. & STEPHENSON, J. D. 1971. Population dynamics of the Red-tailed Hawk (*Buteo jamaicensis*) at Rochester, Alberta. Auk 88: 73–87.

MACARTHUR, R. H. & WILSON, E. O. 1967. The theory of island biogeography. Princeton: University Press.

McCLELLAND, B. R. 1973: Autumn concentrations of Bald Eagles in Glacier National Park. Condor 75: 121–3.

McELROY, H. 1974. Desert hawking. Arizona: Cactus Press.

McGAHAN, J. 1968. Ecology of the Golden Eagle. Auk 85: 1–12.

McGOWAN, J. D. 1975. Distribution, density and productivity of Goshawks in interior Alaska. Rep. of Alaska Dept. Fish & Game.

MACINTYRE, D. 1960. Nature notes of a highland gamekeeper. London: Seeley, Service & Co Ltd.

McINVAILLE, W. B. & KEITH, L. B. 1974. Predator-prey relations and breeding biology of the Great Horned Owl and Red-tailed Hawk in central Alberta. Canad. Field-Nat. 88: 1–20.

McLANE, M. A. & HALL, L. C., 1972. DDE thins Screech Owl eggshells. Bull. Environ. Contam. & Toxicol. 8: 65–8.

MACLEAN, G. L. 1970. The Pygmy Falcon *Polihierax semitorquatus*. Koedoe 13: 1–21.

McMILLAN, J. 1968. Man and the California Condor. New York: Dutton.

McNAB, B. K. 1963. Bioenergetics and the determination of home range size. Am. Nat. 97: 133–40.

MADER, W. J. 1975. Extra adults at Harris' Hawk nests. Condor 77: 482–85.

MADER, W. J. 1975a. Biology of the Harris' Hawk in southern Arizona. Living Bird 14: 59–85.

MAESTRELLI, J. R. & WIEMEYER, S. N. 1975. Breeding Bald Eagles in captivity. Wilson

Bull, 87: 45–53.

MAKATSCH, W. 1953. Der Schwarze Milan. Neue Brehm Bücherei, Leipzig: Wittenburg Lutherstadt.

MALHERBE, A. P. 1963. Notes on birds of prey and some others at Boshoek north of Rustenburg during a rodent plague. Ostrich 34: 95–6.

MARKUS, M. B. 1972. Mortality of vultures caused by electrocution. Nature, Lond. 238: 228.

MARQUISS, M. & NEWTON, I. 1979. Habitat preference in male and female Sparrow-hawks (*Acciptier nisus*). Proc. Hawk Trust Conf., in press.

MATHEW, M. A. 1882. Two Kestrels laying in the same nest. Zoologist 6: 267–8.

MATHEY, W. J. 1966. *Isospora buteonis* in an American Kestrel and a Golden Eagle. Bull. Wildl. Dis. Assoc. 2: 20–2.

MATHISEN, J. E. 1968. Effects of human disturbance on nesting of Bald Eagles. J. Wildl. Manage. 32: 1–6.

MATRAY, P. F. 1974. Broad-winged Hawk nesting and ecology. Auk 91: 307–24.

MAYR, E. 1938. The proportion of sexes in hawks. Auk 55: 522–3.

MEAD, C. J. 1973. Movements of British raptors. Bird Study 20: 259–86.

MEBS, Th. 1964. Zur Biologie und Populationsdynamik des Mäusebussards (*Buteo buteo*). J. Orn. 105: 247–306.

MEBS, Th. 1969. Peregrine Falcon population trends in West Germany. Pp 193–208 in 'Peregrine Falcon populations: their biology and decline', ed. J. J. Hickey. Madison, Milwaukee & London: Univ. Wisconsin Press.

MEBS, Th. 1971. (Death causes and mortality rates of Peregrines (*Falco peregrinus*) calculated by German and Finnish band-recoveries.) Die Vogelwarte 26: 98–105.

MEINERTZHAGEN, R. 1954. The education of young Ospreys. Ibis 96: 153–5.

MEINERTZHAGEN, R. 1959. Pirates and predators. Edinburgh & London: Oliver & Boyd.

MELDE, M. 1956. Die Mäusebussard. Neue Brehm Bucherei, Leipzig: Wittenburg Luther-stadt.

MELQUIST, W. E. & JOHNSON, D. R. 1973. Osprey population status in northern Idaho and north eastern Washington – 1972. Raptor Research Foundation, Raptor Res. Rep. 3: 121–3.

MENDELSSOHN, H. 1972. The impact of pesticides on bird life in Israel. ICBP Bull. 11: 75–104.

MENDELSSOHN, H. & MARDER, U. 1970. Problems of reproduction in birds of prey in captivity. Int. Zoo Yb. 10: 6–11.

MENDELSSOHN, H. & PAZ, U. 1977. Mass mortality of birds of prey caused by azodrin, an organophosphorus insecticide. Biol. Conserv. 11: 163–70.

MENG, H. 1951. The Cooper's Hawk. Ph.D. thesis, Cornell University, Ithaca, New York.

MEYBURG, B–U. 1967. Beobachtungen zur Brutbiologie des Schwarzen Milans (*Milvus migrans*). Vogelwelt 88: 70–85.

MEYBURG, B-U. 1970. Zur biologie des Schreiadlers (*Aquila pomarina*). Deutscher Falkenorden 1969: 32–66.

MEYBURG, B-U. 1971. On the question of the incubation period of the Black Kite *Milvus migrans*. Ibis 113: 530.

MEYBURG, B-U. 1974. Sibling aggression and mortality among nestling eagles. Ibis 116: 224–8.

MEYBURG, B-U. 1975. On the biology of the Spanish Imperial Eagle *Aquila heliaca adalberti*. Ardeola 21: 245–83.

MEYBURG, B-U. 1977. Protective management of eagles by reduction of nestling mortality. Proc. ICBP World Conf. on birds of prey, Vienna, 1975: 387–92.

MIKKOLA, H. 1976. Owls killing and killed by other owls and raptors in Europe. Br. Birds 69: 144–54.

MILLS, G. S. 1975. A winter population study of the American Kestrel in central Ohio. Wilson Bull. 87: 241–7.

MILSTEIN, P. le S., PRESTT, I. & BELL, A. A. 1970. The breeding cycle of the Grey Heron. Ardea 58: 172–257.

MOILANEN, P. 1976. Goshawks and Pheasants. Suomen Luonto 6: 315–18.

MONNERET, R. J. 1977. Project Peregrine. Pp. 64–69 in 'Papers on the veterinary medicine and domestic breeding of diurnal birds of prey', ed. J. E. Cooper & R. E. Kenward. Oxford: British Falconers' Club.

MONTIER, D. 1968. A survey of the breeding distribution of the Kestrel, Barn Owl and Tawny Owl in the London area in 1967. London Bird Report 32: 81–92.

MOORE, N. W. 1957. The past and present status of the Buzzard in the British Isles. Br. Birds 50: 173–97.

MOORE, N. W. 1965. Environmental contamination by pesticides. Pp. 219–37 in 'Ecology and the Industrial Society', ed. G. T. Goodman et al. Oxford.

MOREAU, R. E. 1944. Clutch size: a comparative study, with special reference to African birds. Ibis 86: 286–347.

MOREAU, R. E. 1945. On the Bateleur, especially at the nest. Ibis 87: 224–49.

MOREAU, R. E. 1966. The bird faunas of Africa and its islands. London: Academic Press.

MOREAU, R. 1972. The Palaearctic – African bird migration systems. London: Academic Press.

MORIARTY, F. 1972. Pollutants and food chains. New Scientist 1972: 594–6.

MORIARTY, F. 1975. Pollutants and animals, a factual perspective. London: George Allen & Unwin.

MORITZ, D. & VAUK, G. 1976. Der Zug des Sperbers (Accipiter nisus) auf Helgoland. J. Orn. 117: 317–328.

MORRIS, J. 1972. Peregrine/Saker breeding. Captive breeding of diurnal birds of prey 1 (3): 14–15.

MØRZER BRUYNS, M. F. 1963. Bird mortality in the Netherlands in the spring of 1960, due to the use of pesticides in agriculture. ICBP Bull. 9: 70–75.

MOSHER, J. A. & MATRAY, P. F. 1974. Size dimorphism: a factor in energy saving for Broad-winged Hawks. Auk 91: 325–41.

MOSS, D. 1976. Woodland song-bird populations and growth of nestling Sparrowhawks. Ph. D. thesis, Edinburgh University.

MUELLER, H. C. & BERGER, D. D. 1967. Fall migration of Sharp-shinned Hawks. Wilson Bull. 79: 397–415.

MUELLER, H. C. & BERGER, D. D. 1968. Sex ratios and measurements of migrant Goshawks. Auk 85: 431–6.

MUELLER, H. C. & BERGER, D. D. 1970. Prey preferences in the Sharp-shinned Hawk: the roles of sex, experience, and motivation. Auk 87: 452–57.

MUELLER, H. C., BERGER, D. D. & ALLEZ, G. 1977. The periodic invasions of Goshawks. Auk 94: 652–63.

MULHERN, B. M., REICHEL, W. L., LOCKE, L. N., LAMONT, T. G., BELISLE, A., CROMARTIE, E., BAGLEY, G. E. & PROUTY, R. M. 1970. Organochlorine residues and autopsy data from Bald Eagles 1966–68. Pestic. Monit. J.4: 141–4.

MUNDY, P. J. & LEDGER, J. A. 1976. Griffon Vultures, carnivores and bones. South Afr. J. of Science 72: 106–10.

MURPHY, J. R. 1974. Status of a Golden Eagle population in central Utah, 1967–73. Raptor Research Foundation, Raptor Res. Rep. 3: 91–6.

MURRAY, J. B. 1970. Escaped American Red-tailed Hawk nesting with Buzzard in Midlothian. Scott. Birds 6: 34–7.

NAGY, A. C. 1977. Population trend indices based on 40 years of autumn counts at Hawk Mountain. Proc. ICBP World Conf. on birds of prey, Vienna, 1975: 243–53.

NEILL, D. D., MULLER, H. D. & SHUTZE, J. V. 1971. Pesticide effects on the fecundity of the Gray Partridge. Bull, environ. Contam. & Toxicol 6: 546–51.

NELSON, J. B. 1976. The breeding biology of Frigate Birds. Living Bird 14: 113–155.

NELSON, M. W. & NELSON, P. 1977. Power lines and birds of prey. Proc. ICBP World Conf. on birds of prey, Vienna, 1975: 228–42.

NELSON, R. W. 1970. Some aspects of the breeding behaviour of Peregrine Falcons on Langara Island, B. C. M.Sc. thesis, Calgary University.

NELSON, R. W. 1972. On photoperiod and captive breeding of northern Peregrines. Raptor Research 6: 57–72.

NELSON, R. W. 1972a. The incubation period in Peale's Falcons. Raptor Research 6: 11–15.

NELSON, R. W. 1974. Prairie Falcons: nesting attempt on a building and effects of weather on courtship and incubation. Raptor Research Foundation. Ethology, Information Exchange 1: 10–12.

NELSON, R. W. 1977. On the diagnosis and "cure" of imprinting in falcons which fail to breed in captivity. Pp 39–49 in 'Papers on the veterinary medicine and domestic breeding of diurnal birds of prey', ed. J. E. Cooper & R. E. Kenward. Oxford: British Falconer's Club.

NELSON, R. W. & CAMPBELL, J. A. 1973. Breeding and behaviour of arctic Peregrines in captivity. Hawk Chalk 12 (3): 39–54.

NELSON, R. W. & CAMPBELL, J. A. 1974. Breeding and behaviour of captive arctic Peregrines. Hawk Chalk 13(3): 44–61.

NELSON, R. W. & MYRES, M. T. 1975. Changes in the Peregrine population and its sea bird prey at Langara Island, British Columbia. Raptor Research 3: 13–31.

NETHERSOLE-THOMPSON, C. & D. 1943. Nest-site selection by birds. Br. Birds 37: 108–13.

NEUFELDT, I. A. 1964. Notes on the nidification of the Pied Harrier *Circus melanoleucos* (Pennant), in Amurland, U.S.S.R. J. Bombay Nat. Hist. Soc. 64: 284–306.

NEWTON, I. 1970. Some aspects of the control of birds. Bird Study 17: 177–92.

NEWTON, I. 1972. Birds of prey in Scotland: some conservation problems. Scot. Birds 7: 5–23.

NEWTON, I. 1972a. Finches. London: Collins.

NEWTON, I. 1973. Egg breakage and breeding failure in British Merlins. Bird Study 20: 241–4.

NEWTON, I. 1973a. Studies of Sparrowhawks. Br. Birds 66: 271–8.

NEWTON, I. 1973b. Success of Sparrowhawks in an area of pesticide usage. Bird Study 20: 1–8.

NEWTON, I. 1974. Changes attributed to pesticides in the nesting success of the Sparrowhawk in Britain. J. Appl. Ecol. 11: 95–101.

NEWTON, I. 1975. Movements and mortality of British Sparrowhawks. Bird Study 22: 35–43.

NEWTON, I. 1976. Breeding of Sparrowhawks (*Accipiter nisus*) in different environments. J. anim. Ecol. 45: 831–49.

NEWTON, I. 1976a. Population limitation in diurnal raptors. Canad. Field-Nat. 90: 274–300.

NEWTON I. 1977. Breeding strategies in birds of prey. Living Bird 16: 51–82.

NEWTON, I. 1978. Feeding and development of Sparrowhawk nestlings. J. Zool., Lond. 184: 465–87.

NEWTON, I. & BOGAN, J. 1974. Organochlorine residues, eggshell thinning and hatching success in British Sparrowhawks. Nature, Lond. 249: 582–3.

NEWTON, I. & BOGAN, J. 1978. The role of different organo-chlorine compounds in the breeding of British Sparrowhawks. J. appl. Ecol. 15: 105–16.

NEWTON, I. & MARQUISS, M. 1976. Occupancy and success of Sparrowhawk nesting territories. Raptor Research 10: 65–71.

NEWTON, I. & MARQUISS, M. 1979. Sex ratio among nestlings of the European Sparrowhawk. Amer. Nat. 113: 309–15.

NEWTON, I. MARQUISS, M. & MOSS, D. 1979. Habitat, female age, organo-chlorine compounds and breeding of European Sparrowhawks. J. Appl. Ecol., in press.

NEWTON, I., MARQUISS, M., WEIR, D. N., & MOSS, D. 1977. Spacing of Sparrowhawk nesting territories. J. anim. Ecol. 46: 425–41.

NEWTON, I., MEEK, E. & LITTLE, B. 1978. Breeding ecology of the Merlin in Northumberland. Br. Birds. 71: 376–98.

NEWTON, I. & MOSS, D. 1977. Breeding birds of Scottish pinewoods. Pp. 26–34 in 'Native pinewoods of Scotland', ed. R. G. H. Bunce & J. N. R. Jeffers. N.E.R.C., Cambridge.

NIELSEN, B. P. 1977. Migratory habits and dispersal of Danish Buzzards *Buteo buteo*. Dansk. Orn. Foren. Tidsskr. 71: 1–9.

NIELSEN, B. P. & CHRISTENSEN, S. 1970. Observations on the autumn migration of raptors in the Lebanon. Ornis Scand. 1: 65–73.

NISBET, I. C. T. & SMOUT, T. C. 1957. Autumn observations on the Bosphorus and Dardanelles. Ibis 99: 483–99.

NORDSTROM, G. 1963. Einige Ergebnisse der Vogelberingung in Finnland in der Jahren 1913–1962. Orn. Fenn. 40: 81–124.

ODSJÖ, T. 1971. Klorerade kolväten och äggskalsförtunning hos fiskgjuse. Fauna och Flora 66: 90–100.

ODSJÖ, T. & SONDELL, J. 1976. Reproductive success in Ospreys *Pandion haliaetus* in southern and central Sweden, 1971–73. Ornis Scand. 7: 71–84.

ODSJÖ, T. & SONDELL, J. 1977. Population development and breeding success in the Marsh Harrier *Circus aeruginosus* in relation to levels of DDT, PCB & mercury. Vår Fågelvärld 36: 152–60.

OGDEN, J. C. 1975. Effects of Bald Eagle territoriality on nesting Ospreys. Wilson Bull. 87: 496–505.

OGDEN, V. T. & HORNOCKER, M. G. 1977. Nesting density and success of Prairie Falcons in southwestern Idaho. J. Wildl. Manage. 41: 1–11.

OLENDORFF, R. R. & KOCHERT, M. N. 1977. Land management for the conservation of birds of prey. Proc. ICBP World Conf. on birds of prey, Vienna, 1975: 294–307.

OLENDORFF, R. R. & STODDART, J. W. 1974. The potential for management of raptor populations in western grasslands. Raptor Res. Foundation, Raptor Research Rep. 2: 47–87.

OLSSON, O. 1958. Dispersal, migration, longevity and death causes of *Strix aluco*, *Buteo buteo*, *Ardea cinerea* and *Larus argentatus*. Acta Vertebratica 1: 91–189.

OPDAM, P. 1975. Inter – and intraspecific differentiation with respect to feeding ecology in two sympatric species of the genus *Accipiter*. Ardea 63: 30–54.

ORIANS, G. H. 1961. The ecology of blackbird (*Agelaius*) social systems. Ecol. Monogr. 31: 285–312.

ORIANS, G. 1971. Ecological aspects of behaviour. Pp. 513–46 in 'Avian Biology', Vol. 1, ed. D. S. Farner & J. R. King. New York and London: Academic Press.

ORIANS, G. & KUHLMAN, F. 1956. The Red-tailed Hawk and Great Horned Owl population in Wisconsin. Condor 58: 371–85.

OSBORNE, M. C. 1966. Observaciones y experiencias con aves de presa (Accipitridae). Ardeola 12: 11–18.

ÖSTERLÖF, S. 1951. Fiskgjusens, *Pandion haliaetus* (L.), flyttning. Vår Fågelvärd 10: 1–15.

ÖSTERLÖF, S. 1977. Migration, wintering areas, and site tenacity of the European Osprey *Pandion h. haliaetus* (L). Ornis Scand. 8: 61–78.

OWEN, J. H. 1916–22. Some breeding habits of the Sparrowhawk. Br. Birds 10: 26–37, 50–9, 74–86, 106–115; 12: 61–5, 74–82; 13: 114–24; 15: 74–7.

OWEN, J. H. 1926–7. The eggs of the Sparrowhawk. Br. Birds 20: 114–20.

OWEN, J. H. 1936–7. Further notes on the Sparrowhawk. Br. Birds 30: 22–6.

PARKER, J. W. 1974. Populations of the Mississippi Kite in the Great Plains. Raptor Research Foundation, Raptor Research Report 3: 159–72.

PARSLOW, J. L. F. & JEFFERIES, D. J. 1973. Relationship between organochlorine residues in livers and whole bodies of Guillemots. Environ. Pollut. 5: 87–101.

PEAKALL, D. B. 1970. pp' – DDT: effect on calcium metabolism of extradiol in the blood. Science, N.Y. 168: 592–4.

PEAKALL, D. B. 1971. Effect of polychlorinated biphenyls (PCBs) on the eggshells of ring doves. Bull. environ. Contamin. & Toxicol. 6: 100–1.

PEAKALL, D. B. 1974. DDE: its presence in Peregrine eggs in 1948. Science 183: 673–4.

PEAKALL, D. B. 1975. PCBs and their environmental effects. CRC Critical reviews in Environmental Control, 1975, 469–509.

PEAKALL, D. B., CADE, T. J., WHITE, C. M. & HAUGH, J. R. 1975. Organochlorine residues in Alaskan Peregrines. Pesticides Monit. J. 8: 255–60.

PEAKALL, D. B., LINCER, J. L. & BLOOM, S. E. 1972. Embryonic mortality and chromosomal alterations caused by Aroclor 1254 in ring doves. Environ. Health Perspect. 1: 103–4.

PEAKALL, D. B., LINCER, J. L., RISEBROUGH, R. W., PRITCHARD, J. B. & KINTER, W. B., 1973. DDE-induced egg-shell thinning: structural and physiological effects in three species. Comp. gen. Pharmac. 4: 305–13.

PEAKALL, D. B. & PEAKALL, M. L. 1973. Effects of polychlorinated biphenyl on the reproduction of artificially and naturally incubated dove eggs. J. appl. Ecol. 10: 863–8.

PEAKALL, D. B., REYNOLDS, L. M. & FRENCH, M. C. 1976. DDE in eggs of the Peregrine Falcon. Bird Study 23: 183–6.

PENNYCUICK, C. J. 1976. Breeding of the Lappet-faced and White-headed Vultures (*Torgos tracheliotus* Forster and *Trigonoceps occipitalis* Burchell) on the Serengeti Plains. E. Afr. Wild. J. 15: 67–84.

PERSSON, T. 1975. (Marsh Harrier & Bittern in Lake Tåkern in 1972–1974). Vår Fågelvärd 34: 283–9.

PETERSEN, C. M. 1956. (Studies of the breeding biology of the Kestrel *Falco tinnunculus*. L. in Copenhagen.) Dansk Orn. Foren. Tidsskr 50: 134–59. (Danish, with English summary).

PIANKA, E. R. 1970. On r- and K-selection. Am. Nat. 104: 592–7.

PICKWELL, G. 1930. The White-tailed Kite. Condor 32: 221–39.

PICOZZI, N. 1978. Dispersion, breeding and prey of the Hen Harrier *Circus cyaneus* in Glen Dye, Kincardineshire. Ibis 120: 498–508.

PICOZZI, N. & WEIR, D. N. 1974. Breeding biology of the Buzzard in Speyside. Br. Birds 67: 199–210.

PICOZZI, N. & WEIR, D. N. 1976. Dispersal and causes of death in Buzzards. Br. Birds 69: 193–201.

PIELOWSKI, Z. 1968. Studien über die Beslandsverhältnisse einer Habichtspopulation in Zentralpolen. Schrift. der Land. für Nat. und Land. in Nordrhein – Westfalen 5: 125–36.

PITELKA, F. A. 1957. Some aspects of population structure in the short-term cycle of the Brown Lemming in northern Alaska. Cold Spring Harbor Symp. Quant. Biol. 22: 237–51.

PLATT, J. B. 1976. Gyrfalcon nest site selection and winter activity in the western Canadian Arctic. Canad. Field-Nat. 90: 338–45.

PLATT, J. B. 1976a. Sharp-shinned Hawk nesting and nest site selection in Utah. Condor 78: 102–3.

PLATT, J. B. 1977. The breeding behaviour of wild and captive Gyr Falcons in relation to their environment and human disturbance. Ph.D. thesis, Cornell University.

POMEROY, D. E. 1975. Birds as scavengers of refuse in Uganda. Ibis 117: 69–81.

PORTER, R. D. 1975. Experimental alterations of clutch-size of captive American Kestrels *Falco sparverius*. Ibis 117: 510–15.

PORTER, R. D. & WHITE, C. M. 1973. The Peregrine Falcon in Utah, emphasising ecology and competition with the Prairie Falcon. Brigham Young Univ. Science Bull. 18: 1–74.

PORTER, R. D. & WIEMEYER, S. N. 1969. Dieldrin and DDT: effects on Sparrow Hawk eggshells and reproduction. Science, N.Y. 165: 199–200.

PORTER, R. D. & WIEMEYER, S. N. 1970. Propagation of captive American Kestrels. J. Wildl. Manage. 34: 594–604.

PORTER, R. D. & Wiemeyer, S. N. 1972. Reproductive patterns in captive American Kestrels (Sparrow Hawks). Condor 74: 46–53.

POSTUPALSKY, S. & STACKPOLE, S. M. 1974. Artificial nesting platforms for Ospreys in Michigan. Raptor Res. Foundation, Raptor Res. Rep. 2: 105–17.

POTTS, G. R. 1968. Success of the Shag on the Farne Islands, Northumberland, in relation to their content of dieldrin and p,p' – DDE. Nature, Lond. 217: 1282–4.

PRESTT, I., JEFFERIES, D. J. & MACDONALD, J. W. 1968. Post-mortem examination of four Rough-legged Buzzards. Br. Birds 61: 457–65.

PRESTT, I., JEFFERIES, D. J. & MOORE, N. W. 1970. Polychlorinated biphenyls in wild birds in Britain and their avian toxicity. Environ. Pollut. 1: 3–26.

PRESTT, I. & RATCLIFFE, D. A. 1972 Effects of organochlorine insecticides on European birdlife. Proc. Int. Orn. Congr. 15: 486–513.

PSENNER, H. 1977. The successful breeding of the Bearded Vulture in the Alpenzoo, Innsbruck. Proc. ICBP World Conf. on birds of prey, Vienna, 1975: 370–1.

PURCHASE, D. 1973. Results from banding Brown Goshawks in Australia. Australian Bird Bander 11: 71–5.

RAHN, H., PAGANELLI, C. V. & AR, A. 1975. Relation of avian egg weight to body weight. Auk 92: 750–65.

RATCLIFFE, D. A. 1962. Breeding density of the Peregrine *Falco peregrinus* and Raven *Corvus corax*. Ibis 104: 13–39.

RATCLIFFE, D. A. 1967. Decrease in eggshell weight in certain birds of prey. Nature, Lond., 215: 208–10.

RATCLIFFE, D. A. 1969. Population trends of the Peregrine Falcon in Great Britain. Pp. 239–69 in 'Peregrine Falcon Populations: their biology and decline', ed J. J. Hickey. Madison, Milwaukee & London: University of Wisconsin Press.

RATCLIFFE, D. A. 1970. Changes attributable to pesticides in egg breakage frequency and eggshell thickness in some British birds. J. appl. Ecol. 7: 67–107.

RATCLIFFE, D. A. 1972. The Peregrine population of Great Britain in 1971. Bird Study 19: 117–56.

REDIG, P. T. 1978. Raptor rehabilitation: diagnosis, prognosis and moral issues. Pp. 29–41 in 'Bird of prey management techniques', ed. T. A. Geer. Oxford: British Falconers' Club.

REESE, J. G. 1970. Reproduction in a Chesapeake Bay Osprey population. Auk 87: 747–59.

REICHEL, W. L., CROMARTIE, E., LAMONT, T. G., MULHERN, B. M. & PROUTY, R. M. 1969. Pesticide residues in eagles. Pestic. Monit. J. 3: 142–4.

REICHHOLF, J. 194. Artenreichtum, Haufigkeit und Diversität der Greifvögel in einigen Gebieten von Südamerika. J. Orn. 115: 381–97.

REINDAHL, E. 1941. A story of Marsh Hawks. Nature Mag. April: 191–4.

RENSCH, B. 1950. Die Abhangigkeit der relativen Sexual-differenz von der Körpergrösse. Bonn. Zool. Beitr. 1: 58–69.

RENSSEN, Th.A. 1973. Social roosting of *Circus buffoni* in Surinam. Ardea 61: 188.

RETTIG, N. L. 1978. Breeding behaviour of the Harpy Eagle (*Harpia harpyja*). Auk 95: 629–43.

REYNOLDS, R. T. 1972. Sexual dimorphism in *Accipiter* hawks: a new hypothesis. Condor 74: 191–7.

REYNOLDS, R. T. & WIGHT, H. M. 1978. Distribution, density, and productivity of accipiter hawks breeding in Oregon. Wil. Bull. 90: 182–96.

RHODES, L. I. 1972. Success of Osprey nest structures at Martin National Wildlife Refuge. J. Wildl. Manage. 36: 1296–9.

RICE, J. N. 1969. The decline of the Peregrine Falcon in Pennsylvania. Pp. 155–64 in 'Peregrine Falcon populations: their biology and decline', ed. J. J. Hickey. Madison, Milwaukee & London: Univ. Wisconsin Press.

RICHARDSON, W. J. 1975. Autumn hawk migration in Ontario studied with radar. Proc. N. Amer. Hawk Mig. Conf., Syracuse, 1974: 47–58.

RICHDALE, L. E. 1952. Post-egg period in albatrosses. Biol. Monogr. No. 4. Otago Daily Times and Witness Newspapers, Dunedin.

RICHMOND, W. K. 1959. British birds of prey. London: Lutterworth Press.

RICKLEFS, R. E. 1968. Patterns of growth in birds. Ibis 110: 419–51.

RICKLEFS, R. E. 1973. Fecundity, mortality and avian demography. Pp. 366–435 in 'Breeding biology of birds', ed. by D. S. Farner. National Academy of Sciences, Washington, D.C.

RISEBROUGH, R. W. & ANDERSON, D. 1975. Some effects of DDE and PCB on mallards and their eggs. J. Wildl. Manage. 39: 508–13.

RISEBROUGH, R. W., FLORANT, G. L. & BERGER, D. D. 1970. Organochlorine pollutants in Peregrines and Merlins migrating through Wisconsin. Canad. Field-Nat. 84: 247–53.

RISEBROUGH, R. W., MENZEL, D. B., MARTIN, D. J. & OLCOTT, H. J. 1967. DDT residues in Pacific sea birds: a persistent insecticide in marine food chains. Nature, Lond. 216: 589–90.

RISEBROUGH, R. W., REICHE, P., PEAKALL, D. B., HERMAN, S. G. & KIRVEN, M. N. 1968. Polychlorinated biphenyls in the global ecosystem. Nature, Lond. 220:

1098–102.

ROBBINS, C. S. 1975. A history of North American hawkwatching. Proc. N. Amer. Hawk Mig. Conf., Syracuse, 1974: 29–40.

ROBERTS, A. 1968. The birds of South Africa. London: Witherby.

ROBINSON, W. 1950. Montagu's Harriers. Bird Notes 24: 103–14.

ROCKENBAUCH, D. 1968. Zur Brutbiologie des Turmfalken (*Falco tinnunculus* L.). Anz. orn. Ges. Bayern 8: 267–76.

ROCKENBAUCH, D. 1975. Zwölfjährige Untersuchungen zur Ökologie des Mäusebussards (*Buteo buteo*) auf der Schwabischen Alb. J. Orn. 116: 39–54.

ROEST, A. I. 1957. Notes on the American Sparrowhawk. Auk 74: 1–19.

ROSEN, L. 1966. Rovfågelsträcket vid Falsterbo. Vår Fågelvärld 25: 315–26.

ROSEN, M. N. & MORSE, E. E. 1959. An interspecies chain in a fowl cholera epizootic. Calif. Fish & Game 45: 51–6.

ROWAN, W. 1921–2. Observations on the breeding habits of the Merlin. Br. Birds 15: 122–9, 194–202, 222–31, 246–53.

ROWE, E. G. 1947. The breeding biology of *Aquila verreuxi* Lesson. Ibis 89: 347–410, 576–606.

RUDEBECK, G. 1963. Studies on some Palearctic and arctic birds in their winter quarters in South Africa. 4. Birds of prey (Falconiformes). Pp. 418–53 in 'South African animal life; results of the Lund University Expedition in 1950–51', ed. B. Hanström, P. Brinck & G. Rudebeck. Vol. 9. Stockholm: Almquist & Wiksell.

RUSCH, D. H. & DOERR, P. D. 1972. Broad-winged Hawk nesting and food habits. Auk 89: 139–45.

RYDZEWSKI, W. 1962. Longevity of ringed birds. The Ring 33: 147–54.

RYVES, R. H. 1948. Bird-life in Cornwall. London.

SAFRIEL, U. 1968. Bird migration at Elat, Israel. Ibis 110: 283–320.

ST. JOHN, C. 1849. A tour in Sutherland.

SALMINEN, P. & WIKMAN, M. 1977. Population trend and status for the Peregrine (*Falco peregrinus*) in Finland. Pp. 25–30 in 'Report from a Peregrine Conference', ed. P. Lindberg. Swedish Society for the Conservation of Nature, Stockholm.

SALOMONSEN, F. 1955. The evolutionary significance of bird migration. Dan. Biol. Medd. 22: 1–66.

SANDEMAN, P. W. 1957. The breeding success of Golden Eagles in the southern Grampians. Scott. Nat. 69: 148–52.

SARGENT, W. D. 1938. Nest parasitism of hawks. Auk 55: 82–4.

SAUER, E. G. F. 1973. Notes on the behaviour of Lappet-faced Vultures and Cape Vultures in the Namib Desert of South West Africa. Madoqua, Ser. II, 2: 43–62.

SAUROLA, P. 1976. Mortality of Finnish Goshawks. Suomen Luonto 6: 310–14.

SAUROLA, P. 1978. Artificial nest construction in Europe. Pp. 72–80 in 'Bird of prey management techniques', ed. T. A. Geer, Oxford: British Falconers' Club.

SAXBY, H. L. 1874. The birds of Shetland. Edinburgh: MacLachlan & Stewart.

SCHARF, W. C. & BALFOUR, E. 1971. Growth and development of nestling Hen Harriers. Ibis 113: 323–9.

SCHELDE, O. 1960. Danske Spurvhoges (*Accipiter nisus* (L.)) Traekforhold. Dansk Orn. Foren. Tidsskr. 54: 88–102.

SCHIFFERLI, A. 1964. Lebensdauer, Sterblichkeit und Todesursachen beim Turmfalken, *Falco tinnunculus*. Orn. Beob. 61: 81–9.

SCHIFFERLI, A. 1965. Vom Zugverhalten der in der Schweiz brütenden Turmfalken, *Falco tinnunculus*, nach den Ringfunden. Orn. Beob. 62: 1–13.

SCHIFFERLI, A. 1967. Vom Zug Schweizerischer und Deutscher Schwarzer Milane nach Ringfunden. Orn. Beob. 64: 34–51.

SCHIPPER, W. J. A. 1973. A comparison of prey selection in sympatric harriers, *Circus* sp., in Western Europe. Le Gerfaut 63: 17–120.

SCHMAUS, A. 1938. Der Einfluss der Mäusejahre auf das Brutgeschaft unserer Raubvögel und Eulen. Beitr. 2. Fortpfl. Biol. Vög. 14: 181–4.

SCHMUTZ, J. K. 1977. Relationships between three species of the genus *Buteo* (Aves) coexisting in the prairie-parkland ecotone of southeastern Alberta. M.Sc. thesis,

Edmonton University, Alberta.

SCHMUTZ, J. K. & SCHMUTZ, S. M. 1975. Primary molt in Circus cyaneus in relation to nest brood events. Auk 92: 105–10.

SCHNELL, G. D. 1969. Communal roosts of wintering Rough-legged Hawks (Buteo lagopus). Auk 86: 682–90.

SCHNELL, J. H. 1958. Nesting behaviour and food habits of Goshawks in the Sierra Nevada of California. Condor 60: 377–403.

SCHOENER, T. W. 1968. Sizes of feeding territories among birds. Ecology 49: 123–41.

SCHOENER, T. W. 1969. Models of optimal size for solitary predators. Amer. Nat. 103: 277–313.

SCHÖNWETTER, M. 1960–72. Handbuch der Oologie. Band 1. Berlin: Akademie-Verlag.

SCHRIVER, E. C. 1969. The status of Cooper's Hawks in Western Pennsylvania. Pp 356–59 in 'Peregrine Falcon populations: their biology and decline', ed. J. J. Hickey. Madison, Milwaukie & London: Univ. Wisconsin Press.

SCHUSTER, L. 1940. Langjährige Weederkehr eines Mäusebussards (Buteo buteo) an denselben Uberwinterungsplatz. Vogelzug 11: 86.

SCHUYL, G., TINBERGEN, L. & TINBERGEN, N. 1936. Ethologische Beobachtungen am Baumfalken (Falco s. subbuteo L.). J. Orn. 84: 387–433.

SCIPLE, G. W. 1953. Avian botulism: information on earlier research. U. S. Department Interior, Spec. Sci. Rep.: Wildlife 23: 1–13.

SCOTT. M. L., VADEHRA, D. V., MULLENHOFF, P. A., RUMSEY, G. L. & RICE, R. W. 1971. Results of experiments on the effects of PCB on laying hen performance. Proc. Cornell Nutr. Conf., 56–64.

SEEBOHM, H. 1883. A history of British birds, with coloured illustrations of their eggs, Vol. 1. London.

SEIDENSTICKER, J. C. & REYNOLDS, H. V. 1971. The nesting, reproductive performance and chlorinated hydrocarbon residues in the Red-tailed Hawk and Great Horned Owl in south-central Montana. Wilson Bull. 83: 408–18.

SELANDER, R. K. 1966. Sexual dimorphism and differential niche utilisation in birds. Condor 68: 113–51.

SERVENTY, D. L. 1953. The southern invasion of northern birds during 1952. W. Austr. Nat. 3: 177–96.

SERVENTY, D. L. 1967. Aspects of the population ecology of the Short-tailed Shearwater Puffinus tenuirostris. Proc. Int. Orn. Congr. 14: 165–90.

SERVENTY, D. L. & WHITTELL, H. M. 1951. A handbook of the birds of Western Australia. Perth: University of Western Australia Press.

SERVENTY, D. L. & WHITTELL, H. M. 1976. Birds of Western Australia, Perth: University of Western Australia Press.

SHARROCK, J. T. R. (ed. for British Trust for Ornithology & Irish Wildbird Conservancy) 1976. The atlas of breeding birds in Britain and Ireland. Berkhamsted: Poyser.

SHELFORD, V. E. 1943. The abundance of the Collared Lemming (Dicrostonyx groenlandicus (Tr.) var. Richardsoni Mer) in the Churchill area, 1929 to 1940. Ecology 24: 472–84.

SHELLENBERGER, T. E. & NEWELL, G. W. 1965. Toxicological evaluations of agricultural chemicals with Japanese Quail (Coturnix coturnix japonica). Lab. Anim. Care 15: 119–30.

SHERROD, S. K., WHITE, C. M. & WILLIAMSON, F. S. L. 1977. Biology of the Bald Eagle (Haliaetus leucocephalus alascanus) on Amchitka Island, Alaska. Living Bird 15: 143–82.

SHRUBB, M. 1970. The present status of the Kestrel in Sussex. Bird Study 17: 1–15.

SIEGFRIED, W. R. 1968. Breeding season, clutch and brood sizes in Verreaux's Eagle. Ostrich 39: 139–45.

SIEWERT, H. 1933. Die Brutbiologie des Hühnerhabichts. J. Orn. 81: 44–94.

SIEWERT, H. 1941. Zur Brutbiologie des Fischadlers (Pandion h. haliaetus (L.)). J. Orn. 89: 145–193.

SINCLAIR, J. C. & WALTERS, B. 1976. Lanner Falcons breed in Durban City. Bokmakierie 28: 51–2.

SMEENK, C. 1974. Comparative-ecological studies of some East African birds of prey.

Ardea 62: 1–97.

SMEENK, C. & SMEENK-ENSERINK, N. 1975. Observations on the Pale Chanting Goshawk *Melierax poliopterus*, with comparative notes on the Gabar Goshawk *Micronisus gabar*. Ardea 63: 93–115.

SMITH, C. C. 1968. The adaptive nature of social organisation in the genus of tree squirrels *Tamiasciurus*. Ecol. Monogr. 38: 31–63.

SMITH, D. G. & MURPHY, J. R. 1973. Breeding ecology of raptors in the eastern Great Basin of Utah. Biol. Series, Brigham Young Univ. 18: 1–76.

SMITH, D. G., WILSON, C. R. & FROST, H. H. 1972. The biology of the American Kestrel in Central Utah. Southwest Nat. 17: 73–83.

SMITH, N. 1973. Spectacular *Buteo* migration over Panama Canal Zone, October 1972. American Birds 27: 3–5.

SMITH, S. I., WEBER, C. W. & REID, B. L. 1970. Dietary pesticides and contamination of yolks and abdominal fat of laying hens. Poult. Sci. 49: 233–7.

SNELLING, J. C. 1969. A raptor study in the Kruger National Park. Bokmakierie 21, Suppl: 8–11.

SNELLING, J. C. 1970. Some information obtained from marking large raptors in the Kruger National Park, Republic of South Africa. Ostrich, Suppl. 8: 415–27.

SNELLING, J. C. 1975. Endangered birds of prey: ideas on the management of some African species. J. Sth. Afr. Wildl. Mgmt. Ass. 5: 27–31.

SNOW, D. W. 1968. Movements and mortality of British Kestrels *Falco tinnunculus*. Bird Study 15: 65–83.

SNYDER, N. 1974. Can the Cooper's Hawk survive? Nat. Geog. Mag. 145: 433–42.

SNYDER, N. F. R. 1975. Breeding biology of Swallow-tailed Kites in Florida. Living Bird 13: 73–97.

SNYDER, N. F. R. & SNYDER, H. A. 1973. Experimental study of feeding rates of nesting Cooper's Hawks. Condor 75: 461–3.

SNYDER, N. F. R. & SNYDER, H. A. 1974. Function of eye coloration in North American accipiters. Condor 76: 219–22.

SNYDER, N. F. R. & SNYDER, H. A. 1975. Raptors in range habitat. Pp. 190–209 in 'Proc. Symp. on Manage. of Forest and Range habitats for non-game birds', ed. D. R. Smith. Arizona: Forest Service.

SNYDER, N. F. R., SNYDER, H. A., LINCER, J. L. & REYNOLDS, R. T. 1973. Organo-chlorines, heavy metals and the biology of North American accipiters. BioScience 23: 300–5.

SNYDER, N. F. R. & WILEY, J. W. 1976. Sexual size dimorphism in hawks and owls of North America. Ornithological Monographs 20: 1–96. American Ornithologists Union.

SONDELL, J. 1970. (Nest and hunting territories of Marsh Harrier). Vår Fågelvärld 29: 298–9.

SOUTHERN, H. N. 1970. The natural control of a population of tawny owls (*Strix aluco*). J. Zool., Lond. 162: 197–285.

SOUTHERN, W. E. 1964. Additional observations on winter Bald Eagle populations: including remarks on biotelemetry techniques and immature plumages. Wil. Bull. 76: 121–37.

SPENCER, R. & HUDSON, R. 1977. Report on bird-ringing for 1975. Bird Study Suppl. 24: 8–14.

SPOFFORD, W. R. 1964. The Golden Eagle in the Trans-Pecos and Edwards Plateau of Texas. Audubon Cons. Rep. 1: 1–47.

SPOFFORD, W. 1969. Extra female at a nesting site. Pp. 418–19 in 'Peregrine Falcon populations: their biology and decline', ed. J. J. Hickey. Madison, Milwaukee & London: Univ. Wisconsin Press.

SPRUNT, A. 1963. Bald Eagles aren't producing enough young. Aud. Mag. 65: 32–5.

SPRUNT, A. J. 1969. Population trends of the Bald Eagle in North America. Pp. 347–50 in 'Peregrine Falcon populations: their biology and decline', ed. J. J. Hickey. Madison, Milwaukee & London: Univ. Wisconsin Press.

SPRUNT, A., ROBERSTON, W. B., POSTUPALSKY, S., HENSEL, R. J., KNODER, C. E. & LIGAS, F. J. 1973. Comparative productivity of six Bald Eagle populations. Trans. North Amer. Wildl. & Nat. Res. Conf. 38: 96–106.

STABLER, R. M. 1969. Trichonomas gallinae as a factor in the decline of the Peregrine Falcon. Pp. 435–7 in 'Peregrine Falcon populations: their biology and decline', ed. J. J. Hickey. Madison, Milwaukee & London: Univ. Wisconsin Press.

STAGER, K. E. 1964. The role of olfaction in food location by the Turkey Vulture (Cathartes aura). Los Angeles County Museum. Contributions in Science 81.

STALMASTER, M. V. 1976. Winter ecology and effects of human activity on Bald Eagles in the Nooksack River valley. M.Sc. thesis, Western Washington College, Bellingham.

STEINFATT, O. 1938. Quoted in Meyburg 1970.

STENSRUDE, C. 1965. Observations on a pair of Grey Hawks in Southern Arizona. Condor 67: 319–21.

STEWART, P. A. 1974. A nesting of Black Vultures. Auk 91: 595–600.

STEWART, P. A. 1977. Migratory movements and mortality rate of Turkey Vultures. Bird-banding 48: 122–4.

STEWART, R. E. 1949. Ecology of a nesting Red-shouldered Hawk population. Wilson Bull. 61: 26–35.

STEYN, P. 1972. The Little Sparrowhawk at home. Bokmakierie 24: 13–16.

STEYN, P. 1973. Eagle days. London: Purnell & Sons.

STEYN, P. 1975. Observations on the African Hawk-eagle. Ostrich 46: 75–105.

STICKEL, L. F. 1975. The costs and effects of chronic exposure to low-level pollutants in the environment. Pp. 716–28 in 'Hearings before the sub-committee on the environment and the atmosphere', Committee on Science and Technology, U.S. House of Representatives.

STINSON, C. H. 1977. Familial longevity in Ospreys. Bird-Banding 48: 72–3.

STINSON, C. H. 1977a. Growth and behaviour of young Ospreys Pandion haliaetus. Oikos 28: 299–303.

STONEHOUSE, B. 1960. The King Penguin Aptenodytes patagonica of South Georgia. 1. Breeding behaviour and development. Falkland Is. Dep. Surv. Sci. Rep. 23: 81.

STONEHOUSE, B. & STONEHOUSE, S. 1961. The Frigate Bird Fregata aquila of Ascension Island. Ibis 103b: 409–22.

STORER, R. W. 1966. Sexual dimorphism and food habits in three North American accipiters. Auk 83: 423–36.

SUETENS, V. & GROENENDAEL, P. van. 1966. (On the ecology and the reproduction of the Black Vulture (Aegypius monachus).) Ardeola 12: 19–44.

SUETENS, W. & GROENENDAEL, P. van 1972. (Notes on the ecology and ethology of the Lammergeier Gypaetus barbatus aureus (Hablizl).) Le Gerfaut 62: 203–14.

SULKAVA, S. 1964. Zur Nahrungsbiologie des Habichts, Accipiter gentilis (L.). Aquilo, Ser. Zool. 3: 1–103.

SUMNER, E. L. 1929. Comparative studies on the growth of young raptors. Condor 31: 85–111.

SWARTZ, L. G., WALKER, W., ROSENEAU, D. G. & SPRINGER, A. M. 1974. Populations of Gyrfalcons on the Seward Peninsula, Alaska, 1968–1972. Raptor Research Foundation, Raptor Research Report 3: 71–5.

SYLVÉN, M. 1977. Hybridisation between Red Kite Milvus milvus and Black Kite M. migrans in Sweden 1976. Vår Fågelvärld 36: 38–44.

SYLVÉN, M. 1978. Interspecific relations between sympatrically wintering Common Buzzards Buteo buteo and Rough-legged Buzzards Buteo lagopus. Orn. Scand. 9: 197–206.

TARBOTON, W. 1977. Nesting, territoriality and food habits of Wahlberg's Eagle. Bokmakierie 29: 46–50.

TARBOTON, W. 1978. Hunting and the energy budget in the Black-shouldered Kite. Condor 80: 88–91.

TAYLOR, S. M. 1967. Breeding season status of the Kestrel. Proc. Bristol Nat. Soc. 31: 293–6.

TEMPLE, S. A. 1972. Artificial insemination with imprinted bird of prey. Nature, Lond. 237: 287–8.

TEMPLE, S. A. 1972a. Sex and age characteristics of North American merlins. Bird-Banding 43: 191–6.

TERRASSE, J. F. & M. 1977. Le Balbuzard pêcheur Pandion haliaetus (L.) en Méditerranée occidentale. Distribution, essai de recensement, reproduction, avenir. Nos Oiseaux 34: 111–27.

THIOLLAY, J. M. 1967. Écologie d'une population de rapaces diurnes en Lorraine. La Terre et la Vie 114: 116–83.

THIOLLAY, J. M. 1975. Les rapaces d'une zone de contact Savane-Forêt en Côte-d'Ivoire. Alauda 43: 75–102, 347–416.

THIOLLAY, J. M. 1978. Les migrations de rapaces ens Afrique occidentale: adaptations ecologiques aux fluctuations saisonnières de production des ecosystems. La Terre et la Vie 32: 89–133.

THIOLLAY, J. M. & MEYER, J. A. 1978. Densité, taille des territoires et production dans une population d'Aigles Pêcheurs, *Haliaeetus vocifer* (Daudin). La Terre et la Vie 32: 203–19.

THOMASSON, L. 1947. On the nesting sites of the Peregrine Falcon in the countries around the Baltic. Vår Fågelvärld 6: 72–81. (Swedish, with English summary).

TICEHURST, N. F. 1920. On the former abundance of the Kite, Buzzard and Raven in Kent. Br. Birds 14: 34–7.

TICKELL, W. L. N. 1960. Chick feeding in the Wandering Albatross *Diomedea exulans* Linnaeus. Nature, London, 185: 116–17.

TINBERGEN, L. 1940. Beobachtungen über die Arbeitsteilung des Turmfalken (*Falco tinnunculus*) während der Fortpflanzungszeit. Ardea 29: 63–98.

TINBERGEN, L. 1946. Sperver als Roofvijand van Zangvogels. Ardea 34: 1–123.

TJERNBERG, M. 1977. Individual recognition of Golden Eagles, *Aquila chrysaetos*, in the field and results of winter censuses in southwest Uppland, central Sweden. Vår Fågelvärld 36: 21–32.

TRAINER, D. O. 1969. Diseases in raptors: a review of the literature. Pp. 425–33 in 'Peregrine Falcon Populations: their biology and decline', ed. J. J. Hickey. Madison, Milwaukee & London: Univ. Wisconsin Press.

TRAINER, D. O., FOLZ, S. D. & SAMUEL, W. M. 1968. Capillariasis in the Gyr Falcon. Condor 70: 276–7.

TUBBS, C. R. 1971. Analysis of nest record cards for the Buzzard. Bird Study 19: 96–104.

TUBBS, C. R. 1974. The Buzzard. London: David & Charles.

TUCKER, R. K. & CRABTREE, D. G. 1970. Handbook of toxicity of pesticides to wildlife. Resource publication 84. Washington: Govt. Printing Office.

TUCKER, R. K. & HAEGELE, M. A. 1970. Eggshell thinning as influenced by method of DDT exposure. Bull. environ. Contam. & Toxicol. 5: 191–4.

TYRRELL, W. B. 1936. The Ospreys of Smith's Point, Virginia. Auk 53: 261–8.

ULFSTRAND, S., ROOS, G., ALERSTAM, T. & ÖSTERDAHL, L. 1974. Visible bird migration at Falsterbo, Sweden. Vår Fågelvärld, Suppl. 8: 1–244.

UTTENDÖRFER, O. 1952. Die Ernährung der deutschen Raubvögel und Eulen. Neudamm.

VALVERDE, J. A. 1960. La population d'Aigles Imperiaux (*Aquila heliaca adalberti*) des marismas du Guadalquivir; son évolution depius un siècle. Alauda, 28: 20–26.

VAUGHAN, R. 1961. *Falco eleonorae*. Ibis 103: 114–28.

VAURIE, C. 1965. The birds of the Palearctic fauna. Vol. 2. London: Witherby.

VERMEER, K. & REYNOLDS, L. M. 1970. Organochlorine residues in aquatic birds in the Canadian Prairie Provinces. Canad. Field-Nat. 84: 117–30.

VERMEER, K., RISEBROUGH, R. W., SPAANS, A. L. & REYNOLDS, L. M. 1974. Pesticide effects on fishes and birds in rice fields of Surinam, South America. Environ. Pollut. 7: 217–36.

VERNER, J. 1964. Evolution of polygamy in the Long-billed Marsh Wren. Evolution 18: 252–61.

VERNER, W. 1909. My life among the wild birds in Spain. London.

VILLAGE, A. 1979. The ecology of the Kestrel (*Falco tinnunculus*) in relation to vole abundance at Eskdalemuir, south Scotland. Ph.D. thesis, Edinburgh University.

VISSER, J. 1963. The Black Eagles of Zuurhoek, Jansenville. Afr. Wildlife 17: 191–4.

VOOUS, K. H. 1977. Three lines of thought for consideration and eventual action. Proc. ICBP World Conf. on birds of prey, Vienna, 1975; 343–7.

VRIES DE, Tj. 1975. The breeding biology of the Galapagos Hawk, *Buteo galapagoensis*. Le Gerfaut 65: 29–57.

WALKER, C. H., HAMILTON, G. A. & HARRISON, R. B. 1967. Organochlorine insecticide residues in wild birds in Britain. J. Sci. Food Agric. 18: 123–9.

WALPOLE-BOND, J. 1938. A history of Sussex birds. Vol. 2. London: Witherby.

WALTER, H. 1968. Zur Abhangiekeit des Eleonoren Falken (*Falco eleonorae*) vom Mediterranean Vogelzug. J. Orn 109: 323–65.

WARD, F. P. 1973. A clinical evaluation of parasites in birds of prey. Amer. Assoc. Zoo. Veterinarians Conf. Columbus, Ohio.

WARD, F. P. & BERRY, R. B. 1972. Autumn migrations of Peregrine Falcons on Assateague Island, 1970–1. J. Wildl. Manage. 36: 484–92.

WARD, F. P. & FAIRCHILD, D. G. 1972. Air sac parasites of the genus *Serratospiculum* in falcons. J. Wildl. Dis. 8: 165–8.

WARD, P. & ZAHAVI, A. 1973. The importance of certain assemblages of birds as 'information-centres' for food-finding. Ibis 115: 517–34.

WARHAM, J. 1977. The incidence, functions and ecological significance of petrel stomach oils. Proc. N. Z. Ecol. Soc. 24: 84–93.

WARNCKE, K. & WITTENBERG, J. 1959. Über Siedlungsdichte und Brutbiologie des Mäusebussards (*Buteo buteo*). Vogelwelt 80: 101–8.

WATSON, A. 1957. The breeding success of Golden Eagles in the Northeast Highlands. Scott. Nat. 69: 153–69.

WATSON, A. 1970. Work on Golden Eagles. Pp. 14–18 in 'Research on vertebrate predators in Scotland', Nature Conservancy Progress Report. Edinburgh.

WATSON, A. & MOSS, R. 1970. Dominance, spacing behaviour and aggression in relation to population limitation in vertebrates. Pp. 167–220 in 'Animal populations in relation to their food resources', ed. A. Watson. Oxford: Blackwell Scientific Publications.

WATSON, D. 1977. The Hen Harrier. Berkhamsted: Poyser.

WEAVER, J. D. & CADE, T. J. 1974. Special report on the falcon breeding program at Cornell University. Raptor Research BPIE 90.

WEAVING, A. J. S. 1972. Augur Buzzards at the nest. Bokmakierie, 24: 27–30.

WEBSTER, H. M. 1944. A survey of the Prairie Falcon in Colorado. Auk 61: 609–16.

WEDTS DE SWART, J. C. 1969. Vogelsterfte in de delta. Zeeuws Tijdschr. 19(6): 1–3.

WEIR, D. N. 1971. Mortality of hawks and owls in Speyside. Bird Study 18: 147–54.

WEIR, D. & PICOZZI, N. 1975. Aspects of social behaviour in the Buzzard. Br. Birds 68: 125–41.

WEISS, H. 1923. Life of the harrier in Denmark. London. Wheldon & Wesley.

WENDLAND, V. 1952–53. Populationsstudien an Raubvögeln 1 & 2. J. Orn. 93: 144–53; 94: 103–13.

WENDLAND, V. 1959. Schreiadler und Schelladler. Die Neue Brehm Buchererei, Leipzig: Wittenburg Lutherstadt.

WESTERMARK, T., ODSJÖ, T. & JOHNELS, A. G. 1975. Mercury content of bird feathers before and after Swedish ban on alkyl mercury in agriculture. Ambio 4: 87–92.

WESTON, J. B. 1968. Nesting ecology of the Ferruginous Hawk *Buteo regalis*. M.Sc. thesis, Brigham Young University, Provo. Utah.

WHITE, C. M. 1963. Botulism and myiasis as mortality factors in falcons. Condor 65: 442–3.

WHITE, C. 1968. Diagnosis and relationships of the North American tundra-inhabiting Peregrine Falcons. Auk 85: 179–91.

WHITE, C. M. 1969. Breeding Alaskan and Arctic migrant populations of the Peregrine. Pp. 45–51 in 'Peregrine Falcon populations: their biology and decline', ed. J. J. Hickey. Madison, Milwaukee & London: Univ. Wisconsin Press.

WHITE, C. M. 1969a. Population trends in Utah raptors. Pp. 359–61 in 'Peregrine Falcon populations: their biology and decline', ed. J. J. Hickey. Madison, Milwaukee & London: Univ. Wisconsin Press.

WHITE, C. M. 1974. Current problems and techniques in raptor management and conservation. Trans. North Amer. Wildl. & Nat. Res. Conf. 39: 301–11.

WHITE, C. M. 1975. Studies on Peregrine Falcons in the Aleutian Islands. Raptor Res. Rep. 3: 33–50.

WHITE, C. M. & CADE, T. J. 1971. Cliff-nesting raptors and Ravens along the Colville River in arctic Alaska. Living Bird 10: 107–50.

WHITE, C. M. & CADE, T. J. 1977. Long term trends of Peregrine populations in Alaska. Proc. ICBP World Conf. on Birds of prey, Vienna 1975: 63–72.

WHITE, C. M., EMISON, W. B. & WILLIAMSON, F. S. L. 1973. DDE in a resident Aleutian Island Peregrine population. Condor 75: 306–11.

WHITE, C. M. & ROSENEAU, D. G. 1970. Observations on food, nesting, and winter populations of large North American falcons. Condor 72: 113–15.

WHITFIELD, D. W. A., GERRARD, J. M., MAHER, W. J. & DAVIS, D. W. 1974. Bald Eagle nesting habitat, density and reproduction in central Saskatchewan and Manitoba. Canad. Field-Nat. 88: 399–407.

WHITSON, M. A. & WHITSON, P. D. 1969. Breeding behaviour of Andean Condor (*Vultur gryphus*). Condor 71: 73–5.

WIEMEYER, S. N. & PORTER, R. D. 1970. DDE thins eggshells of captive American Kestrels. Nature, Lond. 227: 737–8.

WIEMEYER, S. N., MULHERN, B. M., LIGAS, F. J., HENSEL, R. J., MATHISEN, J. E., ROBARDS, F. C. & POSTUPALSKY, S. 1972. Residues of organochlorine pesticides, polychlorinated biphenyls and mercury in Bald Eagle eggs and changes in shell thickness – 1969 and 1970. Pestic. Monit. J. 6: 50–5.

WIEMEYER, S. N., SPITZER, P. R., KRANTZ, W. C., LAMONT, T. G. & CROMARTIE, E. 1975. Effects of environmental pollutants on Connecticut and Maryland Ospreys. J. Wildl. Manage. 39: 124–39.

WIESE, I. H., BASSON, N. C. J., van der NYVER, J. H. & van der MERWE, J. H. 1969. Toxicology and dynamics of dieldrin in the Crowned Guinea-fowl *Numida meleagris* (L.). Phytophylactica 1: 161–76.

WIKLUND, C. G. 1977. Breeding success of the Merlin (*Falco columbarius*) in an isolated subalpine birch forest. Vår Fågelvärld 36: 206–65.

WIKMAN, M. 1976. Sex ratio of Finnish nestling Goshawks *Accipiter gentilis* (L). Congr. Int. Union Game Biol. 12.

WIKMAN, M. 1977. Duvhokspredation på Skogsfagel i Sydvästra Finland 1975–6. Pp. 59–72 in 'Foredrag fra Nordisk Skofuglsymposium, 1976, Trondheim.

WILBUR, S. R., CARRIER, W. D. & BORNEMAN, J. C. 1974. Supplemental feeding program for California Condors. J. Wildl. Manage. 38: 343–6.

WILEY, J. M. 1975. The nesting and reproductive success of Red-tailed Hawks and Red-shouldered Hawks in Orange County, California, 1973. Condor 77: 133–9.

WILEY, H. 1974. Evolution of social organisation and life history patterns among grouse. Quart. Rev. Biol. 49: 201–27.

WILLGOHS, J. F. 1961. The White-tailed Eagle *Haliaeetus a. albicilla* (L.) In Norway. Arbok for Universitetet i Bergen, Nat.-Naturv. Serie No. 12. Bergen.

WILLIAMS, G. C. 1966. Natural selection, the costs of reproduction, and a refinement of Lack's principle. Amer. Nat. 100. 687–90.

WILLIAMS, R. B. 1947. Infestation of raptorials by *Ornithodoros aquilae*. Auk 64: 185–8.

WILLOUGHBY, E. J. & CADE, T. J. 1964. Breeding behaviour of the American Kestrel (Sparrow Hawk). Living Bird 3: 75–96.

WINSTANLEY, D., SPENCER R. & WILLIAMSON, K. 1974. Where have all the White-throats gone? Bird Study 21: 1–14.

WITHERBY, H. F., JOURDAIN, F. C. R., TICEHURST, N. F. & TUCKER, B. W. 1938. The handbook of British birds, Vol. 3. London: Witherby.

WITHERINGTON, G. 1910. Rapid re-mating of the Peregrine Falcon. Brit. Birds 3: 263–64.

WOFFINDEN, N. D. & MURPHY, J. R. 1977. Population dynamics of the Ferruginous Hawk during a prey decline. Great Basin Nat. 37: 411–25.

WOOD, M. 1938. Food and measurements of Goshawks. Auk 55: 123–4.

YOCOM, C. F. 1944. Evidence of polygamy among Marsh Hawks. Wilson Bull. 56: 116–17.

ZIMMERMAN, J. L. 1966. Polygyny in the Dickcissel. Auk 83: 534–46.

English and scientific names of raptor species mentioned in this book

Turkey Vulture
Cathartes aura
Black Vulture (New World)
Coragyps atratus
King Vulture
Sarcorhamphus papa
California Condor
Gymnogyps californianus
Andean Condor
Vultur gryphus
Osprey
Pandion haliaetus
African Cuckoo-Falcon
Aviceda cuculoides
Honey Buzzard
Pernis apivorus
Swallow-tailed Kite
Elanoides forficatus
Bat Hawk
Machaerhamphus alcinus
Pearl Kite
Gampsonyx swainsonii
White-tailed Kite
Elanus leucurus
Black-shouldered Kite
Elanus caeruleus
Australian Black-shouldered Kite
Elanus notatus
Letter-winged Kite
Elanus scriptus
African Swallow-tailed Kite
Chelictinia riocourii
Snail or Everglade Kite
Rostrhamus sociabilis
Mississippi Kite
Ictinia misisippiensis
Black Kite
Milvus migrans
Red Kite
Milvus milvus
Whistling Eagle
Haliastur sphenurus
Brahminy Kite or White-headed Sea
Eagle
Haliastur indus

White-bellied Sea Eagle
Haliaeetus leucogaster
African Fish Eagle
Haliaeetus vocifer
Pallas' Sea Eagle
Haliaeetus leucoryphus
Bald Eagle
Haliaeetus leucocephalus
White-tailed or Sea Eagle
Haliaeetus albicilla
Steller's Sea Eagle
Haliaeetus pelagicus
Egyptian Vulture
Neophron percnopterus
Lammergeier or Bearded Vulture
Gypaetus barbatus
Hooded Vulture
Necrosyrtes monachus
Indian White-backed Vulture
Gyps bengalensis
African White-backed Vulture
Gyps africanus
Indian Griffon
Gyps indicus
Rüppell's Griffon
Gyps rueppellii
Griffon Vulture
Gyps fulvus
Cape Vulture
Gyps coprotheres
Lappet-faced Vulture
Torgos tracheliotus
European Black Vulture
Aegypius monachus
White-headed Vulture
Trigonoceps occipitalis
Short-toed or Snake or Black-breasted
Snake Eagle
Circaetus gallicus
Brown Snake Eagle
Circaetus cinereus
Bateleur
Terathopius ecaudatus
Gymnogene or African Harrier Hawk
Polyboroides typus

Marsh Harrier
  Circus aeruginosus
Hen Harrier or Marsh Hawk
  Circus cyaneus
Pallid Harrier
  Circus macrourus
Montagu's Harrier
  Circus pygargus
Pied Harrier
  Circus melanoleucos
Dark Chanting Goshawk
  Melierax metabates
Gabar Goshawk
  Melierax gabar
Goshawk
  Accipiter gentilis
Black Sparrowhawk
  Accipiter melanoleucus
Japanese Lesser Sparrowhawk
  Accipiter gularis
Besra
  Accipiter virgatus
European Sparrowhawk
  Accipiter nisus
Sharp-shinned Hawk
  Accipiter striatus
Red-thighed Sparrowhawk
  Accipiter erythropus
African Little Sparrowhawk
  Accipiter minullus
Australian Goshawk
  Accipiter fasciatus
Grey Frog Hawk
  Accipiter soloensis
Levant Sparrowhawk
  Accipiter brevipes
Shikra
  Accipiter badius
Nicobar Shikra
  Accipiter butleri
Cooper's Hawk
  Accipiter cooperii
Grasshopper Buzzard Eagle
  Butastur rufipennis
White-eyed Buzzard
  Butastur teesa
Lizard Buzzard
  Kaupifalco monogrammicus
Common Black Hawk
  Buteogallus anthracinus
Savannah Hawk
  Heterospizias meridionalis

Grey Eagle Buzzard
  Geranoaetus melanoleucus
Harris' Hawk
  Parabuteo unicinctus
Grey Hawk
  Buteo nitidus
Rufous-thighed Hawk
  Buteo leucorrhous
Red-shouldered Hawk
  Buteo lineatus
Broad-winged Hawk
  Buteo platypterus
Short-tailed Hawk
  Buteo brachyurus
Swainson's Hawk
  Buteo swainsonii
Galapagos Hawk
  Buteo galapagoensis
White-tailed Hawk
  Buteo albicaudatus
Red-backed Buzzard
  Buteo polyosoma
Zone-tailed Hawk
  Buteo albonotatus
Red-tailed Buzzard
  Buteo ventralis
Red-tailed Hawk
  Buteo jamaicensis
Common Buzzard
  Buteo buteo
African Mountain Buzzard
  Buteo oreophilus
Rough-legged Buzzard
  Buteo lagopus
Long-legged Buzzard
  Buteo rufinus
Upland Buzzard
  Buteo hemilasius
Ferruginous Hawk
  Buteo regalis
African Red-tailed Buzzard
  Buteo auguralis
Jackal or Augur Buzzard
  Buteo rufofuscus
Harpy Eagle
  Harpia harpyja
Philippine Monkey-eating Eagle
  Pithecophaga jefferyi
Lesser Spotted Eagle
  Aquila pomarina
Greater Spotted Eagle
  Aquila clanga

Tawny or Steppe Eagle
Aquila rapax
Imperial Eagle
Aquila heliaca
Wahlberg's Eagle
Aquila wahlbergi
Gurney's Eagle
Aquila gurneyi
Golden Eagle
Aquila chrysaetos
Wedge-tailed Eagle
Aquila audax
Verreaux's or Black Eagle
Aquila verreauxi
Bonelli's or African Hawk-eagle
Hieraaetus fasciatus
Booted Eagle
Hieraaetus pennatus
Little Eagle
Hieraaetus morphnoides
Ayres' Hawk-eagle
Hieraaetus dubius
Long-crested Eagle
Lophoaetus occipitalis
Mountain or Feather-toed Hawk-eagle
Spizaetus nipalensis
Crowned Eagle
Stephanoaetus coronatus
Martial Eagle
Polemaetus bellicosus
Secretary Bird
Sagittarius serpentarius
Yellow-throated Caracara
Daptrius ater
Carunculated Caracara
Phalcoboenas carunculatus
Guadalupe Caracara
Polyborus lutosus
Common Caracara
Polyborus plancus
African Pigmy Falcon
Polihierax semitorquatus
Lesser Kestrel
Falco naumanni
Greater or White-eyed Kestrel
Falco rupicoloides
American Kestrel
Falco sparverius
European Kestrel
Falco tinnunculus
Madagascar or Aldabra Kestrel
Falco newtoni

Mauritius Kestrel
Falco punctatus
Seychelles Kestrel
Falco araea
Grey Kestrel
Falco ardosiaceus
Dickinson's Kestrel
Falco dickinsoni
Madagascar Banded Kestrel
Falco zoniventris
Australian or Nankeen Kestrel
Falco cenchroides
Red-footed Falcon
Falco vespertinus
Eastern Red-footed Falcon
Falco amurensis
Red-headed Falcon
Falco chiquera
Merlin
Falco columbarius
Hobby
Falco subbuteo
Eleonora's Falcon
Falco eleonorae
Sooty Falcon
Falco concolor
Bat Falcon
Falco rufigularis
Aplomado Falcon
Falco femoralis
Black Falcon
Falco subniger
Lanner Falcon
Falco biarmicus
Prairie Falcon
Falco mexicanus
Laggar Falcon
Falco jugger
Saker Falcon
Falco cherrug
Gyrfalcon
Falco rusticolus
Orange-breasted Falcon
Falco deiroleucus
Taita Falcon
Falco fasciinucha
Kleinschmidt's Falcon
Falco kreyenborgi
Peregrine
Falco peregrinus

Tables 1–68

Differences in body weight and mean prey weight in some strongly dimorphic raptors. On the data available, the Montagu's Harrier was the only species which did not show a prey-weight difference between the sexes. From Storer 1966, Opdam 1975, Schipper 1973

	Mean body weight of raptor	Mean body weight of prey
North American accipiters		
Sharp-shinned Hawk, male	99	18
Sharp-shinned Hawk, female	171	28
Cooper's Hawk, male	295	38
Cooper's Hawk, female	441	51
Goshawk, male	818	397
Goshawk, female	1137	522
European accipiters		
Sparrowhawk, male	149	25
Sparrowhawk, female	258	66
Goshawk, male	716	277
Goshawk, female	1127	605
European harriers		
Montagu's Harrier, male	265	37
Montagu's Harrier, female	345	36
Hen Harrier, male	340	69
Hen Harrier, female	500	122
Marsh Harrier, male	530	134
Marsh Harrier, female	720	204

TABLE 2

Sex ratios of eggs and well-grown nestling raptors. Those marked with an asterisk are significantly different from equal (P<0·05)

Locality		No. of broods	No. of males	No. of females	Ratio females/males	Size dimorphism[1]	Reference
Among eggs[2]							
European Sparrowhawk	Britain	109	260	250	0·96	1·19	Newton & Marquiss 1979
American Kestrel	East U.S.A.	6	19	19	1·00	1·07	Porter & Wiemeyer 1972
Peregrine	Britain	23	49	36	0·73	1·16	R. Mearns, D. N. Weir, I. Newton
Among fledglings							
Hen Harrier	Scotland, Orkney	?	492	569	1·16*	1·09	Balfour & Cadbury 1979
	Scotland, SW	14	17	25	1·47	1·09	Watson 1977
Goshawk[3]	Finland	378	568	526	0·93	1·13	Wikman 1976
European Sparrowhawk	Britain	651	1102	1061	0·96	1·19	Newton & Marquiss 1979
	Denmark	?	147	121	0·82	1·19	Shelde 1960
Cooper's Hawk	East U.S.A.	?	35	36	1·03	1·13	Meng 1951
Harris' Hawk	Arizona	?	56	51	0·91	1·08	Mader 1976
American Kestrel	East U.S.A.	?	?	?	1·00	1·07	Heintzelman & Nagy 1968
	East U.S.A.	19	37	34	0·92	1·07	Porter & Wiemeyer 1972
	Utah	22	19	31	1·63	1·07	Smith et al 1972
	California	?	61	66	1·08	1·07	Balgooyen 1976
Hobby	Germany	?	29	43	1·48	1·04	Fiuczynski 1978
Prairie Falcon	Colorado	?	45	46	1·02	1·15	Enderson 1964
Peregrine	East U.S.A.	?	23	39	1·70*	1·16	Hickey 1942
	Scotland, S	73	47	53	1·13	1·16	R. Mearns
	Scotland, N	45	82	80	0·98	1·16	D. N. Weir
	Britain, other areas	41	47	54	1·15	1·16	I. Newton

[1] As judged by female wing length/male wing length.
[2] Sexed retrospectively from nests in which all eggs laid gave rise to fledged young.
[3] Excludes young from broods of one, in which sexing was thought to be uncertain.
The Table excludes the extensive samples of Cavé (1968) for the European Kestrel in which some young were wrongly sexed.

TABLE 3

Sex ratios in full-grown raptors. Those marked with asterisks are significantly different from equal; *P<0·05, **P<0·01, ***P<0·001

	Location and method (shot (s), trapped (t), observed (o))	Migrating (m) or wintering (w)	First winter (f.w.) or adult (ad)	No. of males	No. of females	Ratio female/male	Reference
Goshawk	Pennsylvania (s)	w	f.w.	11	8	0·73	Wood 1938
	Pennsylvania (s)	w	ad	52	92	1·77***	Hagen 1942
	Norway (t)	w	f.w. & ad	87	39	0·45***	Höglund 1964
	Fennoscandia (t)[1]	w	f.w. & ad	540	354	0·66***	Haukioja & Haukioja 1970
	Fennoscandia (t)[1]	w	f.w.	36	39	1·03	Haukioja & Haukioja 1970
	Fennoscandia (t)[1]	w	ad	18	5	0·28**	Haukioja & Haukioja 1970
	Wisconsin (t)	m	f.w.	39	18	0·46**	Mueller & Berger 1968
	Wisconsin (t)	m	ad	27	20	0·74	Mueller & Berger 1968
Sparrowhawk	Heligoland (t)	m	f.w.	217	136	0·63***	Moritz & Vauk 1976
	Heligoland (t)	m	ad	37	21	0·57*	Moritz & Vauk 1976
	Scotland, forest (t)[2]	w	f.w. & ad	123	43	0·35***	Marquiss & Newton 1979
	Scotland, mixed woods & farms[2]	w	f.w. & ad	23	14	0·61	Marquiss & Newton 1979
	Scotland, open farmland[2]	w	f.w. & ad	23	34	1·48	Marquiss & Newton 1979
Sharp-shinned Hawk	Pennsylvania (s)	m	f.w. & ad	93	20	0·22***	Brown & Amadon 1968
Brown Goshawk	Australia (t)	w	f.w. & ad	46	38	0·83	Purchase 1973
Harris' Hawk	Arizona (t)[3]	w	f.w.	9	3	0·33	Mader 1976
	Arizona (t)[3]	w	ad	22	8	0·36**	Mader 1976

[1] Based on ring recoveries, mainly trapped.
[2] Mid-August to October in a non-migratory population.
[3] On breeding range; a polyandrous population.

TABLE 3 (cont.)

	Location and method (shot (s), trapped (t), observed (o))	Migrating (m) or wintering (w)	First winter (f.w.) or adult (ad)	No. of males	No. of females	Ratio female/male	Reference
American Kestrel	California (o)	w	f.w. & ad	277	451	1·63***	Willoughby & Cade 1964
	California, orchards etc.	w	f.w. & ad	85	35	0·41***	Koplin 1973
	California, open fields	w	f.w. & ad	111	442	3·98***	Koplin 1973
	Pennsylvania (o)	w	f.w. & ad	67	40	0·60**	Roest 1957
	Pennsylvania (o)	m	f.w. & ad	107	70	0·65**	Heintzelman & Nagy 1968
	Ohio (o)	w	f.w. & ad	86	203	2·36***	Mills 1975
European Kestrel	Netherlands (t)	w	f.w. & ad	195	341	1·75***	Cavé 1968
Prairie Falcon	Western America (t,o)	w	f.w.& ad	24	48	2·00**	Enderson 1964
Gyr Falcon	Middle America (o)	w	f.w. & ad	6	20	3·33**	Platt 1976
Peregrine	Maryland Coast (t)	m	f.w.	162	377	2·33***	Ward & Berry 1972
	Maryland Coast (t)	m	ad	8	92	11·50***	Ward & Berry 1972
	Texas Coast (t)	m	f.w.	50	164	3·28***	Hunt et al 1975
	Texas Coast (t)	m	ad	0	36	—	Hunt et al 1975
	Texas Coast (t)	m	f.w. & ad	2	42	21·00***	Enderson 1965

TABLE 4
Instances of polygyny in raptors

Species	Same nest (Type A)*	Separate nests close together in an area that would normally be occupied by one pair (Type B)*	Separate nests far apart in areas that would normally be occupied by separate pairs (Type C)*	Reference
Osprey	+	+		D. N. Weir
Marsh Harrier		+	+	In Table 5
Hen Harrier	+	+	+	In Table 5
Montagu's Harrier		+	+	In Table 5
Sparrowhawk	+		+	Jourdain 1928, Newton 1973a
Shikra		+		I. R. Taylor
Red-tailed Hawk**	+			Wiley 1975
Buzzard		+		Picozzi & Weir 1974, G. Shaw
Kestrel	+	+		Mathew 1882, Witherby et al 1938, Petersen 1956
Peregrine***	+	+		Spofford 1969, D. N. Weir
Merlin		+		J. Roberts

In Type A the evidence consisted of two females seen or shot together at a nest containing a double clutch and which only one male was known. A double clutch alone is insufficient because one female might have placed another, or the same female might have laid both clutches (Balfour 1962). In Types B and C the evidence based on one male consistently visiting more than one nest at which no other male was seen.

Both females were fed by the male, brooded and fed the young, but only one laid.

* Only the larger of two females was fed by the male and only she laid; the smaller female sat on the nest only hen the larger one was absent. In the other case, the two females used alternative cliffs in the same home range d each laid.

TABLE 5
Incidence of polygyny in harriers

	Number of males with following number of females							
	1	2	3	4	5	6	*Locality*	*Reference*
Marsh Harrier	2	1	1	—	—	—	SE England	Buxton 1947–48
Marsh Harrier	24	1	—	—	—	—	Skåne, Sweden	Bengston 1967
Marsh Harrier	30	8	1	—	—	—	SE England	Axell, in Brown 1976a
Hen Harrier	3	1	—	—	—	—	United States	Reindahl 1941
Hen Harrier	9	1	—	—	—	—	Manitoba	Hecht 1951
Hen Harrier	82	7	1	—	—	—	Wisconsin	Hamerstrom 1969
Hen Harrier	53	6	—	—	—	—	NE Scotland	Picozzi 1978
Hen Harrier	45	0	2	—	—	—	SW Scotland	Watson 1977
Hen Harrier,								
yearling males	31	1	—	—	—	—		
older males	51	58	30	9	2	1	Orkney, Scotland	Balfour & Cadbury 1979

In this table, records for different years have been combined, with each year recorded as a separate incide even though some of the same birds may have been involved.

Isolated records of polygyny were also recorded in Marsh Harrier by Hosking 1943 (3 females), Benson 1958 females), Hildén & Kalinainen 1966 (2 females); in Hen Harrier by Yocum 1944 (2 females), Kraan & Strien 19 (7 females); and in Montagu's Harrier by Dent 1939 (2 females) and Hosking 1943 (2 females).

The following studies recorded no polygyny, but it was not always certain whether the author checked: Marsh Harrier see Colling & Brown 1946 (1 nest), Sondell 1970 (6 nests), Johannesson 1975 (6 nests), Perss 1975 (169 nests); for Hen Harrier see Errington 1930 (3 nests), Breckenridge 1935 (10 nests), Hall 1947 (5 nest Hagen 1969 (35 nests), Sondell 1970 (6–7 nests); for Montagu's Harrier see Hosking 1943 (1 nest), Colling Brown 1946 (1 nest), Buxton 1947–48 (4 nests), Robinson 1950 (3 nests).

TABLE 6

rformance of individual Hen Harriers in breeding groups of different sizes, Scotland.
the males the trend from 1–6 is statistically significant, as is the difference between
ɔnogamous and polygynous females mated to older males. Calculated from details in
Balfour & Cadbury 1979

	One female mated to yearling male	Number of females mated to older males					
		1	2	3	4	5	6
. of nests	27	28	92	84	36	13	6
. (%) of nests successful	17(63)	23(82)	60(65)	45(54)	14(39)	5(38)	3(50)
ung produced per male	1·6	2·3	3·0	3·9	4·0	6·5	—
ung produced per female	1·6	2·3	1·5	1·3	1·0	1·3	—

TABLE 7

Performance of individual Harris' Hawks in monogamous and polyandrous
breeding groups, Arizona. Calculated from details in Mader 1976

	One female mated to one male	One female mated to two males
No. of nests	27*	23
No. (%) successful	16(59)	18(78)
Young produced per male	1·3	1·0
Young produced per female	1·3	2·0

* Includes three re-nests; the total number of females studied was 47.

TABLE 8

Stability in breeding populations of raptors with fairly stable food sources. Data from populations not significantly affected by human influence

Honey Buzzard

(1) 5–6 pairs in 1938–42, no record 1939, Denmark (Holstein 1944).

(2) 4, 4, 4, 4 and 3 pairs in 1940–44, 3 pairs in 1948, and 2 pairs in 1950 and 1951, Germany (Wendland 1953).

Black Kite

(1) 7, 8, 8, 8 and 7 pairs in 1940–44, and 7 pairs in 1950 and 1951, Germany (Wendland 1953).

Red Kite

(1) 1, 1, 2, 3, 0 pairs in 1940–44, 2 pairs in 1948, and 1 pair in 1950 and 1951, Germany (Wendland 1952/53).

African Fish Eagle

(1) 56 pairs in 1968–69 and again in 1970–71 on Lake Naivasha, Kenya, despite a marked increase in total population (Brown & Hopcraft 1973).

Goshawk

(1) 5 pairs in 10 years, 1956–65, Poland (Pielowski 1968).

Sparrowhawk

(1) 13 and 15 pairs in 1940–41, Germany (Wendland 1953).

(2) 6 pairs in 1941–1943, Netherlands (Tinbergen 1946).

Hen Harrier

(1) About 24 males in 1970–74, northeast Scotland, females varied in number with amount of polygyny (N. Picozzi).

Red-tailed Hawk

(1) 23, 22, 21, 21 and 20 pairs in 1967–71, Alberta (McInvaille & Keith 1974).

(2) 27, 33, 27 pairs in 1953–1955, Wisconsin (Orians & Kuhlman 1956).

(3) 64 and 73 pairs in 1971 and 1972, Montana (Johnson 1975).

Buzzard

(1) 28, 29, 30 and 29 pairs in 1941–44, 29 and 30 pairs in 1950–51, Germany (Wendland 1952).

(2) 33–37 pairs in ten years in 1962–71, England (Tubbs 1974).

(3) 25 pairs in 1956 and 1958, 27 pairs in 1959, Germany (Mebs 1964).

(4) 12, 14 and 14 pairs in 1957–59, England (Dare 1961).

(5) 31, 32, 38 and 34 pairs in 1969–73, Scotland (Picozzi & Weir 1974).

Golden Eagle

(1) 10–13 pairs in 26 years, northeast Scotland, 1944–69 (Watson 1970).

(2) Change by one pair more or less over 10 years in four Scottish areas holding 16, 13, 12, and 8 pairs (Brown & Watson 1964).

(3)   31 and 30 pairs in 1963–64, Montana (McGahan 1968).

(4)   4 pairs 1947–58, 3 pairs in 1959–76, 4 pairs in 1977, Rhum, Scotland (M. Ball, pers. comm.).

Black Eagle

(1)   52–59 pairs in 1964–76, Matopos Hills, Rhodesia (Gargett 1977).

Bonelli's Eagle

(1)   16 pairs in 1973–76, Israel (Leshem 1976).

Greater Kestrel

(1)   10 pairs in 1975–77, Transvaal (Kemp 1978).

Kestrel

(1)   11, 11, 10, 12 pairs in 1941–44, 11 pairs in 1950, 13 pairs in 1951, Germany (Wendland 1953).

(2)   52, 48, 50 pairs in 1964–66, England (Taylor 1967).

Merlin

(1)   3–4 pairs for about 19 years from 1898, England (Rowan 1921–22).

(2)   6–8 pairs in 1971–75, Sweden (Wiklund 1977).

Hobby

(1)   10, 9, 9, 10 pairs in 1941–44, 12 pairs in 1950–51, Germany (Wendland 1953).

Prairie Falcon

(1)   6, 7 and 7 pairs in 1966–68 in one area, 13 and 14 pairs in 1967–68 in another area, central Canada* (Fyfe *et al* 1969).

Peregrine*

(1)   Four populations of up to 6, 12, 18 and 25 pairs fluctuated by no more than 8% of mean, 1945–60, Britain (Ratcliffe 1962).

(2)   10–13 pairs in 1935–42 and 1947, Massachusetts (Hagar 1969).

(3)   17–20 pairs in 1954–61, East Germany (Kleinstauber 1969).

(4)   10, 8, 11 and 9 pairs in 1956–59, Colville River, Alaska: on a larger stretch of river 32 pairs in 1952, 36 in 1959 plus 5 unmated adults (Cade 1960), 34, 32, 33 and 31 pairs in 1967–69 and 1971, after which population declined from pesticides (White & Cade 1977).

(5)   18 pairs + 1 unmated bird in 1939, 19 pairs in 1940, around New York (Hickey 1942).

(6)   1 pair per 10 miles of Yukon River, Alaska, in 1899, 1 pair per 9·3 miles in 1951, and 1 pair per 10·5 miles in 1966 (Cade *et al* 1968).

(7)   3 pairs in 1940–44, after which population declined from pesticides, Germany (Wendland 1953).

(8)   17, 19, 22 and 19 pairs in 1969–72, Amchitka, Aleutian Islands, Alaska (16 pairs in 1973, an abnormal year in which some pairs had probably deserted before the first survey) (White 1975).

* Excludes other areas where populations had declined from pesticide poisoning.

TABLE 9

Stability in the total raptor population of particular areas. From Wendland 1952–53*, and Craighead & Craighead 1956.

	(a) Area 1, near Berlin, Germany					
	1941	1942	1943	1944	1950	1951
Buzzard	28	29	30	29	29	30
Kestrel	11	11	10	12	11	13
Hobby	10	9	9	10	12	12
Black Kite	8	8	8	7	7	7
Honey Buzzard	4	4	4	3	2	2
Red Kite	1	2	3	0	1	1
Totals	62	63	64	61	62	65

	(b) Area 2, Michigan			(c) Area 3, Michigan	
	1942	1948	1949	1947	1948
Red-shouldered Hawk	19	16	14	17	18
Red-tailed Hawk	2	5	6	9	12
Cooper's Hawk	8	7	7	6	6
Hen Harrier	7	9	8	5	6
Swainson's Hawk	0	0	0	0	0
Goshawk	0	0	0	0	0
Kestrel	0	0	0	0	0
Prairie Falcon	0	0	0	0	0
Osprey	0	0	0	0	0
Totals	36	37	35	37	42

* Excludes Peregrine which disappeared in this period associated with DDT usage, and Sparrowhawk which was not covered every year.

TABLE 10

Instances of replacement of lost mates in the same season.

	Sex replaced	Time period	References
ɔrey	Both	?	St John 1849; Bent 1938
ɪite-tailed Eagle	Female	Within 1 week	Saxby 1874
ɪite-backed Vulture	?	Within 5 days	Ali & Ripley 1968
eleur	Female	?	Brown 1952–1953
ntagu's Harrier	Both	?	Mayr 1938
shawk	Male	Within 2 weeks	Holstein 1942
ɪrrowhawk	Both	Within 1 week	Owen 1936–1937; I. Newton & M. Marquiss
ɪrp-shinned Hawk	?	?	Brooks 1927
kra	?	?	Dementiev & Gladkov 1954
ɔper's Hawk	Female	Within 'some days'	Brooks 1927; Bent 1938; Schriver 1969
l-shouldered Hawk	?	'Promptly'	Bent 1938
ainson's Hawk	Male	3 instances, within 1–4 days	Schmutz 1977
nmon Buzzard	Both	?	Dementiev & Gladkov 1954; Nethersole-Thompson 1943
ican Red-tailed Buzzard	Female	Within 1 day	Brown & Amadon 1968
ɔerial Eagle	Both	?	Valverde 1960
lden Eagle	Female	Within 10 weeks	Dixon 1937
ck Eagle	Female	Within 3 weeks	Visser 1963
ser Kestrel	Female	8 birds in 'quick succession'	Lucanus 1936
ɪerican Kestrel	Both	One female within 5 days	Mayr 1938; Enderson 1960
ɔpean Kestrel	Both	Within 1 day	Frere 1886; Village 1979
rlin	Both	?	Seebohm 1883; Dementiev & Gladkov 1954
bby	?	?	Dementiev & Gladkov 1954
irie Falcon	?	?	Webster 1944
er Falcon	?	?	Dementiev & Gladkov 1954
r Falcon	Female	?	Bent 1938
egrine	Both	Several cases within 1 day	Brooks 1927; Walpole-Bond 1938; Witherington 1910; Hickey 1942; Nethersole-Thompson 1943; Ferguson-Lees 1951; Hall 1955; Macintyre 1960; Hagar 1969; Ratcliffe 1969; Kumari 1974

TABLE 11

Buzzard densities in relation to rabbit numbers in different areas of Britain.

Good rabbit areas	No. pairs	Pairs/km^2
Skomer Island, Pembroke	7	2·4
Monmouth	12	1·5
Devon A	21	1·0
Devon B	9	0·9
Argyll	7	1·0
Poor rabbit areas		
Sedbergh, Yorkshire	5	0·1
New Forest, Hampshire	33–37	0·1

Densities from Moore (1957), Holdsworth (1972), Tubbs (1974); division of rabbit areas, I. Newton.

TABLE 12

Annual variation in breeding populations of raptors with greatly fluctuating food sources.

A. *Species that eat rodents (approximately 4-year cycles)*

Rough-legged Buzzard
*(1) 0–9 pairs during 9 years, North Norway (Hagen 1969).
(2) 51, 46 and 61 nests during 3 years, Colville River, Alaska (White & Cade 1971).
(3) 10–82 pairs during 5 years, Seward Peninsula, Alaska (Swartz et al 1974).

Hen Harrier
(1) 10–24 females in 33 km^2 during 22 years, Orkney, Scotland (Balfour, in Hamerstrom, 1969).
*(2) 13–25 females in 160 km^2 during 5 years, Wisconsin (Hamerstrom 1969)**.
*(3) 0–9 pairs in 6 years between 1938 and 1946 (Hagen 1969).

European Kestrel
*(1) 35–109 clutches in 4 years; 109 clutches at vole density index 24, 97 at index 13, 35 at index 9, 50 at index 4, Netherlands (Cavé 1968).
*(2) Approximately 20-fold fluctuation in index of number of broods ringed in Britain over 42 years, with peaks every 4–5 years (Snow 1968).
*(3) 1–14 pairs during 5 years, North Norway (Hagen 1969).
*(4) 1–16 nests during 12 years, Swabian Alps (Rockenbauch 1969).
*(5) 26, 28 and 38 pairs during 3 years, south Scotland (Village 1969).

Black-shouldered Kite   *(1) Increase from 1 to 8 nests in one year, associated with rodent plague, South Africa (Malherbe 1963).

B. *Species that eat gallinaceous birds or hares (4 year or 10-year cycles)*

Ferruginous Hawk   *(1) 5–16 pairs during 8 years in one area, 1–8 pairs during 3 years in another area, Utah (Woffinden & Murphy 1977).

Goshawk   (1) 0–4 nests in 100 km^2 during 13 years; 2–9 nests in 200 km^2 during seven years, in two areas of Sweden (Höglund 1964).
    *(2) 1–9 nests in 372 km^2 during 4 years, Alaska (McGowan 1975).

Gyr Falcon   *(1) 13–49 pairs during 5 years, Seward Peninsula, Alaska (Swartz *et al* 1974).
    *(2) 19–31 occupied cliffs and 12–29 successful nests during 4 years, Alaska (Platt 1979).

* Pre population also assessed and related to raptor numbers.
** Excluding one year when population dropped from DDT poisoning.

TABLE 13

Winter raptor numbers in relation to rodent numbers in 96 km^2 of southern Michigan. From Craighead & Craighead 1956.

	Index of mouse numbers	Total raptor numbers	Raptors per km^2	Unidentified Buteo sp	Red-tailed Hawk	Red-shouldered Hawk	Rough-legged Hawk	Kestrel	Hen Harrier	Cooper's Hawk
41–42	303	96	1·0	13	17	1	13	5	37	10
47–48	75	27	0·3	0	12	3	0	5	1	6

TABLE 14
Nesting places of raptor genera[1]

	Number of genera using:				
	Trees only	Cliffs only	Ground only	Trees or cliffs	Trees, cliffs or ground
Genera that do not build nests	2	0	0	2	4
Genera that do build nests	50[2]	2[3]	1[4]	9	6

[1] The table excludes the New World Falconidae genera *Spiziapteryx* and *Micrastur* because I can find no description of a nest, or of whether or not they build.
[2] Includes seven forest genera whose nests have apparently not been described, but may be assumed to use trees and, from their relationships, to build nests. They include *Leptodon*, *Henicopernis*, *Dryotriorchis*, *Eutriorchis*, *Urotriorchis*, *Harpyhaliaetus* and *Spizastur*.
[3] Includes *Phalcoboenas* caracaras which also nest on the ground, and Lammergeier *Gypaetus barbatus* which does not.
[4] The ten *Circus* harriers which include one Australian species, *C. assimilis*, that nests in trees.
The range of situations recorded for some species may well increase with further study.

TABLE 15
Genera in which species differ in choice of nesting place

	Number of species using:			
	Trees only	Cliffs only (and occasionally ground)	Trees and cliffs (and occasionally ground)	Unknown
Gyps	2	4	1[1]	0
Buteo	11	2	11	1[2]
Aquila	4	0	4	1[3]
Hieraaetus	3	0	2	0
Falco	15	7	13	2[4]

[1] The Indian Griffon *Gyps indicus* uses cliffs in one part of its range (subspecies *G. i. indicus*) and trees in another (subspecies *G. i. nudiceps*).
[2] Rufous-thighed Buzzard *B. leucorrhous* from South America.
[3] Gurney's Eagle *A. gurneyi* from New Guinea and neighbouring islands.
[4] Madagascar Banded Kestrel *F. zoniventris*, Kleinschmidt's Falcon *F. kreyenborgi* from South America.

TABLE 16
Raptors that have nested on buildings or other man-made structures

Cathartes:	Turkey Vulture (in barns and sheds, Brown & Amadon 1968).
Coragyps:	Black Vulture (in a barn, Stewart 1974).
Pandion:	Osprey (telegraph and power poles, duck blinds, channel markers, ruined buildings and nest-platforms, Witherby *et al* 1938, Henny *et al* 1974).
Elanus:	Black-shouldered Kite (two records on telegraph poles, Broekhuysen 1974).
Milvus:	Black Kite (on old buildings, Brown & Amadon 1968).
Haliaeetus:	Bald Eagle (nest platforms on pylons, Nelson 1977).
Neophron:	Egyptian Vulture (on buildings, Brown & Amadon 1968).
Buteo:	Swainson's Hawk (telegraph and power poles, Dunckle 1977), Red-backed Buzzard, *B. polyosoma* (telegraph poles, Brown & Amadon 1968), Red-tailed Hawk (power poles, H. Snyder), Buzzard (haystacks, deserted huts, Mebs 1964), Upland Buzzard (haystacks, Dementiev & Gladkov 1951), Ferruginous Hawk (pylons, Nelson 1977).
Aquila:	Steppe Eagle (ruined buildings, haystacks, Dementiev & Gladkov 1954), Golden Eagle (power poles, M. Nelson; remote ruined buildings, Dementiev & Gladkov 1954).
Hieraaetus:	Bonelli's Eagle (old buildings, Brown & Amadon 1968).
Polemaetus:	Martial Eagle (pylon tops, A. C. Kemp).
Falco:	Lesser Kestrel (buildings and walls, Witherby *et al* 1938), Greater Kestrel (pylons, in old crow nests, A. C. Kemp), American Kestrel (buildings, Brown & Amadon 1968), European Kestrel (buildings, walls, bridges, cranes, Witherby *et al* 1938), Madagascar Kestrel *F. newtoni* (buildings, Brown & Amadon 1968), Seychelles Kestrel *F. araea* (buildings, Brown & Amadon 1968), Australian Kestrel *F. cenchroides* (buildings, Brown & Amadon 1968), Merlin (one on a radio tower, in old crow nest, D. Okill), Bat Falcon (crane, Brown & Amadon 1968), Lanner Falcon (buildings, pylons, in old crow nests, Kemp 1972, Sinclair & Walters 1976), Prairie Falcon (one attempt on building, Nelson 1974), Lagger Falcon (buildings, Vaurie 1965), Gyr Falcon (old cranes and dredgers, White & Roseneau 1970), Orange-breasted Falcon (buildings, Brown & Amadon 1968), Peregrine (buildings and other structures, Hickey & Anderson 1969).

TABLE 17

Laying seasons of raptors in Zambia (Northern Rhodesia). Figures show the number of clutches started in different months, excluding species for which less than five records were available. Summer rains occur October–April, and the remaining months are dry. From Benson et al 1971.

	J	F	M	A	M	J	J	A	S	O	N	D
Black-shouldered Kite				5	1		1	3				
Yellow-billed Kite								12	9	3	1	
African Fish Eagle		2	5	10	7	9	7	2		1		
Hooded Vulture					1	2	2					
White-backed Vulture	1			3	9	11	5	3				
Lappet-faced Vulture			1		6	4	1					
White-headed Vulture					1	2	3	5				
Black-breasted Snake Eagle					2	4						2
Bateleur	4	6	3	2								
Dark Chanting Goshawk								2	6	1		
Gabar Goshawk								2	3			
Black Sparrowhawk								1	2	5	1	
Little Sparrowhawk								1	2	2		
Shikra									7	6	2	2
Lizard Buzzard								1	3	11	1	1
Tawny Eagle				2	2	4	4	2	2			
Wahlberg's Eagle									16	10	2	1
African Hawk-eagle					2	1	3					
Long-crested Eagle								3	2			
Martial Eagle			3			1	1					
Secretary Bird			1	1			1		1		3	1
Lanner Falcon								5	3			
Taita Falcon							2	1	2			

TABLE 18

Breeding parameters of Nearctic and Palaearctic raptors. Data mainly from references given in right-hand column, otherwise from Brown & Amadon (1968), Dementiev & Gladkov (1954), Schoenwetter (1960–72) for egg weights, Snyder & Wiley (1976) for some female weights

	Range[1]	Main food[2]	Female weight (g)	Egg weight (g)	Egg wt as % of female wt	Usual clutch size	Interval between eggs (days)	Incubation period (days)	Nestling period (days)	Post-fledging period (weeks)	Age adult plumage acquired (years)	Additional references
Turkey Vulture	N	C	1440	80	5·6	2		38–41	70–80+			Stewart 1974
Black Vulture	N	C	1560	103	6·6	2	2	38	80			Koford 1953
California Condor	N	C	11120	280	2·5	1		55–56	220	30	5–8	Osterlof 1951, Green 1976, Stinson 1977
Osprey	N P	F	1600	72	4·5	2–3		37–38	44–59	6–7	2	
Honey Buzzard	P	I	875	49	5·6	2	3–5	37	40–44	2+	2	Holstein 1944
Swallow-tailed Kite	N	ARI	460	37	7·9	2		28	36–42	2+		Snyder 1975
White-tailed Kite	N	Ro	307	24	7·6	4–5		30	35–40	4+	1[3]	Pickwell 1930
Black-shouldered Kite	P	Ro	258	21	8·1	2–4	2	30–32	32–36	10	1[3]	J. Mendelsohn
Snail Kite	N S	S	395	32	8·1	2–3			ca 30		2	Haverschmidt 1970
Mississippi Kite	N	I	310	27	8·7	1–2		30	34			
Black Kite	P	MFC	885	56	6·3	2–3	2–3	31–37	42–50	2–3+	2	Makatsch 1953, Meyburg 1971
Red Kite	P	MBC	1030	63	6·1	2–3	3	31–32	50–60	4–10	2	Davies & Davis 1973
Pallas' Sea Eagle	P	FBC	3270	117	3·6	2	2–3	35	70–105		5	
Bald Eagle	N	FBC	5000	117	2·3	2–3	2–3	34–36	70–77		5	Maestrelli & Wiemeyer 1975
White-tailed Eagle	P	FBC	6000	140	2·7	1–3	2–3	38	70–84	5–6+	5	Fentzloff 1975
Steller's Sea Eagle	P	FBC									5	
Egyptian Vulture	P	C	1890	94	4·9	2	2–4	40–42	95	4+	5	
Lammergeier	P	C	5100	216	4·2	1–2	4–5	53–60	107–117		5	Glutz von Blotzheim et al 1971
Indian White-backed Vulture	P	C	5250	204	3·9	1						
Griffon Vulture	P	C	6900	252	3·6	1		52–59	110–115	13–23	5–6	Mendelssohn & Marder 1970
Lappet-faced Vulture	P	C	6800	238	3·5	1		54–56	100–125		5–6	Anthony 1976, Pennycuick 1976
Black Vulture	P	C	10000	243	2·4	1		52–55	90–120	7–9	6	Suetens & Groenendael 1966
Short-toed Eagle	P	R	2070	136	6·6	1		47	70–75	3	3	Boudoint et al 1953
Marsh Harrier	P	BM	760	40	5·3	4–6	2–3	32–38	42	3–3·5	2(♀),3(♂)	Hilden 1966
Hen Harrier	N P	RoB	530	31	5·8	4–5	2–3	29–33	35–38	3–3·5	2(♀),3(♂)	Breckenridge 1935, Balfour 1957

[1] N – Nearctic, P—Palaearctic

[2] A—amphibia, B—birds, C—carrion, F—fish, I—insects, M—mammals except rodents, Ro—rodents, R—reptiles, S—snails.

[3] Achieved by autumn moult.

Since almost all the breeding parameters in this table are variable, I chose for regression analyses (in Figures 19–20), a single value from the middle of the range of values given. Where a choice was available, I took a mean weight from near the centre of a species breeding range. Hence, almost all the values used in the regression analyses could be refined to some extent, but this could not alter the overall conclusions.

TABLE 18 (cont.)

	Range[1]	Main food[2]	Female weight (g)	Egg weight (g)	Egg wt as % of female wt	Usual clutch size	Interval between eggs (days)	Incubation period (days)	Nestling period (days)	Post-fledging period (weeks)	Age adult plumage acquired (years)	Additional references
Pallid Harrier	P	RoI	380	28	7.4	4–5	2–3	30	35	2–3	2	Laszlo 1941
Montagu's Harrier	P	RoBR	340	24	7.1	4–5	2–4	28–30	35–40	2–3	2(♀),3(♂)	Neufeldt 1964
Pied Harrier	P	RoB	455	28	6.2	4–5	2	31	40		2	
Dark Chanting Goshawk	P	RIM										
Goshawk	N P	BM	1200	64	5.3	3–4	2–3	35–38	40–43	4.5–7.5	2	Holstein 1942, Reynolds & Wight 1978
Japanese Lesser Sparrowhawk												
Besra	P	B	192	18	9.5	2–4						
European Sparrowhawk	P	BRI	300	23	7.5	4–6	2–3	32–35	26–30	3–4	2	Newton 1976, 1978
Sharp-shinned Hawk	N	B	180	19	10.7	4–5	2	30–34	24–27	2.5–3.5	2	Platt 1976, Snyder & Wiley 1976
Grey Frog Hawk	P	A		18		3–5					2	
Levant Sparrowhawk	P	BI		22		2–4		32–35		2+	2	
Shikra	P	Re	270	21	7.6			33–35	30		2	
Cooper's Hawk	N	BMR	560	40	7.1	4–5	2	35–36	30–34	5.5–6	2	Meng 1951, Snyder & Wiley 1976
White-eyed Buzzard	P	IAR										
Grey-faced Buzzard Eagle	P	RAI										
Black Hawk	N	AFR	945	65	6.9	1–2		38–39	43–50	6–8	2	J. Schnell
Harris' Hawk	N	MB	1050	50	4.8	2–4		33–36	40	8–12+	2	Mader 1976
Grey Hawk	N	R	635	43	6.8	2–3		32	42			R. Glinski, unpublished
Red-shouldered Hawk	N	RoRA	700	57	8.1	2–4	2–3	33	42	8–10	2	Snyder & Wiley 1976, P. Glasier, unpublished
Broad-winged Hawk	N	MBA	490	42	8.6	2–3		30–38	35–42	3–4	2	Matray 1974, Fitch 1974
Short-tailed Hawk	N	BR	425	48	11.3	2–3						
Swainson's Hawk	N	RoMI	1100	61	5.7	2–3		34–35	42–44	4–4.5	2	Fitzner 1978
White-tailed Hawk	N	M		70		2						
Zone-tailed Hawk	N	R		58		2		35	35–42			
Red-tailed Hawk	N	RoMR	1220	64	5.2	2–4		32–35	43–48	5–10	2	Fitch et al 1946, Johnson 1975, Wiley 1975
Buzzard	P	RoMB	940	60	6.4	2–4	2–4	33–35	42–49	9–11	2	Mebs 1964
Rough-legged Buzzard	N P	Ro	1040	60	5.8	3–6		31	41	3–6	2	
Long-legged Buzzard	P	RoR	1350	73	5.4	2–3		32–34	45–48	6	2	I. Sela
Upland Buzzard	P	Ro	1800	79		2–4			45	4–6	2	
Ferruginous Hawk	N	RoM		80	6.5	3–5		33–36	44–48		2	Angell 1968, Schmutz 1977, P. Glasier
Lesser Spotted Eagle	P	MB	1500	88	5.9	2	3–4	42–43	58	3–4	5	Meyburg 1970, Brooke et al 1972

TABLE 18 (cont.)

	Range[1]	Main food[2]	Female weight (g)	Egg weight (g)	Egg wt as % of female wt	Usual clutch size	Interval between eggs (days)	Incubation period (days)	Nestling period (days)	Post-fledging period (weeks)	Age adult plumage acquired (years)	Additional references
Greater Spotted Eagle	P	MBA	2580	110	4·3	2		42–44	60–65	4+	5	Wendland 1959
Steppe Eagle	P	M	3560	117	3·3	2		45	55–70		7	Brooke et al 1972
Imperial Eagle	P	MRB	3500	132	3·8	2–3	2–3	43	65		9	Valverde 1960, Brooke et al 1972
Golden Eagle	N P	MB	4690	142	3·0	2	3–4	43–45	65–77	12+	5	Gordon 1955
Bonelli's Eagle	P	BM	2500–	112	4·5	2	2–3	37–39	65	8	3	Leshem 1976
Booted Eagle	P	BM	700	62	8·9	2	2–4		63			
Mountain Hawk Eagle	P	MB										
Common Caracara	N	CI	1500	76	5·1	2–3		32	42+			
Lesser Kestrel	P	IRo	168	16	9·4	4–5		28	26–28			P. Glasier
American Kestrel	N	RoI	140	16	11·4	4–6	2	27–31	28–30	3	1[3]	Porter & Wiemeyer 1970, 1972
European Kestrel	P	Ro	220	21	9·5	4–6	2	28	28–32	3·5–4	2	Tinbergen 1940
Red-footed Falcon	P	IA	150	17	11·6	3–4	2	25–28	26–28	1–3	2	
Red-headed Falcon	P	B	220	24	10·9	3–4		34–35	37–38			Koehler 1970
Merlin	N P	B	215	22	10·2	3–5	2	28–32	26–32	2–3	2	Campbell & Nelson 1975, Newton et al 1978
Hobby	P	BI	230	24	10·4	2–3	2–3	28	28–32	2–3	2	Walter 1968
Eleonora's Falcon	P	IB	210	26	10·5	2–3		28	37			
Sooty Falcon	P	BI	430	22		2–3						
Aplomado Falcon	N	BRoRI	580	47	8·1	3–4		32	35	4		D. Hector
Lanner Falcon	P	BM	865	49	5·7	4–5		32–35	44–46	4–6	2	Dalling 1974
Prairie Falcon	N	BM	755	43	5·7	3–4		29–33	36–41		2	Webster 1944, Enderson 1964
Laggar Falcon	P	BM						28–33			2	
Saker Falcon	P	MB	1050	53	5·0	3–5			40–45	4–6	2	
Gyr Falcon	N P	B	1750	70	4·0	3–5	2–3	34–35	46–49	4	2	Platt 1977
Peregrine	N P	B	950	50	5·3	3–4	2	32–34	35–40	5–6	2	Walpole-Bond 1938, Nelson 1972a

[1]N—Nearctic, P—Palaearctic

[2]A—amphibia, B—birds, C—carrion, F—fish, I—insects, M—mammals except rodents, Ro—rodents, R—reptiles, S—snails.

[3]Achieved by autumn moult.

Since almost all the breeding parameters in this table are variable, I chose for regression analyses (in Figures 19–20), a single value from the middle of the range of values given. Where a choice was available, I took a mean weight from near the centre of a species breeding range. Hence, almost all the values used in the regression analyses could be refined to some extent, but this could not alter the overall conclusions.

TABLE 19

Breeding parameters of some large raptors with long breeding cycles. Periods are in days and are approximate.

	Clutch size	Incubation period	Nestling period	Post-fledging period	Whole breeding cycle exceeds one year	Year adult plumage acquired	Reference
King Vulture	1	50–53	80–90	150	No		Brown & Amadon 1968, Cuneo 1968
California Condor	1	55–56	220	210	Yes	5–8	Koford 1953
Andean Condor	1	54–58	180	180+	Yes	5+	Brown & Amadon 1968
Lappet-faced Vulture	1	54–56	100–125	95–160	No	5–6	Anthony 1976, Pennycuick 1976
White-headed Vulture	1	55	100+	180+	No		Pennycuick 1976
Bateleur	1	52–55	90–125	180+	No	7	Brown & Cade 1972, A. C. Kemp, Steyn 1973
Harpy Eagle	2	54–56	141–148		Probably	5	Fowler & Cope 1964, Rettig 1978
Philippine Monkey-eating Eagle	1	60	105		Probably		Brown & Amadon 1968
Crowned Eagle	1–2	48–49	105–116	330–350**	Yes**		Brown 1966
Martial Eagle	1	44–51	100	180	No	7	Brown & Amadon 1968, Steyn 1973

Note ** Breeding every second year, with long post-fledging periods, has been recorded in Kenya (Brown 1966), but in southern Africa some pairs breed every year, with short post-fledging periods (Steyn 1973, Fannin & Webb 1975). For the Harpy and Philippine Monkey-eating Eagle the evidence for annual breeding is as yet inconclusive.

TABLE 20
Raptors recorded as paired or breeding in sub-adult plumage

	Sex	Paired	Eggs laid	Young raised	Reference
Snail Kite	Both	+	+	?	Haverschmidt 1970
Mississippi Kite	?	+	+	+	Parker 1974
Bald Eagle	Both	+	+	+	Bent 1938
Egyptian Vulture	?	+	?	?	Brown & Amadon 1968
Marsh Harrier	Both	+	+	+	Colling & Brown 1946
Hen Harrier	Both	+	+	+	Balfour & Cadbury 1974, 1978
Pallid Harrier	Both	+	+	+	Lundevall & Rosenberg 1955
Goshawk	Both	+	+	+	Hoglund 1964, Haukioja & Haukioja 1970, McGowan 1975
Sparrowhawk	Both	+	+	+	Newton 1976
Sharp-shinned Hawk	♀	+	?	?	Brooks 1927
Red-thighed Sparrowhawk	♀	+	+	+	Brown & Amadon 1968
Little Sparrowhawk	♀	+	+	+	Steyn 1972
Cooper's Hawk	♀	+	+	+	Meng 1951
Red-shouldered Hawk	♀	+	+	+	Henny et al 1973, Wiley 1975
Broad-winged Hawk	♀	+			Burns 1911
Swainson's Hawk	Both	+	+	+	Schmutz 1977
Red-tailed Hawk	♀	+	+	+	Luttich et al 1971
Common Buzzard	♀	+	?	−	D. N. Weir, unpublished
Imperial Eagle	♀	+	+	+	Valverde 1960
Golden Eagle	Both	+	+	+	Sandeman 1957, Bates 1976
Black Eagle	♀	+	−	−	I. Sela, unpublished
Bonelli's Eagle	Both	+	−	−	Leshem, 1976
European Kestrel	Both	+	+	+	Village 1979
Merlin	Both	+	+	+	Temple 1972a, Hodson 1975, Newton et al 1978
Hobby	♀	+	+	+	Fiuczynski 1978
Prairie Falcon	Both	+	+	+	Webster 1944, R, Fyfe, Platt 1978
Gyrfalcon	Both	+	+	?	Dementiev & Gladkov 1954
Peregrine	♀	+	+	−	Hickey 1942, I. Hopkins, C. White

TABLE 21

Proportions of immature-plumaged birds in breeding populations. Different years are combined

| | Males | | Females | | Unspecified sex | | |
	Total	% in immature plumage	Total	% in immature plumage	Total	% in immature plumage	Reference
Mississippi Kite					146	8	Parker 1974
Goshawk, good food year	17	0	13	38			McGowan 1975
Goshawk, poor food year			34	0			
			93	10			Glutz et al 1971
			70	0			Reynolds & Wight 1978
Sparrowhawk, good food habitat	16	44	131	20			Newton 1976
Sparrowhawk, poor food habitat	29	14	73	5			
Cooper's Hawk			36	6			Meng 1951
			34	6			Reynolds & Wight 1978
Red-shouldered Hawk	57	0	57	0			Craighead & Craighead 1956*
	74	0	74	1			Henny et al 1973
	29	0	29	10			Wiley 1975
Red-tailed Hawk	36	0	36	0			Craighead & Craighead 1956**
	24	0	24	0			Hagar 1957**
	67	0	67	0			Oriens & Kuhlman 1956**
Golden Eagle, not persecuted					132	1	Luttich et al 1971,
					61	1	Gates 1972
Golden Eagle, persecuted					62	1	Sandeman 1957
					94	4	
Bonelli's Eagle	61	3***	60	4			Leshem 1976
Hen Harrier	201	16					Balfour & Cadbury 1979
	46	7	66	24			Schmutz & Schmutz 1975
	47	9					Watson 1977

TABLE 21 (cont.)

	Males Total	Males % in immature plumage	Females Total	Females % in immature plumage	Unspecified sex Total	Unspecified sex % in immature plumage	Reference
European Kestrel, good food year	37	35	38	55			Village 1979
European Kestrel, poor food year	25	4	22	18			
Merlin	20	3					Temple 1972a, Newton et al 1978, & unpub.
Peregrine	38	13			250	0	Cade 1960
			34	3			Hickey 1942
	107	0	99	6			Hagar 1969
	143	0	142	1			Rice 1969

* Cited in Henny 1972;   ** Cited in Luttich et al 1971;   *** In one case sex not certain.

In all species mentioned, one or more immature-plumaged birds raised young, except for Golden Eagle and Peregrine (though elsewhere records are available for these species).

TABLE 22

Incubation and nestling periods (in days) of temperate zone raptors compared with those of closely-related tropical forms. (Partly after Brown 1976, otherwise references for temperate zone species as in Table 18 and tropical species as listed)

Temperate Zone			Tropical			
Species (wt in gm)	Incubation	Nestling	Species (wt in gm)	Incubation	Nestling	
White-tailed Eagle (6000)	38	70–84	African Fish Eagle (2400)	43	70–75	Steyn 1973
Snake Eagle (2070)	47	70–75	Black-breasted Snake Eagle (2200)	51–52	90–100	
Common Buzzard (940)	33–35	42–49	Augur Buzzard (1200)	38	52–60	Laing 1965, Weaving 1972
Red-tailed Hawk (1220)	32–35	43–48	Galapagos Hawk (1350)	37–38	50–60	de Vries 1975
Lesser Spotted Eagle (1500)	43	58	Wahlberg's Eagle (1000)	45–46	72–80	Brown 1952–53, 1955
Golden Eagle (4700)	43	65–77	Black Eagle (5000)	44–47	91–98	Gargett 1971
Bonelli's Eagle (2500)	37–39	65	African Hawk Eagle (1500)	42–44	61–71	Steyn 1975
Booted Eagle (700)		63	Ayre's Hawk Eagle (900)	45	73–75	Brown 1952–53, 1955
Merlin (215)	28–32	26–32	Red-headed Falcon (220)	34–35	37–38	Koehler 1970

TABLE 23
Breeding success of raptors not seriously affected by pesticides

Species	Location (years)	Number of pairs on territory (years combined)	Number (%) of nests in which at least 1 egg laid	Number of nests in which at least 1 egg hatched	Number (%) of nests from which at least 1 young flew	Mean clutch size	Mean brood size at hatch in successful nests	Mean brood size at fledging in successful nests	Mean number of young per clutch started	Mean number of young per territorial pair	Reference
Osprey	Virginia (1934)	176	47		37 (79)			2·0	1·6		Tyrrell 1936
	Washington State (1972)	404	151 (86)		97 (64)	2·5		2·0	1·3	1·1	Melquist & Johnson 1973
	Florida (1968–74)	45	336 (83)		221 (66)			1·2	0·8	0·7	Ogden 1975
	Scotland (1954–71)	329	35 (78)		21 (60)			2·2	1·3	1·1	Weir, in Brown 1976a
	Sweden (1971–73)	22	300 (91)		214 (71)	2·9	2·5	2·1	1·5	1·4	Odsjö & Sondell 1976
Honey Buzzard	Denmark (1938–42)		14 (64)		10 (63)	2·0		1·6	1·0		Holstein 1944
	Germany (1940–51)		16					1·9			Wendland 1953
Black Kite	Germany (1940–51)	173			112					1·2	Fiuczynski & Wendland 1968
Red Kite	France (1966)	27	23 (85)		20 (87)			1·1	1·1	0·9	Thiollay 1967
	Wales (1951–72)	395	321 (81)		156 (49)	2·1		0·7	0·7	0·5	Davies & Davis 1973
Bald Eagle	Florida (1939–46)	392	317 (81)		283 (89)			1·7	1·4	1·0	Broley 1947
	Saskatchewan (1969)	54	51 (94)		36 (71)			1·9		1·3	Whitfield et al 1974
Marsh Harrier	France (1966)	20	11 (55)		7 (64)			3·7	2·4	1·3	Thiollay 1967
	England (1953–71)		49		42 (86)			2·6	2·3		Axell, in Brown 1976a
	Finland (1963–64)		79		50 (63)			3·2	1·5		Hilden & Kalinainen 1966
	Sweden (1956–59)		26		18 (69)			3·3	2·2		Bengston 1967
Hen Harrier	North Dakota (1937–39)	16	60					1·0	1·0		Hammond & Hendry 1949
	Michigan (1942, 1948)		15 (94)					1·2	1·2	1·1	Craighead & Craighead 1956
	Scotland, Orkney (1944–56)		299			4·0		1·3	1·3		Balfour 1957
	Norway (1938–46)		35			4·6		2·0	2·0		Hagen 1969
	Wisconsin (1959–65)		80	61	46 (58)	4·6		1·9	1·9		Hamerstrom 1969
	Scotland, SW (1959–75)		51	41	31 (61)	4·7		2·7	1·7		Watson 1977
	Scotland, NE (1970–70)		39	28	22 (56)	4·7		3·1	1·7		Picozzi 1978
Goshawk	Germany (1940–51)		14		10 (71)			2·3	1·6		Wendland (1953)
	Germany (1942–70)		220		160 (73)	3·2		2·7	2·0		Kramer 1973
Sparrowhawk	Alaska (1971–73)	18	40	29	29 (73)		2·9	3·2	1·7		McGowan 1975
	Netherlands (1941–43)		18 (100)		10 (55)			3·0	2·0	1·7	Tinbergen 1946
	Germany (1916–70)		911		604 (66)						Kramer 1973
Cooper's Hawk	Michigan/Wyoming (1942, 1947–48)	18	14 (78)						2·9		Craighead & Craighead 1956
	Pennsylvania (1947–59)	24	23 (96)		20 (87)	4·2		3·3	2·9	2·3	Schriver 1969[1]
	Eastern U.S. (1880–1945)		36		29 (81)	4·2		3·5	2·8	2·8	Henny & Wight 1970

[1] Four territories (1947–59) considered here from eight (1947–64) listed by Schriver.

TABLE 23 (cont.)

	Location (years)	Number of pairs on territory (years combined)	Number (%) of nests in which at least 1 egg laid	Number nests in which at least 1 egg hatched	Number (%) of nests from which at least 1 young flew	Mean clutch size	Mean brood size at hatch in successful nests	Mean brood size at fledging in successful nests	Mean brood of young per clutch started	Mean number of young per territorial pair	Reference
Harris' Hawk	Arizona (1969–73)		50		34 (68)	3·0		2·3	1·6		Mader 1975a
	Texas (1975)	19	18 (95)		8 (44)	2·9		1·9	0·8	0·8	Griffin 1976
Red-shouldered Hawk	Michigan (1942, 1948)	40	39 (98)			3·5			1·8	1·8	Craighead & Craighead 1956
	Maryland (1960–71)	74			50			2·3		1·6	Henny *et al* 1973
	California (1973)		29	27	19 (66)	2·7	2·1	2·1	1·3		Wiley 1975
Swainson's Hawk	Wyoming (1964)	55	49 (89)	34	33 (67)	2·6	2·2	2·1	1·4	1·2	Dunkle 1977
Red-tailed Hawk	California (1939–40)		26	19	14 (54)			2·3	1·2		Fitch *et al* 1946
	Michigan/Wyoming (1942, 1947–48)	19	16 (84)			2·2			1·4	1·2	Craighead & Craighead 1956
	Wisconsin (1954–55)	66	60 (91)		40 (67)			1·9	1·3	1·2	Orians & Kuhlman 1956
	New York State (1951–52)	37			22			1·9			Hagar 1957
	Alberta (1967–68)	66	57 (86)		47 (82)	2·0	1·9	1·7	1·4	1·2	Luttich *et al* 1971
	Montana (1966–67)		54		27 (50)	2·9		1·7	0·9		Seidensticker & Reynolds 1971
	Wisconsin (1962–64)		31		20 (65)			1·8	1·1		Gates 1972
	Montana (1971–72)	137	121 (88)	78	73 (60)		2·6	2·6	1·6	1·4	Johnson 1975
	California (1973)		51	49	39 (76)	2·6	2·2	2·2	1·6		Wiley 1975
	Utah (1967–70)	25	20 (80)		14 (70)	2·9	2·3	1·7	1·2	1·0	Smith & Murphy 1973
Common Buzzard	Germany (1940–51)		170		117 (69)			1·4	1·0		Wendland 1952
	West England (1956–58)	40	29 (73)	21	14 (48)	2·4		1·3	0·6	0·5	Dare 1961
	Wales (1954–64)	51	40 (78)	28	26 (65)			1·7	1·1	0·9	Davis & Saunders 1965
	Germany (1955–60)		95	85	82 (86)	2·6		1·9	1·7		Mebs 1964
	England (1962–71)	215	178 (83)		150 (84)			1·5	1·2	1·0	Tubbs 1974
	All Britain (1937–69)		645		499 (77)	2·6		1·8	1·4		Tubbs 1971
	Scotland (1969–72)	135			80	3·0		2·6		1·6	Picozzi & Weir 1974
	Germany (1960–71)		1740		1129 (65)			1·6	1·1		Rockenbauch 1975
Rough-legged Buzzard	Norway (1938–46)		43						1·8		Hagen 1969
Ferruginous Hawk	Utah/Idaho (1972–73)	97			57	2·8				1·5	Howard & Wolfe 1976
	Utah (1967–70)	40	36 (90)		27 (75)	3·2	2·3	2·0	1·5	1·4	Smith & Murphy 1973
	South Dakota (1973–74)	31	27 (87)		16 (59)	4·3		2·9	1·8	1·5	Lokemoen & Duebbert 1976
Lesser Spotted Eagle	West Russia (1950's)		35	25	21 (60)	1·8		1·0	0·6		Goluduschko 1961
	Slovakia (1968–69)		17		13 (76)	1·8		1·0	0·8		Meyburg 1970
Golden Eagle	Scotland (1950–56)	67			18			1·4		0·4	Sandeman 1957
	Scotland (1945–57)	64	55 (86)	40	40 (73)			1·3	0·9	0·8	Watson 1957
	Scotland (1964–68)	489	395 (81)		247 (63)[2]			1·2[2]	0·7[2]	0·6[2]	Everett 1971
	Utah (1967–73)	112	61 (54)		50 (82)			1·6	1·3	0·7	Murphy 1973
	Idaho (1967–69)		146		93 (64)	2·1	1·8	1·6	1·1		Beecham & Kochert 1975
	Scotland (1964–71)	167	105 (63)		63 (60)			1·1	0·7	0·4	Weir, in Brown 1976a
	Scotland (?)	114	84 (74)		51 (61)			1·1	0·6	0·5	Merrie, in Brown 1976a
Wedge-tailed Eagle	Australia (1964)	15	13 (87)		10 (77)	2·0		1·2	0·9	0·8	Leopold & Wolfe 1970
[illegible]	[illegible]	[illegible]	[illegible]	[illegible]	[illegible]	[illegible]		[illegible]	[illegible]	[illegible]	[illegible]

	Species / Locality (years)								Source
European Kestrel	Utah (1967–70)	9		6 (67)		2·8	3·4	4·7	Smith et al 1972
	Utah (1967–70)	22		9 (100)		3·0	1·9	5·2	Smith & Murphy 1973
	Germany (1940–51)		35	298 (79)		4·1	2·0	5·0	Wendland 1953
	Netherlands (1960–64)[3]		375	11 (73)		3·2		4·0	Cavé 1968
	England (1964–67)		15	59 (69)		2·8	1·8	5·0	Shrubb 1970
	Scotland (1976–78)	94	86 (91)	19 (63)		4·0	2·7	3·5	Village 1979
Red-footed Falcon	Hungary (1953)		30	17 (53)		2·5	1·6		Horvath 1955
Hobby	Germany (1941–51)		32			2·1	1·1		Wendland 1953
	Germany (1956–75)	358		277		2·4			Fiuczynski 1978
Prairie Falcon	Colorado/Wyoming (1960–62)	77	67 (87)	32 (48)	2·7	2·9	2·5	4·5	Enderson 1964
	Idaho (1970–72)	128	110 (86)	91 (83)	4·5	3·7	3·1	4·3	Ogden & Hornocker 1977
Peregrine	New York area (1939–40)	38	29 (76)	16 (55)	4·3	2·5	1·4		Hickey 1942
	England (1939–40)	34		24		2·4	1·7		Ryves 1948
	Germany (1940–51)		17	4 (24)		2·0	0·5		Wendland 1952
	Colville River, Alaska (1950–58)	25		14		2·5	1·4		Cade 1960
	Yukon River, Alaska (1950–58)	20		14		1·9	0·8		Cade 1960
	Yukon River, Alaska (1966)	17		12		2·3	1·8		Cade et al 1968
	Hudson River area, eastern U.S.A. (1931–49)	72	62 (86)	28 (45)	3·1	2·4	1·1	0·9	Herbert & Herbert 1969
	Pennsylvania (1939–46)	65	44 (68)	35 (80)	2·1	2·3	1·8	1·3	Rice 1969
	Scotland (1964–71)	125	94 (75)	70 (74)		2·4	1·7	1·3	Weir, in Brown 1976a
	Aleutian Is. (1970–72)	57		38		2·7		1·8	White 1975
Tropical and sub-tropical regions									
Black-shouldered Kite	South Africa	27	18 (67)	7 (39)	3·1	2·7	1·1	0·7	W. Tarboton
Fish Eagle	Tanzania (1953–55)	21	12 (57)	9 (75)	2·1	1·3	1·0	0·6	Brown 1960
	Kenya (1968–71)	112	71 (64)	41 (58)		1·3	0·7	0·5	Brown & Hopcroft 1973
	Uganda (1976)	100		37		1·1	1·1	0·4	Thiollay & Meyer 1978
Lappet-faced Vulture	Zimbabwe Rhodesia (1975)		18	12 (67)	1·0	1·0	1·0	0·7	Anthony 1976
	Tanzania (1972–73)	31	55	ca 23 (42)	1·0	1·0	0·4		Pennycuick 1976
Brown Snake Eagle	Kenya (1949–53)	10	6 (60)	3 (50)	1·0	1·0	1·0	0·5	Brown 1955
Bateleur	Kenya (1949–53)	8	7 (88)	4 (57)	1·0	1·0	1·0	0·6	Brown 1955
	Zimbabwe Rhodesia	14	14 (100)	11 (79)	1·0	1·0	1·0	0·8	Brown et al 1977
Tawny Eagle	Zimbabwe Rhodesia (1963–71)	19	16 (84)	12 (75)	2·0	1·0	1·0	0·8	Steyn 1973
	Kenya (1970–72)	15		7		1·0		0·6	Smeenk 1974
Wahlberg's Eagle	Kenya (1949–53)	32	26 (81)	20 (77)	1·0	1·0	1·0	0·8	Brown 1955
	Kenya (1970–72)	12	9 (75)	7 (78)	1·0	1·0	1·0	0·6	Smeenk 1974
	South Africa (1974–76)	33	29 (88)	13 (45)		1·0	0·5	0·6	Tarboton 1977
	Zimbabwe Rhodesia	63	55 (87)	35 (64)		1·0	0·9	0·4	Brown et al 1977
Black Eagle	Zimbabwe Rhodesia (1964–76)	652	442 (68)	339 (77)	1·7	1·0	1·2	0·5	Gargett 1977
African Hawk Eagle	Kenya (1949–65)	23	16 (70)	11 (69)		1·0	1·0	0·6	Brown 1966
	Kenya (1970–72)	11	2 (18)	2 (100)		1·0	1·0	0·1	Smeenk 1974
	Zimbabwe Rhodesia (1961–72)	21	17 (81)	10 (59)	1·8	1·0	1·0		Steyn 1975
Ayre's Hawk Eagle	Kenya (1949–65)	16	13 (81)	10 (77)		1·0	1·0	0·8	Brown 1966
Crowned Eagle	Kenya (1949–65)	34	18 (53)	15 (53)	1·3	1·0	0·8	0·8	Brown 1966
Martial Eagle	Kenya (1949–65)	31	18 (58)	13 (73)	1·0	1·0	1·0	0·7	Brown 1966
Galapagos Hawk	Galapagos (1969–70)		34		1·9	1·0		0·7	de Vries 1975

[2] adjusted to account for pairs with unknown success; [3] calculated from tables 27 and 29 in Cavé.

In this table, statistics have sometimes been calculated and presented differently from those in the original paper, so as to make them comparable with one another.

TABLE 24

Breeding of Rough-legged Buzzard in relation to vole population levels at Dovr
Norway, 1938—40. From Hagen 1969

Vole population level	Stage of breeding reached	Usual number young reared per pair
Extreme low	No territorial behaviour or pair formation.	0
Low	Territory and pair formation, some nest building, no eggs laid.	0
Below average	Eggs laid. Clutches of 2–4 eggs, soon deserted.	0
Average	Incubation, eggs deserted or hatched, young normally die.	0
Above average (a) Increasing (b) Decreasing	Eggs hatched (a) Few young reared. (b) Young reared.	0–1
High	Young reared. Clutches of 3–6 eggs, broods of 2–4 young.	2–4
Extreme high	Young reared. Clutches of 4–7 eggs, broods of 4–5 young.	4–5

TABLE 25

Numbers and breeding success of four species of raptor in relation to vole population levels, Dovre, Norway. Species arranged in order of importance of voles in diet. From Hagen 1969

Vole population	Rough-legged Buzzard		European Kestrel		Hen Harrier		Merlin	
	Mean No. nesting pairs/year	No. young reared per pair	Mean No. nesting pairs/year	No. young reared per pair	Mean No. nesting pairs/year	No. young reared per pair	Mean No. nesting pairs/year	No. young reared per pair
Low (1939, 1943)	0	0	1·0	(3)	0·5	(3)	4	1·5
Medium (1944, 1946)	5	0·3	4·5	1·8	6·0	1·2	3·5	2·6
High (1938, 1941, 1942, 1945)	8·3	2·2	11·0	3·6	7·3	2·1	7	3·3

TABLE 26

Nesting success of Common Buzzards in relation to differing vole population densities, near Castell, Germany. Vole densities in spring varied by about 5-fold, with densities of 3·5 animals, per 1,000 m² in a low year, 6·5 in a medium year, and 18·3 in a high year. From Mebs 1964

Vole population	No. clutches	No. in which at least 1 egg hatched	No. in which at least 1 young fledged	Mean laying date	Mean clutch-size	Mean brood-size at fledging per successful nest	Mean number young fledged per nest
Low (1956 and 1960)	31	29	26	14 Apr	2·1	1·6	1·4
Medium (1955 and 1958)	34	31	31	13 Apr	2·4	1·9	1·8
High (1950 and 1959)	30	25	25	30 Mar	3·2	2·4	2·0

TABLE 27

Variations in breeding output of species that are subject to fluctuating food supplies

A. *Species that eat rodents*

Rough-legged Buzzard	*(1)	0·3–2·8 young reared per pair, during 8 years, North Norway (Hagen 1969).
Common Buzzard	(1)	0·6–1·1 young reared per pair during 10 years, Germany (Wendland 1952).
	*(2)	1·4–2·1 young reared per nest during 6 years, Germany (Mebs 1964).
	*(3)	0·4–2·1 young reared per nest during 12 years, Germany (Rockenbauch 1975).
Hen Harrier	*(1)	1·2–3·0 young reared per pair during 7 years, North Norway (Hagen 1969).
	*(2)	1·6–3·3 young reared per nest during 10 years, Wisconsin (Hamerstrom 1969).
European Kestrel	*(1)	1·4–3·7 young reared per pair during 6 years, North Norway (Hagen 1969).
	*(2)	0·8–3·7 young reared per pair during 3 years, South Scotland (Village 1979).

B. *Species that eat game birds or hares*

Goshawk	(1)	0·5–3·2 young reared per nest during 7 years, Sweden (Höglund 1964).
	*(2)	1·8–2·5 young reared per nest during 3 years, Alaska (McGowan 1975).
Gyr Falcon	*(1)	0·9–2·9 young reared per nest during 5 years, Alaska (Swartz *et al* 1975).
Golden Eagle	*(1)	0·3–1·1 young reared per pair during 7 years, Utah (Murphy 1974).
Ferruginous Hawk	*(1)	1·0 young per pair in a poor food year, 1·9 in a good one, Utah/Idaho (Howard & Wolfe 1976).
	*(2)	0·6–3·0 young per pair during 8 years, Utah (Woffinden & Murphy 1977).

* Prey population also assessed and related to raptor breeding success.

TABLE 29

Breeding success of Buzzards in territories with contrasting food supply; difference significant at 10 % level. From Dare 1961

	No. nests	Total chicks		Mean number young
		hatched	fledged	fledged/nest
Rabbits abundant	9	21	12	1·33
Rabbits scarce	10	24	6	0·60

TABLE 28

*eding parameters of Sparrowhawks in four areas in Southern Scotland, 1971–73. All
ects of breeding were better on low ground, where prey were plentiful, than on high
ground, where prey were scarce. From Newton 1976*

	Esk valley woods	Annan valley woods	Annan hillside forests	Annan large hill-forest
nests found	85	205	46	59
nests laid in	79–81	176–182	33–36	47–48
nests in which at least egg hatched	59	133	24	30
nests in which at least young fledged*	55 (65%)	120 (59%)	23 (50%)	26 (44%)
an laying date***	7 May	9 May	12 May	16 May
an clutch-size***	5·3	4·8	4·5	4·5
an brood-size*	3·6	3·3	3·1	2·7
of young per nest built	2·4	1·9	1·6	1·2
yearlings among breeding ₃males*	25	22	13	5

as are arranged in order of declining breeding success, all years combined. Asterisks show the number of
·s out of three in which the area differences were significant statistically. The same trends were apparent in all
·s.

TABLE 29 (*see facing page*)

TABLE 30

*Effects on numbers and breeding of Buzzards of elimination of rabbits (through
myxomatosis epidemic), Avon Valley, Devon. Note the complete lack of
success in 1955. From Dare 1961*

	Pre-myxomatosis 1954	Post-myxomatosis	
		1955	1956
Pairs resident in April	21	14	12
building nest	19	13	9
laying eggs	19	1	8
hatching eggs	17	0	5
fledging young	17	0	5
No. young reared	28–33	0	7
No. young per territorial pair	1·3–1·7	0	0·6
No. young per breeding pair	1·5–1·9	0	0·9

TABLE 31

Response of certain raptors to variations in food supply. Number of asterisks indicate strength of response and contribution to total breeding output. A dash indicates an apparent lack of response, and a question mark a lack of information

Response in good food years compared to poor ones	Species that eat rodents					Species that eat game birds or hares		
	Rough-legged Buzzard	Common Buzzard	Hen Harrier	European Kestrel	Golden Eagle	Goshawk	Gyr Falcon
Greater density of territorial pairs	***	*	**	**	—	**	**
Greater proportion of 'young' birds among breeders	?	?	**	**	?	**	?
Greater proportion of pairs producing clutches	***	**	*	*	***	*	**
Greater proportion of clutches producing young	***	*	***	***	—	**	**
Earlier laying dates	**	**	*	**	?	*	**
Larger clutches or broods	***	*	**	—	*	**	**
Better parental care	*	?	?	?	?	?	?
More repeat laying by failed pairs	?	*	?	?	—	—	—

Based on Hagen (1969) for Rough-legged Buzzard, Mebs (1964) and Holdsworth (1971) for Buzzard, Hagen (1969), Hamerstrom (1969) and Balfour (1957, 1962) for Hen Harrier, Snow (1968), Hagen (1969), Cavé (1968) and Village (1979) for Kestrel, Murphy (1974) for Golden Eagle, Sulkava (1964), Höglund (1964) and McGowan (1975) for Goshawk, Swartz et al 1974 for Gyr Falcon.

TABLE 32

Nest success in relation to situation and vulnerability to mammalian predators. In each case, the more accessible nests were less successful

(a) Merlin, northern England (Newton, Meek & Little 1978)

	No. nests	No. successful
Ground nests	112	70 (63%)
Tree nests	31	29 (94%)
Significance of difference		$P < 0.001$

(b) Osprey, Connecticut (Ames & Mersereau 1964)

	No. nests	Total eggs laid	Total young fledged
Ground nests[1]	33	78	1 (1%)
Raised nests[2]	125	362	35 (10%)
Significance of difference		$P < 0.5$	

(c) Prairie Falcon, Idaho (Ogden & Hornocker 1977)

	No. nests	No. successful
Cliff nests accessible to mammals	17	9 (53%)
Cliff nests inaccessible to mammals	93	82 (88%)
Significance of difference		$P < 0.1$

The success of the Osprey population was already very low from pesticide contamination.
[1] Includes nests on flat ground or on fallen logs.
[2] Includes nests on duck blinds, artificial nest platforms and trees.

TABLE 33

Influence of human predation on nesting success of Red-tailed and Red-shouldered Hawks in California. From Wiley 1975

	No. of nests	No. nests hatched	No. nests fledged (%)	No. young/nest started (±s.e.)	Causes of failure			
					Human disturbance[1]	Human predation[2]	Natural predation	Others and unknown
Red-tailed Hawk								
Disturbed area*	20	17	8 (40%)	0·95±1·23	2	9	0	1
Undisturbed area*	33	32	31 (94%)	2·06±0·93	0	0	1	1
Red-shouldered Hawk								
Disturbed area*	18	16	10 (56%)	1·06±1·11	2	5	0	1
Undisturbed area*	11	11	9 (82%)	1·82±1·08	0	0	2	0

* Disturbed area was within 0·25 mile of road; undisturbed more distant.
Difference between areas significant in Red-tailed Hawk only (P < 0·001).
[1] Caused by land development.
[2] Spike marks in tree; young taken by falconers.

TABLE 34

Presumed causes of nest failure in some raptors. Some references expressed findings in terms of clutches, others in terms of individual eggs

	Red-footed Falcon	Sparrowhawk	Prairie Falcon	Buzzard	Red Kite	Bald Eagle
Number of eggs or clutches	105 eggs	335 clutches	302 eggs	645 clutches	231 clutches	619 clutches
Number (%) failed	57 (54)	111 (33)	117 (39)	146 (23)	137 (59)	171 (28)
Eggs addled or infertile[1]	14	16	34	32	15	25
Desertion of eggs	3	16	—	—	—	—
Desertion of young		1				

	Horvath 1955	Newton 1976	Ogden & Hornocker 1977	Tubbs 1971	Davies & Davis 1973	Broley 1947
Predation of young[2]	23	4	6	—	—	1
Eggs broken[3]	—	26	9	21	—	—
Nestlings died	—	—	4	7	—	3
Interaction with other species[4]	—	—	3	—	2	31
Adult died naturally	—	—	—	—	3	1
Adverse weather[5]	—	—	—	1	4	45
Collapse of nest tree or nest	—	2	—	1	—	8
Human predation or deliberate disturbance[6]	—	23	—	49	20	15
Nest tree felled	—	4	—	2	—	11
Felling or other incidental disturbance nearby	—	—	—	—	6	10
Unknown[7]	—	16	42	33	87	19

[1] Incubated for full time, but not hatched.
[2] In Red-footed Falcon mainly by neighbouring Rooks *Corvus frugilegus* at egg stage and by Goshawks at nestling stage; in Sparrowhawk mainly by Tawny Owls *Strix aluco* at nestling stage and in Prairie Falcon mainly by mammalian predators.
[3] In Prairie Falcon mainly accidental, and in Sparrowhawk and Buzzard associated with shell-thinning and DDE contamination.
[4] In Prairie Falcon with Barn Owls *Tyto alba*, in Kite with Ravens *Corvus corax*, and in Bald Eagle with Great Horned Owls *Bubo virginianus*.
[5] In Buzzard nest-flooded, in Kite prolonged heavy rain in nestling period, and in Bald Eagle a hurricane which was followed by much non-breeding and failure of eggs to hatch.
[6] Includes taking of eggs and young, or shooting of adults.
[7] Includes mainly instances where nest-contents disappeared without trace.
The table excludes non-laying which influenced about 12% of all territorial pairs of Sparrowhawks, about 14% of Prairie Falcons, 16% of Kites, 13% of Bald Eagles, and an unknown % of Red-footed Falcons and Buzzards.

TABLE 35

Number of young fledged from Kestrel broods of differing sizes in Holland. Early broods were from eggs laid up to 30 April, and late broods from eggs laid after 30 April. From Cavé 1968

	Early Broods					Late Broods			
Initial brood-size	1–3	4	5	6	7	1–3	4	5	6
Frequency	9	14	47	53	11	25	44	61	34
% young fledged	71	89	90	83	75	82	85	85	75
No. fledged per brood	1·7	3·6	4·5	5·0	5·3	2·1	3·4	4·2	4·5

The difference in survival between early and late young is significant only in broods of 6.

TABLE 36

Use of alternative nests by Common Buzzards and Red Kites according to breeding performance in the previous year. In all three cases, birds were more likely to return for another season in the same nest after a successful attempt than after a failure, though the overall tendency to return differed between studies. From Tubbs 1974, Holdsworth 1971 and Davies & Davis 1973

	Buzzard, New Forest		Buzzard, Sedbergh area		Red Kite, central Wales	
	Nest retention	Nest change	Nest retention	Nest change	Nest retention	Nest change
After success	54	76	8	8	52	13
After failure	2	21	2	9	27	41

New Forest Buzzard figures significant at 1% level, Red Kite figures significant at 0·1% level.

TABLE 37

Distance from birthplace of Black Kites recovered in the breeding season (20 April–21 July). From Schifferli 1967

Year of age	Recoveries	Distance from birthplace (km)					Mean distance (km)
		0–100	101–200	201–500	501–1000	1001–2000	
1	24	2(8%)	2	8	4	8	705
2	25	7(28%)	3	11	1	3	350
3	17	8(47%)	2	2	5	0	251
4+	11	10(91%)	0	1	0	0	66

TABLE 38

Dispersal of Sparrowhawks, based on results from a local study in southern Scotland. Longer movements are under-represented because of the failure to record birds that settle outside the area but, within the area, results are comparable between the different sex and age groups. From I. Newton & M. Marquiss

Distance away (km)	Recoveries in later breeding seasons of birds ringed as:		
	Adult males	Adult females	Nestlings*
Same territory	10	68	0
1–5	13	53	25
6–10	1	7	22
11–15	0	1	15
16–20	1	0	4
21–25	0	0	3
26–30	0	1	4
31–35	0	0	4
36–40	0	0	1
41–45	0	0	0
46–50	0	0	1

*Includes first recapture only; one bird bred in its natal territory after first breeding elsewhere.

TABLE 39

Dispersal of various raptors from birthplace to presumed breeding place

Species (region, number)	% of all breeding season ring recoveries within following distances (km) from birthplace					Reference
	20	50	100	200	500	
Osprey (Fennoscandia, 79)	?	43	59	67	77	Österlöf 1977
Black Kite (Switzerland, 11)	?	?	91	?	100	Schifferli 1967
Goshawk (Sweden, 55)	31	71	84	93	98	Hoglund 1964
Sparrowhawk (Britain, 138)	76	98	99	100	100	B.T.O. records
Buzzard (Britain, 15)	73	93	100	100	100	B.T.O. records
(Germany, 45)	?	62	?	98	100	Mead 1973
(Denmark 10)	30	60	80	100	100	Nielsen 1977
(Fennoscandia, 15)	67	95	100	100	100	Olsson 1958
Kestrel (Britain 18)	61	89	89	100	100	Snow 1968
(Holland 150)	?	63	80	93	100	Cavé 1968
(Switzerland 35)	?	43	46	51	91	Schifferli 1965
Merlin (Britain, 16)	38	75	81	100	100	Mead 1973**
Hobby (Germany, 12)*	92	100	100	100	100	Fiucyynski 1978

* Probably all males, females disperse further; **amplified by letter.

TABLE 40

Mean distance (km ± s.e.) between birthplace and presumed breeding place in various raptor species. From Galushin (1974), based on recoveries from Europe and Russia of birds ringed as nestlings and recovered at least two years later. Most recoveries refer to areas east of those in Table 39; sample sizes not given.

Hobby	39±14
Sparrowhawk	41±21
Black Kite	49±36
Goshawk	60±16
Osprey*	82±53
Buzzard	90±21
Marsh Harrier	174±137
Kestrel	192±43
Montagu's Harrier	199±58
Honey Buzzard	997±375
Rough-legged Buzzard	1955±1079

* Excludes one bird which moved exceptionally far from Sweden to Scotland.

TABLE 41

Difference in the dispersal distances of male and female Sparrowhawks between birthplace and presumed place of first breeding. From Newton, unpublished, based on British ring recoveries

	Dispersal distance	
	Less than 10 km	More than 10 km
Males	20	11
Females	9	20
Significance of difference, $P < 0.01$		

TABLE 42

Fidelity of female Sparrowhawks to nesting territory of previous year. Comparing successive years, females more often stayed on the same territory in good habitat than in poor (P < 0·05), and within habitats, they more often stayed after a successful breeding attempt than after a failed one (P < 0·001 and P < 0·05 in intermediate and poor habitat). From I. Newton and M. Marquiss, unpublished

| | In second year, female in | |
	same territory	different territory
Good habitat		
After success	11	0
After failure	1	0
Intermediate habitat		
After success	26	9
After failure	2	12
Poor habitat		
After success	16	11
After failure	2	11

TABLE 43

Four-year cycles of two prey-predator pairs in South Norway, 1870–1912. From Johnsen 1929, Elton 1942, Lack 1951, all summarised by Lack 1954

| | Years of peak population | | |
Lemming Lemmus lemmus	Rough-legged Buzzard	Willow Grouse	Goshawk
1871–72	1872	1872	1871–73
1875–76	1875	1876	1876
1879–80	1879–80	1880	1880
1883–84	1883–84	1883	1883–84
1887–88	1887–88	1887	1889
1890–91	1891	1891	1891
1894–95	1894–95	1895	1895
1897	1898–99	1897	1899
1902–03	1902	1903	1901–02
1906	1906	1906	1905–06
1909–10	1909–10	1908–09	—
		1911–12	1911–12

TABLE 44

Years of recorded invasions into eastern North America of four predatory birds. In t
ten-year cycle, Great Horned Owls usually appeared in greatest numbers a year lo
than Goshawks. From Spiers 1939, Gross 1947, Keith 1963, Hanson 1971, Mueller et
1977

1. Approximately four-year cycle of species which eat small rodents		2. Approximately ten-year cycle of species which eat hares and game bird	
Rough-legged Buzzard	Snowy Owl	Goshawk	Horned Owl
?	1882	1886	1887
?	1886	1896	1897
?	1889	1906	1907
?	1892	?	1916
?	1896	1926	1927
?	1901	1935	1936
?	1905	1945	?
?	1909	1954	?
?	1912	1962	?
1917	1917	1972	?
?	1921		
1926	1926		
1930	1930		
1934	1934		
1937	1937		
?	1941		
?	1945		
?	1949		
?	1953		
?	?		
?	1960		
?	1963		
?	1966		

TABLE 45

Wintering areas of European raptors (west of 45°E). A—amphibia, B—birds, C—carrion and refuse, F—fish, I—insects and other invertebrates, M—mammals, R—reptiles

Species	Main Foods	Wintering Areas North of Sahara	Wintering Areas South of Sahara
Osprey*	F	+	+
Honey Buzzard*	I		+
Black-shouldered Kite	M	+	
Black Kite*	FCIM		+
Red Kite	MCB	+	
White-tailed Eagle	FBC	+	
Egyptian Vulture*	CR		+
Lammergeier	C	+	
Griffon Vulture*	C	+	
Lappet-faced Vulture	C	+	
Black Vulture	C	+	
Short-toed Eagle*	R		+
Marsh Harrier*	BM	+	+
Hen Harrier	MB	+	
Pallid Harrier*	MRI	+	+
Montagu's Harrier*	RMA		+
Goshawk	B	+	
Sparrowhawk*	B	+	
Levant Sparrowhawk	BRI		+
Common Buzzard*	MB	+	+
Rough-legged Buzzard	M	+	
Long-legged Buzzard*	M	+	+
Lesser Spotted Eagle*	MIAR		+
Greater Spotted Eagle*	MIAR	+	+
Steppe Eagle*	MI	+	+
Imperial Eagle*	M	+	+
Golden Eagle	MB	+	
Bonelli's Eagle	BM	+	
Booted Eagle*	BM		+
Lesser Kestrel*	I		+
Kestrel*	MI	+	+
Red-footed Falcon*	I		+
Merlin	B	+	
Hobby*	IB		+
Eleonora's Falcon	IB		+
Sooty Falcon*	IB		+
Lanner Falcon*	BM	+	
Saker Falcon*	BM	+	+
Gyrfalcon	B	+	
Peregrine*	B	+	

* In all these species, at least some individuals from more eastern Palearctic breeding populations are known to reach Africa south of the Sahara in winter.
Details mainly from Moreau 1972.

TABLE 46

Counts of Palearctic raptors on migration. From Bernis et al 1975 (Gibraltar), Beaman et al 1979 (Bosphorus), Beaman et al 1979 (eastern Black Sea) and Christensen et al 1979 (northern Red Sea)

	Gibraltar 1 Aug–14 Oct 1972	Bosphorus 13 Aug–8 Oct 1971	Eastern Black Sea 17 Aug–10 Oct 1976	Northern Red Sea (Elat) 20 Feb–17 May 1977
Osprey		10	20	120
Honey Buzzard	117000	26000	138000	226000
Black Kite	39000	2700	5800	27000
Red Kite		<10		
Egyptian Vulture	3700	550	<10	800
Griffon Vulture		120	<10	20
Snake Eagle	8800	2300	240	220
Marsh Harrier	360	<10	390	120
Hen Harrier		<10	<10	<10
Pallid Harrier		<10	130	<10
Montagu's Harrier	1700	10	120	<10
Goshawk		<10	20	<10
Sparrowhawk	920	130	690	150
Levant Sparrowhawk		5700	290	6000
Buzzard	2600	19000	205000	316000
Long-legged Buzzard			<10	30
Lesser Spotted Eagle		19000	740	70
Greater Spotted Eagle		<10	20	<10
Steppe Eagle			270	19000
Imperial Eagle		<10	<10	100
Booted Eagle	14400	530	470	180
Lesser Kestrel	540	10	50	30
Kestrel	1200	30	30	10
Red-footed Falcon		10	20	<10
Hobby	220	220	190	<10
Eleonora's Falcon		<10		<10
Saker Falcon		10	10	
Unidentified	4000	750	27000	9000*
Approximate totals	194000	77000	38000	605000

* Almost all eagles.

TABLE 47

Longevity records of some raptors in captivity and in the wild. Figures show the year in which they died

In captivity

King Vulture Sarcorhamphus papa 30, Andean Condor 52, Black Kite 22, Red Kite 38, Whistling Eagle Haliastur sphenurus 21, Brahminy Kite H. indus 19, White-tailed Eagle 42, Bald Eagle 36, Steller's Sea Eagle Haliaeetus pelagicus 32, Egyptian Vulture 23, Lammergeier 26, Griffon Vulture 37, Lappet-faced Vulture 25, European Black Vulture 25, Bateleur 55 (still alive), Goshawk 19, Savannah Hawk Heterospizias meridionalis 19, Grey Eagle-buzzard Geranoaetus melanoleucus 18, Red-backed Buzzard Buteo polyosoma 23, Buzzard 30, Golden Eagle 48, Bonelli's Eagle 20, Yellow-throated Caracara Daptrius ater 26, Carunculated Caracara Phalcoboenus carunculatus 20, Common Caracara 37, European Kestrel 12, Red-footed Falcon 13, Merlin 8.

In the wild

Osprey 25, Honey Buzzard 16, Black Kite 24, Red Kite 26, Short-toed Eagle 18, Marsh Harrier 17, Hen Harrier 14 (Europe), 18 (North America), Goshawk 19, European Sparrowhawk 15, Red-shouldered Hawk 13, Swainson's Hawk 13, Red-tailed Hawk 18, Buzzard 26, Ferruginous Hawk 19, American Kestrel 11, European Kestrel 17, Merlin 8, Hobby 13, Prairie Falcon 14, Peregrine 17.

Details from Flower 1923, 1938, Rydzewski 1962, Brown & Amadon 1968, Lucas 1970, Österlöf 1977, and references in Table 49.

TABLE 48

Life table of Kestrels ringed as nestlings in the Netherlands between 1911 and 1955, and recovered up to 1965. From Cavé 1968

Year class	Number reported dead	Number alive	Annual mortality	Mean expectation of further life (years)
1	126	245	51%	1·5
2	49	119	41%	
3	34	70	49%	
4	14	36		
5	8	22		
6	4	14		42% 1·9
7	5	10	39%	
8	1	5		
9	2	4		
10	2	2		

Total recoveries—245; oldest bird 9·98 years; initial date 20 June, final date 31 December 1965.

TABLE 49
Annual mortalities of raptors calculated from ring recoveries

Species	Region of ringing	Minimum % killed by man	Oldest bird (year of death)	First-year % Mortality	First-year Mean expectation of further life[1]	Second-year % Mortality	Second-year Mean expectation of further life[1]	Later years % mortality	Later years Mean expectation of further life[1]	Reference
Osprey	E. United States	40	18	57	1·3			18	5·1	Henny & Wight 1969
Black Kite	Switzerland	80	24					30	2·8	Schifferli 1967
Hen Harrier	United States	90	17					30	2·8	Hickey 1952
	Britain		13	62	1·2			30	2·8	Watson 1977[2]
Goshawk[3]	Finland, Sweden	92	10	63	1·1	33	2·5	19[4]	4·8	Haukioja & Haukioja 1970
	Germany		19	58	1·1	38	2·1	38[5]	2·1	Kramer 1973[2]
	Finland			64	1·1	35	2·4	18[5]	5·1	Saurola 1976
European Sparrowhawk	Germany		12	62	1·1			42	1·9	Kramer 1973[2]
	W Europe		8	68	1·0			51	1·5	Tinbergen 1946
	Denmark	71	15	63	1·1			40	2·0	Shelde 1960
	Britain	48	8	70	0·9			57	1·3	Newton 1975
Cooper's Hawk	E United States (1925–40)	75	7	83	0·7			44	1·8	Henny & Wight 1972
	E United States (1941–57)	73	8	78	0·8			34	2·4	Henny & Wight 1972
Red-shouldered Hawk	E United States		13	58	1·2			31	2·7	Henny 1972
Red-tailed Hawk	N America, above 42°N	71[6]	18	62	1·1			21	4·3	Henny & Wight 1972
	N America, below 42°N		13	67	1·0			24	3·7	Henny & Wight 1972
Common Buzzard	Fennoscandia	62	18	57	1·3	30	2·8	20	4·5	Olsson 1958
	Germany	50–80	26	51	1·5	32	2·6	19	4·8	Mebs 1964
American Kestrel	United States	48	11	65	1·0			49	1·5	Roest 1957
	United States	18	15	69	1·1			47	1·6	Henny 1972
European Kestrel	Britain	23	10	60	1·2			34	2·4	Snow 1968
	Netherlands		10	51	1·5	47	1·6	42	1·9	Cavé 1968
	Switzerland	65	17	50	1·5			35	2·4	Schifferli 1964
	Finland		10		1·3			44	1·8	Nordstrom 1963
Prairie Falcon	W United States	62	14	74	0·8			25	3·5	Enderson 1969
Peregrine	Sweden	48	17	59	1·2			32	2·6	Lindberg 1977
	United States	45	14	70	0·9			25	3·5	Enderson 1969
	Finland	78	16	71	0·9			19	4·8	Mebs 1971
	Germany	43	14	56	1·3			28	3·1	Mebs 1971

[1] Calculated by the formula 2-m/2m, where m is the annual mortality as a fraction of unity.
[2] Calculated by me.
[3] The two Fennoscandian estimates are not independent, because they use some of the same recoveries.
[4] 19% in 3rd year, 17% in 4th, and 11% in later years.
[5] 18% in 3rd year, 15% in later years.
[6] First-year only.

TABLE 50

Causes of mortality in 231 Bald Eagles found dead in U.S.A., 1966–74. From Reichel et al 1969, Mulhern et al 1970, Belisle et al 1972, Cromartie et al 1975, Prouty et al 1977

	Numbers	% of total
Direct persecution		
Shot	83	36 ⎫
Trapped	4	2 ⎬ 43
Poisoned[1]	11	5 ⎭
Pesticide poisoning[2]		
Dieldrin	20	9 ⎫ 10
DDE/DDT	2	1 ⎭
Accidental		
Impact	26	11 ⎫
Electrocuted	11	5 ⎬ 21
Choked	2	1 ⎟
Drowned[3]	9	4 ⎭
Natural		
Disease[4]	19	8 ⎫ 12
Other natural[5]	10	4 ⎭
Uncertain	34	14

[1] Included 9 cases involving thallium, 1 strychnine and 1 lead.
[2] In other birds organo-chlorines were present, but judged insufficient to have caused death.
[3] Caught in mammal traps first.
[4] Included nephrosis (4 cases, one probably caused by mercury poisoning), Arteriosclerosis (1), streptococcal infection (1), aspergillosis (1), avian cholera (2), coccidiosis (6), pneumonia (1), visceral gout (1), and other infectious disease (2).
[5] Seven emaciated, two haemorrhage and one from faecal impaction.

TABLE 51

Raptors in the diets of other raptors and owls in Central Europe. From Uttendörfer 1952

Prey-species	Predator-species				
	Goshawk	Sparrowhawk	Peregrine	Eagle Owl	Tawny Owl
Total prey items (approximate)	9,000	59,500	6,400	5,800	58,500
Osprey				1	
Honey Buzzard	2			2	
Black Kite	3			2	
Red Kite				1	
Goshawk	14			9	
Sparrowhawk	87	46 (young)	3	4	6
Common Buzzard	16			51	
Rough-legged Buzzard				1	
Kestrel	113	2	2	28	5
Merlin	3				
Hobby	3		1	3	1
Peregrine	2		2 (young)	15	
Barn Owl	5			2	
Eagle Owl				12 (young)	
Pygmy Owl		1		1	
Little Owl	18	1		5	14
Tawny Owl	46	1 (young)	1	39	5 (young)
Long-eared Owl	179	2	1	69	3
Short-eared Owl	42		2	6	
Tengmalms Owl	1			5	

Other European raptors and owls took few or no raptors in their prey.
Scientific names of owls in order mentioned on left: Tyto alba, Bubo bubo, Glaucidium passerinum, Athene noctua, Strix aluco, Asio otus, Asio flammeus, Aegolius funereus.

TABLE 52
Reported causes of death (%) from ring recoveries of two British raptors in different periods. From Glue 1971

| | Sparrowhawk (N=166 & 71) | | Kestrel (N=175 & 457) | |
	1911–54	1955–69	1911–54	1955–69
Killed	60	16	41	10
Found dead	25	46	30	52
Collisions	13	32	17	28
Others	2	6	12	10

TABLE 53
Official Austrian game statistics on birds of prey killed between 1948 and 1968. From Bijleveld (1974), derived from Oesterreichisches Statistisches Zentralamt

	'Accipiters'	'Harriers'	'Buzzards'	Total
1948	10,943	386	3,482	14,811
1949	11,406	821	3,757	15,984
1950	13,181	973	5,252	19,406
1951	13,385	674	4,770	18,829
1952	13,533	757	4,152	18,442
1953	13,788	902	5,479	20,169
1954	12,567	695	5,157	18,419
1955	12,024	694	4,757	18,475
1956	12,952	1,068	5,922	19,942
1957	11,967	838	5,071	17,876
1958	11,518	1,033	5,315	17,866
1959	11,886	1,045	5,606	18,537
1960	12,558	908	6,173	19,639
1961	12,983	921	6,825	20,729
1962	13,838	944	7,590	22,372
1963	11,399	911	7,293	19,603
1964	10,755	879	5,872	17,506
1965	7,109	615	4,826	12,550
1966	7,162	695	4,661	12,518
1967	8,922	596	5,292	14,810
1968	9,262	569	5,823	15,654

TABLE 54

Effects of human persecution on Golden Eagles, South Grampians, Scotland 1950–56. From Sandeman 1957

	Number of territory years	Territory with only one bird	One member of pair immature	Mean brood-size in successful nests	Mean brood-size in all nests
Deer areas (no persecution)	35	0	0	1·4	0·6
Sheep & Grouse areas (much persecution)	63	8	4	1·4	0·3

TABLE 55

Percentage of ringed raptors recovered, compared with some other birds in the British ringing scheme. From Spencer & Hudson 1977

Raptors		Recognised pest-species	
Osprey	11	Crow	7
Red Kite	11	Magpie	6
Marsh Harrier	11	Bullfinch	2
Hen Harrier	9	Cormorant	20
Montagu's Harrier	14	Recognised quarry-species	
Sparrowhawk	10	Red Grouse	11
Common Buzzard	7	Partridge	5
Golden Eagle	8	Pheasant	7
Kestrel	12	Mallard	17
Merlin	11	Teal	18
Peregrine	8	Wigeon	16
Owls		Snipe	5
Barn Owl	17	Woodcock	8
Little Owl	9	Woodpigeon	10
Tawny Owl	9	Songbirds	
Long-eared Owl	7	Resident species	1–4
		Migrant species	<1

The table excludes swans and geese in which the recovery rates have been inflated by detailed studies; swans are also especially likely to be found after death.

TABLE 56

Percentages of recovered British-ringed raptors reported deliberately killed before and after 1954, when protective legislation was enacted. Figures are minima, and it is not known what were the true proportions of recovered birds that were killed. Sparrowhawk and Kestrel to 1969, other species to 1976–77

	Up to 1954		After 1954	
	Total recovered	Reported as killed	Total recovered	Reported as killed
Hen Harrier	25	20%	148	10%
Montagu's Harrier	19	68%	24	50%
Sparrowhawk	166	60%	71	16%
Common Buzzard	33	48%	173	14%
Golden Eagle	—	—	18	28%
Kestrel	175	41%	457	10%
Merlin	73	52%	107	16%
Peregrine	16	56%	55	22%

TABLE 57

Recoveries of birds of prey ringed in Finland. From Nordstrom 1963

	Number ringed 1913–62	% recovered	% recovered birds shot or trapped
Goshawk	1,006	27	78
Sparrowhawk	1,724	17	49
Marsh Harrier	208	11	65
Rough-legged Buzzard	164	13	67
Honey Buzzard	242	13	45
Common Buzzard	803	9	49
Peregrine	195	22	62
Kestrel	2,135	6	53
Osprey	800	4	67

TABLE 58

First-year recoveries of three raptors in eastern North America in different periods. From Henny 1972, Henny & Wight 1972

Species	Years	Numbers ringed	% recovered	% recovered birds reported to have been killed
Cooper's Hawk	1929–40	153	21	75
	1941–57	382	10	73
American Kestrel	1920–40	267	6	41
	1941–60	626	3	21
Red-tailed Hawk	1926–40	196	9	78
	1941–60	1,115	10	62
	1961–64	1,216	5	46

TABLE 59

Toxicity of various common pesticides and poisons. Figures show the single oral dose (mg/kg) that would be lethal to 50% of a population of the test species. From Tucker & Crabtree 1970

	Mallard	Pheasant
Highly poisonous		
Strychnine*	3	25
Endrin	6	2
Phosdrin	5	1
Parathion	2	12
Ten-eighty*	9	6
Moderately poisonous		
Aldrin	520	17
Dieldrin	381	79
Toxaphene	71	40
Mildly poisonous		
Chlordane	1,200	?
DDT	>2,240	1,296
Heptachlor	>2,000	?
Lindane	>2,000	?
Malathion	1,485	?

* The figure for Golden Eagle is 5 mg/kg for strychnine, and 1·3–3·0 for ten-eighty.

TABLE 60

Numbers of breeding raptors in forest subject to two levels of disturbance; the Netherlands sector subjected to more human disturbance than the German sector. From Voous 1977, based on research by P. F. M. Opdam

	Numbers of nesting pairs in	
	Netherlands sector	*German sector*
Goshawk, 1969	0	6
Goshawk, 1974	2	14
Common Buzzard, 1969	3	25
Honey Buzzard, 1969	1	6

Other species—no data.

TABLE 61

Organo-chlorine compounds commonly found in tissues and eggs of raptors

A. Persistent organo-chlorine insecticides and their persistent metabolites:
    (1) DDT (mostly pp' DDT)—an insecticide.
    (2) DDE—a metabolite of DDT.
    (3) TDE (DDD)—a metabolite of DDT and an insecticide in its own right.
    (4) Dieldrin (HEOD)—a metabolite of the insecticide aldrin and an insecticide in its own right.
    (5) Heptachlor epoxide—a metabolite of the insecticide heptachlor.
    (6) BHC, generally in the form of the gamma-isomer (lindane or gammexane)—an insecticide.
    All these substances are used in agriculture: some, notably DDT and dieldrin, are used in the control of disease vectors; DDT and dieldrin are also used industrially for moth proofing woollen goods. Other organo-chlorine pesticides include endrin, toxaphene and chlordane.

B. Polychlorinated biphenyls (PCBs):
    These include a large number of related compounds and metabolites used in the plastics, engineering and electrical industries. It has been calculated that up to 210 different compounds could theoretically be present in a PCB mixture.
    The PCBs were first synthesised in 1929 and the organo-chlorine insecticides in the 1940s and 1950s, and their use has greatly increased since those times, so that today some or all of them are probably used in all countries of the world. PCBs are now manufactured by Monsanto in the United States (trade name Aroclor), Prodelée in France (Phenoclor) and Bayer in Germany (Colphen). Other manufacturers are located in Japan and the Soviet Union.

TABLE 62

Shell-thinning and population trends in various raptor populations following the advent of DDT in agriculture.
From a large literature, I have selected only those based on at least 20 post-DDT clutches. Diet: A—amphibia, B—birds, F—fish,
I—insects, M—mammals, R—reptiles

Species	Diet	Area, Years	Mean % decrease in shell-index	Population declining in period concerned	Reference on shell-index
Osprey	F	Scandinavia, 1948–67	12	No	Anderson & Hickey 1974
		Fennoscandia, 1947–53	8	No	Odsjö 1971
		Fennoscandia, 1954–59	12	No	Odsjö 1971
		Fennoscandia, 1960–67	16	?	Odsjö 1971
		North America, 1949–60	18	Yes	Anderson & Hickey 1974
Bald Eagle	FB	North America, 1947–62	16	Yes	Anderson & Hickey 1974
White-tailed Eagle	FB	Sweden, 1953–69	16	Yes	Anderson & Hickey 1974
Marsh Harrier	BM	Europe, NW, 1949–65	7	No	Anderson & Hickey 1974
		Sweden, centre, 1965–76	14	No	Odsjö & Sondell 1977
Hen Harrier	MB	North America, 1947–69	15	No	Anderson & Hickey 1974
Montagu's Harrier	MRB	Europe, NW, 1948–69	3	?	Anderson & Hickey 1974
Goshawk	BM	Europe, NW, 1948–61	8	No	Anderson & Hickey 1974
		California, 1947–64	8	No	Anderson & Hickey 1972
European Sparrowhawk	B	Britain, SE, 1947–66	21	Yes	Ratcliffe 1970
		Britain, NW, 1947–67	14	Yes	Ratcliffe 1970
		Northern Ireland, 1947–69	15	Yes	Ratcliffe 1970
		Scotland, South, 1971–73	18	Yes	Newton 1976
		Europe, NW, 1947–53	10	No	Anderson & Hickey 1974
		Netherlands, 1945–71	16	Yes	Koeman et al 1972
Cooper's Hawk	BMR	California, 1947–48	0	No	Anderson & Hickey 1972
		California, 1951–52	6	No	Anderson & Hickey 1972
		California, 1956–59	6	No	Anderson & Hickey 1972
		New Mexico, Utah 1952–56	0	No	Anderson & Hickey 1972

Species	Code	Location, years	No.	DDE	Reference
Red-shouldered Hawk	MA	California, 1950–55	0	No	Anderson & Hickey 1972
		California, 1961–67	5	No	Anderson & Hickey 1972
		Florida, 1950–59	4	No	Anderson & Hickey 1972
		Florida 1960–68	0	No	Anderson & Hickey 1972
		Texas, 1949–51	9	No	Anderson & Hickey 1972
		Texas, 1952–54	4	No	Anderson & Hickey 1972
Red-tailed Hawk	MR	California, 1950–59	5	No	Anderson & Hickey 1972
		Ontario, 1949–68	0	No	Anderson & Hickey 1972
		Texas, 1948–67	5	No	Anderson & Hickey 1972
		Florida, 1949–67	3	No	Anderson & Hickey 1972
Common Buzzard	M	Britain, 1948–66	0	No	Ratcliffe 1970
Golden Eagle	MB	Scotland, west, 1951–65	10	No	Ratcliffe 1970
		California, 1953–55	9	No	Anderson & Hickey 1972
		New Mexico, 1951–55	0	No	Anderson & Hickey 1972
American Kestrel	MI	North America, 1947–67	3	No	Anderson & Hickey 1972
European Kestrel	M	Britain, 1946–66	5	No	Ratcliffe 1970
Hobby	BI	England, 1952–64	5	No	Ratcliffe 1970
Prairie Falcon	MB	North America, 1947–67	12	No	Anderson & Hickey 1974
		California, 1948–51	13	No	Anderson & Hickey 1974
Peregrine	B	England, north, 1947–69	20	Yes	Ratcliffe 1970
		England, south, 1947–59	18	Yes	Ratcliffe 1970
		Canada, Queen Charlotte Is, 1947–53	0	No	Anderson & Hickey 1974
		United States, 1947–54	16	Yes	Anderson & Hickey 1974
		Northern Canada, Alaska 1947–69	17	Yes	Anderson & Hickey 1974

TABLE 63

Relationship between clutch success and organo-chlorine levels in eggs. Asterisks denote a difference in levels from those in successful clutches significant at 5% (*) or 1% (**) levels; n.s.—not significant

	Failed, cause unspecified			Broken eggs			Addled eggs			Deserted eggs		
	DDE	PCB	HEOD	DDE	PCB	HEOD	DDE	PCB	HEOD	DDE	PCB	HEOD
Sparrowhawk	**			**	n.s.	n.s.	**	**	n.s.	n.s.	n.s.	n.s.
Cooper's Hawk	*			*						*		
Golden Eagle			**									
Brown Pelican	**	n.s.	*									
Shag	n.s.		*									

Details from Newton & Bogan 1978, Snyder et al 1973, Lockie et al 1969, Blus et al 1974, Potts 1968.

TABLE 64

Effects of organo-chlorine compounds on captive birds. +, a measureable effect; −, no measurable effect; ±, effect, at one concentration but not at a lower one; gaps mean that the parameter concerned was not reported on

	Behavioural changes	Delayed egg-laying	Reduced egg-production	Low weight eggs	Shell-thinning	Egg breakage	Reduced hatchability	Infertility	Increased deaths of young	References
DDE/DDT										
American Kestrel			−		+	+	+	−	+	Porter & Wiemeyer 1969
Screech Owl					+	+	+			McLane & Hall 1972.
Mallard/Black Duck			−		+	+	+	±	±	Cooke 1975, Heath et al 1969, Longcore et al 1971, Tucker & Haegele 1970.
Domestic Chicken					±		±	−		Brown et al 1965, Dunachie & Fletcher 1966, 1969, Lillie et al 1972, Sauter et al 1972, Smith et al 1970, Stephen et al 1971,

Species							Reference
Pheasant				+		±	Azevedo et al 1965, Dewitt 1956.
Bobwhite Quail	+			+			James & Davis 1965, Tucker & Haegele 1970.
Japanese Quail	−	+	±	+	+	±	Bitman et al 1969, De Witt 1955, 1956, Kenny et al 1972, Kreitzer & Heinz 1974.
Ring Dove	+	+	+	+	+		Haegele & Hudson 1973, Peakall 1970a.
Bengalese Finch[1]	+	+	−	+	+	+	Jefferies 1967, 1969, 1971
HEOD (Dieldrin)							
Mallard				+	+	+	Lehner & Egbert 1969
Domestic Chicken			−			+	Brown et al 1965, Graves et al 1969.
Pheasant	+	±	−	−	±	±	Atkins & Linder 1967, Dahlgren et al 1970, Dewitt 1956, Genelly & Rudd 1956.
							Baxter et al 1969, 1972.
Japanese Quail	+	−		−	±	±	Dewitt 1956, Kreitzer & Heinz 1974, Shellesberger et al 1965.
Partridge				−		−	Neill et al 1971.
Crowned Guinea Fowl[2]				+	+	−	Weise et al 1969.
Ring Dove	+	+					Peakall 1970a. Peakall & Peakall 1973.
PCB							
American Kestrel				−			Lincer 1972.
Mallard	−			−		−	Heath et al 1972.
Domestic Chicken	±			−	±	±	Briggs & Harris 1973. Britton & Huston 1973, Lillie et al 1974, Scott et al 1971.
Pheasant	+			+	+	+	Dahlgren & Linder 1971.
	−			−	−		Heath et al 1972.
Japanese Quail	+			+	−	+	Cecil et al 1973, Call et al 1973. Kreitzer & Heinz 1974.
Ring Dove	+			−		+	Peakall 1971, Peakall et al 1972, Peakall & Peakall 1973.

[1] Shells thickened rather than thinned, and incubation and nestling periods were lengthened. Thickening of shells was also noted in Moorhens under the influence of DDE (Graves et al 1971).

[2] Egg production increased.

TABLE 65

Mortalities (%) of three North American raptors estimated from ring recoveries, before and during the time when organo-chlorine pesticides were in widespread use. From Henny & Wight 1972.

		1924–45*	1946–65*
Red-shouldered Hawk	First year	58	59
	Second and later years	31·3±2·0	29·7±2·4
	Overall	42·4±1·9	41·8±2·2
American Kestrel	First year	69	60·7
	Second and later years	46·8±5·0	46·0±4·6
	Overall	60·2±3·5	53·9±3·0
Osprey	First year	57·3	53·3
	Second and later years	18·5±1·8	19·6±2·0
	Overall	30·2±1·8	29·6±2·0

*The only significant difference between periods is for first year Kestrels.

TABLE 66

Changes in the status of raptors in Israel during 1949 to 1965, associated with the use of thallium sulphate for rodent control.
The total number of species was 28, but each may be represented more than once if different populations were involved. Excluding passage migrants, most of the species for which no change was noted were already very rare. From Mendelssohn 1972

	Number of species showing:		
	Disappearance or near-disappearance	Marked decline	No change
Residents	5	4	3
Summer visitors	5	1	2
Winter visitors	24	0	1
Passage migrants	0	1	16

TABLE 67

Changes in the abundance and diversity of South American raptors in relation to cultivation and human settlement. From Reichholf 1974.

| | Extent of human influence on landscape | | | |
	Low	Medium	Great	Very great
Number of species	19	16	12	4
Number of individuals				
per unit time	91	35	152	452
Species diversity index*	2·05	1·65	1·13	0·05

* Takes into account both number of species and the proportion each forms of the total. The increase in individuals in highly modified areas was due almost entirely to scavenging vultures and caracaras.

TABLE 68

Influence of age at capture on breeding performance of captive Peregrines. From Cade and Temple 1977

	Male and female both nestlings	Mixed pairs	Male and female both post-nestlings
Number of pairs	35	9	18
Total number of matings	97	16	45
Number of laying females	24	3	1
Number with fertile eggs	12	2	0
Number rearing young	9	2	0

Index